Nuclear and New Energy Technology

Nuclear and New Energy Technology

Editor

Roman Vadimovich Davydov

MDPI

Basel • Beijing • Wuhan • Barcelona • Belgrade • Novi Sad • Cluj • Manchester

Editor
Roman Vadimovich Davydov
Institute of Cybersecurity and
Computer Science
Peter the Great St.Petersburg
Polytechnic University
St. Petersburg
Russia

Editorial Office
MDPI
St. Alban-Anlage 66
4052 Basel, Switzerland

This is a reprint of articles from the Special Issue published online in the open access journal *Energies* (ISSN 1996-1073) (available at: www.mdpi.com/journal/energies/special_issues/LOC69J0R29).

For citation purposes, cite each article independently as indicated on the article page online and as indicated below:

Lastname, A.A.; Lastname, B.B. Article Title. *Journal Name* **Year**, *Volume Number*, Page Range.

ISBN 978-3-7258-0496-2 (Hbk)
ISBN 978-3-7258-0495-5 (PDF)
doi.org/10.3390/books978-3-7258-0495-5

Contents

About the Editor

Roman Vadimovich Davydov

Dr. Roman Davydov is currently an Associate Professor at the Peter the Great St. Petersburg Polytechnic University, Institute of Cybersecurity and Computer Science, Higher School of Cyberphysical Systems & Control. Additionally, he is head of the Department of Bioinformatics and Mathematical Biology at Alferov University. His research interests include the mathematical modeling of physical processes, nuclear magnetic resonance, nuclear power plants, the express control of flowing media, computer science, and artificial intelligence. He has published more than 80 research papers in scientific journals and conference proceedings, and is an active reviewer for various MDPI and Elsevier journals.

Preface

The demand for electricity is growing globally, making nuclear energy a promising area for research and advancement. Unlike other power sources reliant on factors such as climate change or fuel availability, nuclear power plants offer consistent electricity production. Over the years, nuclear energy has gained traction as a clean, efficient, and low-carbon energy source. However, it is crucial to note that nuclear power can complement other electricity generation methods or be applied independently in various scenarios.

Efforts are needed to enhance the efficiency of electricity generation in nuclear plants and other renewable sources such as solar and wind energy. Alongside efficiency improvements, stringent environmental safety measures must be upheld at power facilities. All of this necessitates continuous research, technological development, and modernization based on data analysis, mathematical models, and experimental studies.

This Special Issue addresses various aspects of technology development and enhancement of nuclear energy systems, including design, subsystems, operational safety, and efficiency improvement. Additionally, it contains research results of energy storage and controlling systems and technologies for solar power plants, reflecting the broader scope of innovation in the energy sector. High-quality articles were prepared by scientists working at universities from different countries, including China, Hungary, Russia, Japan, Korea, and the UK. This confirms the relevance and importance of creating new technologies for energy production and storage systems.

Thanks to all the authors, the Editorial Office, and peer reviewers for their contributions to the Special Issue.

Roman Vadimovich Davydov
Editor

energies

Editorial

Nuclear and New Energy Technology

Roman Davydov

Institute of Physics and Mechanics, Peter the Great St. Petersburg Polytechnic University, 195251 St. Petersburg, Russia; davydovrv@spbstu.ru

Due to the growing demand for electrical energy generation worldwide [1], nuclear energy seems to be one of the most promising areas for research and new developments. Unlike other types of power plants, nuclear power plants do not depend on climate change (which is critical in the operation of hydroelectric power plants, solar installations, or wind turbines) or on the availability of large fuel reserves (which is critical in the operation of thermal power plants) to produce electricity. For several decades, nuclear energy has been being increasingly developed worldwide as clean, low-carbon, efficient primary load energy. However, other options for the generation of electricity can be successfully used with nuclear power and used separately in different situations.

In this regard, there is a need to create new technologies (and improve existing ones) with the aim to increase the efficiency of electricity generation in nuclear power plants, as well as other (solar or wind) power plants [2]. At the same time, it is essential to ensure that strict requirements are in place regarding environmental safety at electric power facilities [3]. During the operation of nuclear power plants, special attention should be paid to the processing and storage of spent nuclear fuel [4]. To do this, it is necessary to constantly conduct research and develop and modernize new technologies based on data analysis regarding the operation of energy facilities, mathematical modeling, and experimental studies.

Topics of interest in this Special Issue include the development of new and the improvement of existing technologies in:

- Nuclear energy systems design development;
- Different subsystems of nuclear power plant;
- Nuclear fuel cycle strategies and spent nuclear fuel management;
- Increasing the efficiency of electrical energy production;
- Energy storage systems;
- Wind and solar power plants.

check for updates

Citation: Davydov, R. Nuclear and New Energy Technology. *Energies* **2022**, *15*, 6046. https://doi.org/10.3390/en15166046

Received: 25 July 2022
Accepted: 18 August 2022
Published: 20 August 2022

Funding: This research received no external funding.

Acknowledgments: I would like to express special thanks to all the authors who submitted their work to this Special Issue.

Conflicts of Interest: The authors declare no conflict of interest.

References

1. Kober, T.; Schiffer, H.-W.; Densing, M.; Panos, E. Global energy perspectives to 2060—WEC's world energy scenarios 2019. *Energy Strategy Rev.* **2020**, *31*, 100523. [CrossRef]
2. Zhan, L.; Bo, Y.; Lin, T.; Fan, Z. Development and outlook of advanced nuclear energy technology. *Energy Strategy Rev.* **2021**, *34*, 100630. [CrossRef]
3. Di Maio, F.; Pedroni, N.; Tóth, B.; Burgazzi, L.; Zio, E. Reliability assessment of passive safety systems for nuclear energy applications: State-of-the-art and open issues. *Energies* **2021**, *14*, 4688. [CrossRef]
4. Alwaeli, M.; Mannheim, V. Investigation into the current state of nuclear energy and nuclear waste management—A state-of-the-art review. *Energies* **2022**, *15*, 4275. [CrossRef]

Short Biography of Authors

Dr. Roman Davydov is currently an Associate Professor at the Peter the Great St. Petersburg Polytechnic University, Institute of Physics and Mechanics, Department of Higher Mathematics. He received a B.S. degree in Applied Mathematics and Informatics, an M.S. degree in Applied Mathematics and Informatics, and a PhD degree in Physics and Mathematics from the Peter the Great St. Petersburg Polytechnic University in 2013, 2015, and 2021, respectively. His research interests include the mathematical modeling of physical processes, nuclear magnetic resonance, nuclear power plants, the express control of flowing media, computer science, and artificial intelligence. He has published more than 70 research papers in scientific journals and conference proceedings.

energies MDPI

Review

Review of the Requirements for Load Following of Small Modular Reactors

Choong-koo Chang * and **Harold Chisano Oyando**

Department of Nuclear Power Plant Engineering, KEPCO International Nuclear Graduate School (KINGS), 658-91 Haemaji-ro, Seosang-myeon, Ulju-gun, Ulsan 45014, Korea
* Correspondence: ckchang@kings.ac.kr; Tel.: +82-52-712-7303

Abstract: CO_2 net neutralization by 2050 is a global target. Renewable energy and nuclear power generation are emerging as major power sources for CO_2 net neutralization. Therefore, this paper comprehensively reviews the load-following operation method of nuclear power plants as a method to compensate for intermittency, which is the biggest weakness of renewable energy. First, this paper looks at the types of SMRs and elaborates the concept and necessity of load following. The comprehensive requirements for the load-following operation of an SMR, i.e., planned operation, automatic generation control, governor-free operation, cooperative control of the reactor and turbine generator, unit control of a multiple-module SMR, cogeneration, etc., are studied. Finally, the interaction between an SMR and the power grid during load-following operation and other technical issues are also reviewed. This paper can be used as a guide for load-following operations or as a guide for requirement analysis when developing a comprehensive control system of load following in SMR fleets.

Keywords: small modular reactor (SMR); load following; power grid; variable renewable energy (VRE); nuclear power plant (NPP)

Citation: Chang, C.-k.; Oyando, H.C. Review of the Requirements for Load Following of Small Modular Reactors. *Energies* **2022**, *15*, 6327. https://doi.org/10.3390/en15176327

Academic Editor: Roman Vadimovich Davydov

Received: 20 July 2022
Accepted: 26 August 2022
Published: 30 August 2022

1. Introduction

Most generating stations are connected to the grid, except generators that are installed in remote areas and conducting island operations. Grid-connected generating stations are classified as constant-output and load-following generating stations. In practice, large-capacity steam turbine power plants, such as nuclear power plants and thermal power plants, operate at base load. In addition, power plants that start and stop quickly, such as liquefied natural gas (LNG) power plants and hydroelectric power plants are used for demand control. However, as the share of renewable energy increased, it became difficult to demand control. Because it is impossible to maintain or arbitrarily control the output of solar or wind power plants, the gap between the maximum and minimum of the power generation output becomes very large. Inevitably, as the share of renewable energy increases, the share of flexible power generation such as LNG power generation must also increase, which leads to an increase in power generation costs.

In order to solve such a problem, a load-following operation that adjusts the output of the existing large-capacity thermal power plant or nuclear power plant according to the increase or decrease in the load is required. While this is not technically impossible, it is technically and economically undesirable. The next alternative that comes to mind is a small modular reactor (SMR). This is because, by operating multiple SMRs of less than 300 MW together, as much as necessary, it is easy to carry out a load-following operation, and the safety of an SMR is much higher than that of existing large scale nuclear power plants.

The purpose of this paper is to analyze the basic characteristics of SMRs and to suggest the load-following methods and countermeasures required for an SMR to perform demand control when the renewable energy generation capacity connected to the same grid and demands fluctuate together.

2. Overview of Small Modular Reactors

Small modular reactors (SMRs) are generally defined as nuclear reactors with power outputs between 10 megawatts electric (MWe) and 300 MWe. Reference [1] introduces the SMR types. There is growing interest in SMRs and their applications. An international conference on climate change and the role of nuclear power in September 2019 revealed that SMRs are being considered by many member states as a potential viable nuclear option that can contribute to climate change mitigation [1]. Most SMRs have their own safety features and can be designed as single- or multi-module plants.

SMRs are in development for all major reactor lines and can be categorized by moderator, reaction type, and advanced generation IV reactors. Recent publications introduce a number of advanced SMR design concepts that are currently being developed worldwide (in most cases, not yet deployed) [2]. Among the above SMRs, some SMRs have recently been connected to a grid or are currently under construction. In most cases, these designs are significantly supported by the government and are being built as prototype facilities or first of a kind (FOAK) demonstrations and commercial facilities. Some SMRs are likely to be deployed in the next 10 years after design, testing, and R&D work. These designs reflect technological advances and receive significant government or private sector support [3]. They are more likely to use traditional technologies to shorten development timelines with evolutionary change but may not be as efficient as more advanced technologies.

2.1. Classified by Moderator

A light-water reactor is a type of thermal reactor that uses "light water" (plain water) as a neutron moderator or coolant. Light-water reactors are the most commonly used among thermal reactors [4]. Heavy-water reactors (HWR) use heavy water as a neutron moderator. Heavy water is deuterium oxide, D_2O. The neutrons in a nuclear reactor that use uranium are fast moving and must be slowed down to initiate further fission. Gas-cooled reactors (GCRs) mostly use carbon dioxide and recently use helium as a coolant to transfer heat to turbines and graphite as a moderator. As with heavy water, a graphite moderator allows the use of natural uranium (GCR), or slightly enriched uranium (AGR) can be used as a fuel [4].

2.2. Classified by Reaction Type

Fast neutron spectrum reactors can use very different coolants, including, but not limited to, liquid sodium, lead, lead–bismuth eutectic, molten salt, and helium, which might significantly challenge the structural integrity of the fuel and other reactor components [5]. SMR technology with a fast neutron spectrum could be an attractive alternative for developed countries seeking to reduce plutonium stockpiles [3].

2.3. Advanced Generation IV Reactors

Some other types of SMRs are likely to be deployed within 20 years. These designs are currently planning R&D programs to test new approaches and materials and may require significant testing and operational experience in areas such as fuel safety performance, corrosion resistance, aging mechanisms, and component reliability [4]. High-temperature gas-cooled reactors (HTGR), molten salt reactors (MSR), sodium-cooled fast reactors (SCFR), and lead-cooled fast reactors (LCFR) are typical generation IV reactors.

3. Needs and Definitions of Load-Following Operation

3.1. Increase in Variable Generation in Power Grids

In the most countries, the share of renewable energy is increasing, and most of the new power plants are solar and wind power, as shown in Table 1 [6]. As of 2020, the global renewable energy capacity was about 35.7% of the total power electricity capacity, including hydroelectric power generation plants, and about 20% of the total electricity capacity excluding hydroelectric power plant capacity [7].

Table 1. World Total Renewable Energy Capacity (excerpted).

Capacity (GW)	2012	2013	2014	2015	2016	2017	2018	2019	2020	2021
World	1444	1566	1698	1852	2014	2185	2357	2542	2807	3064
China	302.1	359.5	414.7	479.1	541.0	620.9	695.5	758.8	899.6	1020
India	60.5	63.6	71.9	78.6	90.4	105.2	118.2	128.4	134.4	147.1
Japan	39.0	46.1	56.1	67.4	76.2	84.2	91.3	99.3	106.9	111.9
Korea	3.7	4.3	5.7	7.2	9.4	11.4	14.0	18.0	20.4	24.3

However, the solar and wind turbine generation plants have a peak load contribution of up to 10–20% and cannot control output. This increases the need for flexibility and backup resources in other parts of the power system [8]. There are several ways to increase flexibility, but flexible power plants, especially those already in operation, will provide the necessary flexible backups in the short term, and other flexibility measures such as demand-side engagement should also be implemented. Coping with power ramps, that is, sudden and massive active power (MW) variation control, will become increasingly important. Demand was always variable, and supply flexibility was always needed. However, the increase in net demand variation due to increased renewable energy will significantly change the way the power system operates by adding a new dimension on top of the traditional variable demand control. Table 2 [9] shows the flexibility of conventional power generation technologies.

Table 2. Flexibility of Conventional Power Generation Technologies.

Description	NPP	HC	Lign	CCG	PS
Start-up Time "Cold"	~40H	~6H	~10H	<2H	~0.1H
Start-up Time "warm"	~40H	~3H	~6H	<1.5H	~0.1H
Load Gradient (up) "nominal output"	~5%/M	~2%/M	~2%/M	~4%/M	>40%/M
Load Gradient (down) "nominal output"	~5%/M	~2%/M	~2%/M	~4%/M	>40%/M
Minimal Shutdown Time	No	No	No	No	~10H
Minimal Possible Load	50%	40%	40%	<50%	~15%

Abbreviations: NPP: nuclear power plants, HC: hard coal-fired power plants, Lign: ignited-fire power plants, CCG: combined-cycle gas-fired power plants, PS: pumped storage power plants, H: hour, M: minute.

3.2. *Definitions and Types of Load Following*

(1) Load Following

Until now, large-scale nuclear power plants operated at maximum output in principle. This is because the capital cost is high and the fuel cost is low. Therefore, the nuclear power plants should be operated at the highest safely achievable power. However, France, which has a high proportion of nuclear power, is known to operate a load-following operation, and in recent years, as the share of renewable energy, which is an intermittent power source, has increased, the load following of nuclear power plants has emerged as an important concern. There are two ways to regulate the thermal output of nuclear power plants. The first is a primary loop control method for controlling the fuel rod (reactor following turbine), and the second is a method for controlling the amount of main steam supplied to the turbine (turbine following reactor), as shown in Figure 1 [10].

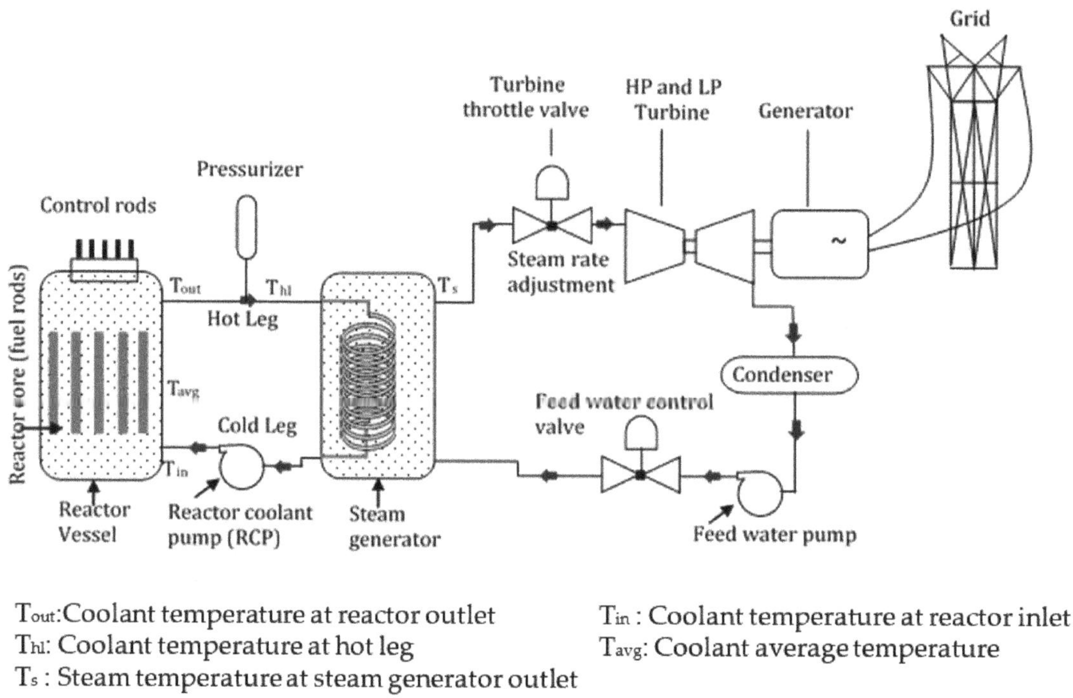

Figure 1. Schematic of PWR Nuclear Power Plants.

(2) Frequency Control

Power demand is difficult to accurately predict in advance. Therefore, fluctuations in demand result in frequency fluctuations. In order to keep the plant frequency stable at the rated frequency, the frequency of the grid must be monitored and the generation level must be adjusted immediately (primary control), as shown in Figure 2 [11]. The variation in the frequency, Δf, would require a change in the power of the plant of:

$$\frac{\Delta p}{P_0} = \frac{1}{s}\frac{\Delta f}{f_0} \rightarrow \Delta p = k\Delta f, \text{ with } k = \frac{1}{s}\frac{P_0}{f_0} \tag{1}$$

where f_0 is the target frequency (e.g., 60 Hz in Korea), P_0 is the power level of the plant (as a % of the rated power, Pr), Δp is the power change, and S is the droop measured in %.

The primary frequency control, also known as governor-free (GF) control, means short-term adjustment in the time frame of 2 to 30 s after deviations in power generation and demand are observed. This is so-called governor-free control. The secondary frequency control method operates in longer time units (e.g., seconds to minutes) and calculates the average frequency deviation over a period of time to restore the rated frequency. The secondary frequency control is called AGC (automatic generation control), and in AGC mode, the transmission system operator (TSO) gives control commands to the generator. In general, secondary control is especially important when the national grid is interconnected with the grid of another country [11].

Table 3 [12] shows some of the key features of the German nuclear power plant design for flexible operation. These design features have been incorporated into the design since the early years of the operation, when flexible operation was not expected.

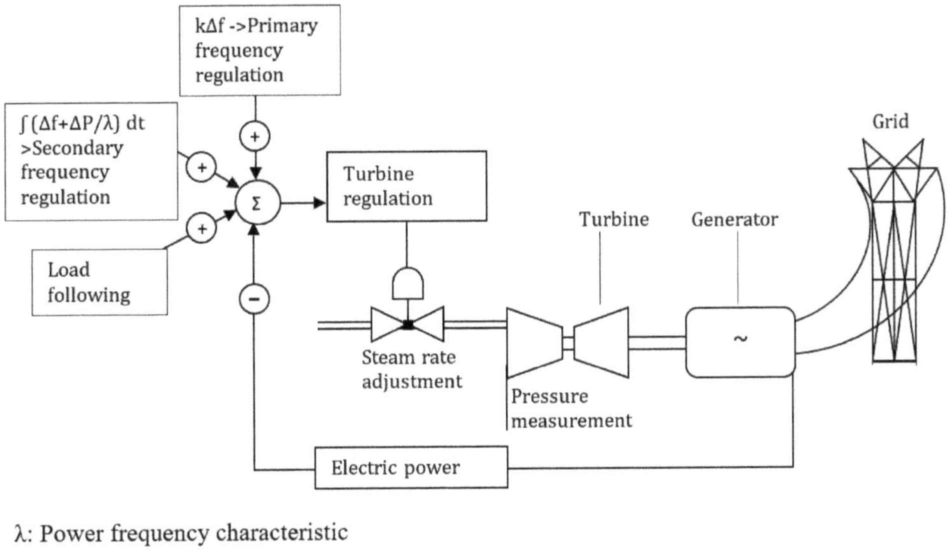

Figure 2. Frequency control in a power plant.

Table 3. Design Philosophy for Flexible Operation in German Light Water Reactor.

Response Base	Response Mode	Parameter and Properties
Predicted daily demand variations	Load following	Low-power period (power level and duration) Power change rate (slow, fast) Time in cycle (beginning, end)
Spontaneous limited demand variations	Frequency control	Local frequency control: frequency deviation (Δf) converted into power change (Δp) (Δp amplitude, slope of change) Remote frequency control: signal from the dispatcher (Δp amplitude, slope of change) Superimposition of local and remote frequency control
Grid disturbances	Spinning reserve	Ramp (amplitude, slope, from minimum power level) Steps (amplitude, from minimum power level) House load capability (loss of off-site power without reactor trip) Fast (e.g., 5% rated thermal power/minute) return to full power without advance notice
Longer-term forecasted demand	Extended low-power operation	Reduced power level during extended period (number of occurrences, duration)

(3) Voltage stability

The frequency and voltage stability of the power grid are maintained by active power and reactive power control, respectively. Voltage stability is the ability of the power system that maintains the voltage within a predetermined range on all buses after a fault or failure to prevent power outages. Once a generator is synchronized to the power grid, the active power of the generator can be controlled through the shaft torque, and the reactive power can be controlled by the field current [8]. However, as the share of renewable energy, such as solar and wind power, which do not have the ability to supply or absorb reactive power, increases, the problem of maintaining voltage stability is becoming a challenge.

Currently, most commercial photovoltaic (PV) inverters operate as grid-following (GFL) sources that regulate their power output by measuring the angle of the grid voltage using a phase-locked loop [13]. Hence, they merely follow the grid angle/frequency and do not actively control their frequency output. In contrast, a grid-forming source (GFM) actively controls its frequency and voltage output and has been extensively used in microgrid configurations [14]. However, given that power electronics inverters typically have many times smaller power ratings compared to synchronous machines, this means that the system load of inverter-based infrastructure must be satisfied with a much larger number of inverters. For large power grids, this translates into the need to install millions of inverter interface variable renewable energy (VRE) units over a large geographical area [8].

Consequently, in power systems with a high share of renewable energy, the reactive power control capability of SMRs can contribute significantly to voltage stability. This is because the SMRs can control reactive power output to maintain voltage stability.

(4) European Utilities Requirements (EUR) for Load Following

The EUR cover a wide range of conditions for nuclear power plants to operate efficiently and safely [11]. It states that modern nuclear reactors must implement significant maneuverability and, in particular, be able to operate in load-following mode. Four transients—excessive load increase, uncontrolled control rod withdrawal, the uncontrolled dilution of boric acid, and uncontrolled control rod drop—can cause clad/pellet interaction and rupture the clad [15]. Only the first of these affects load-following regulations. The EUR have defined common requirements applicable to new LWRs, as shown in Table 4 [15].

Table 4. Power Margins of European Light Water Reactor NPPs.

Common Requirements Applicable to NEW LWRs	**Parameter**
Continuous operation range (mandatory)	50%~100% Pn
Down to minimum (option)	20%
Primary control (mandatory); 2~30 s after deviation observed	±2% Pn/min
Higher values by agreement between system operator and plant operator	± 5% Pn/min
Activating total primary range of control requested	Within 30 s
Secondary control (option); several seconds to several minutes	±10%
Load-Following Capabilities (option)	
Load-following capability until ( ) % of whole fuel cycle	90%
Load-following range	100% Pn~minimum
From full power to minimum load and back to full power operation	2 per day, 5 per week Cumulatively 200 per year
Emergency load variation (by agreement between grid operator and unit operator); amplitude down to minimum load of the unit.	20% of Pn/min.

Note: Pn denotes rated power of the power plant.

4. Requirements for the Load-Following Operation of SMRs

Since an SMR has a smaller unit capacity compared to a large nuclear reactor, thermal power control is relatively easier and less risky in terms of nuclear safety. In addition, because start up and shut down are faster, bulk output power control is possible through unit control.

Load following means changing the power generation as closely as possible to the expected power demand. Load-following generation can match demand by the output changes in a planned manner or in response to instructions or signals from the grid control center or transmission system operator (TSO). Changes in output can be large or small and frequent or infrequent [9]. The following are typical requirements for each SMR load-following mode based on the above review.

4.1. Power Change Dependent on Grid Plans

(1) Planned Operation

This refers to the planned control of the power plant output between 20% and 80% based on the power supply and demand plan, and the output control timescale is hours or days. In some cases, the reactor output is adjusted to the level of 20% to 80% for the repair or recovery of the reactor, and the timescale in this case is hours. The power output maneuvering range is a function of time. During the first 65% of the fuel cycle, output power is controllable between 100% of nominal power and around 25% of nominal power. Then, the power control range is gradually reduced from 25% to 80% of nominal power because of excess reactivity and low boron concentrations. Nuclear power plants can operate at a minimum power level of 10%. However, the minimum output is around 20%, as for many conventional power plants [16].

(2) Unplanned Operation

This refers to the case where the power plant output is unplanned and adjusted according to the power grid conditions, and the load-following operation is performed by controlling the output of the nuclear reactor between 20% and 80%. The timescale for TSO instruction operation is minutes.

4.2. Power Change Dependent on Frequency

(1) Automatic Generation Control (AGC)

This refers to automatically reducing or increasing plant output within a limited range according to signals from the transmission system operator (TSO). This type of operation is also referred to as the 'Automatic Generation Control' (AGC). The power change is typically within 20~40% of the RTP [10] or ±10% of the rated thermal power (RTP) [12], and the timescale is minutes.

(2) Governor-Free (GF) Control

This refers to controlling frequencies outside a specified frequency range, either by reducing the generator output by the turbine governor when the system frequency exceeds the maximum limit or increasing the output when the system frequency falls below the minimum limit. The plant operator, in response to the frequency deviation, can initiate frequency control automatically or manually. The power change is typically within 20~40% of the RTP [10] or ±10% of the RTP [12], and the timescale is seconds.

4.3. Coordinated Rapid Load Following

A coordinated control approach accomplishes a rapid load-following operation by wisely combining the 'reactor following turbine mode' and 'turbine following reactor mode' along with the satisfaction of the reactivity constraints. The coordinated mode of power variation can be explained as follows:

The reactor is assumed to operate normally in the 'reactor following turbine mode' and at a certain power output level of $P_{th} = P_{tha}$, much lower than 100% full power (FP). In that condition, the output is increased to a high value of P_{thb} within a short time interval "T" and held to that value for the rest of the time. The coordinated mode may be due to the nature of the daily load curve. For example, where the reactor's power has to match a growing demand, a predefined instant consumption suddenly increases [10].

4.4. Multi-Module Unit Operation

If the plant has multiple SMRs, the entire power output of the plant can be adjusted by the disconnection of some reactor modules during periods of low demand, for scheduled maintenance, or when significant, high-priority energy becomes available from intermittent renewable energy system (RES) [17]. The performance data of the SMRs are not yet verified by commercial operation, but for German nuclear power plants, at least 3 h of downtime and at least 1 h of operation have been established, taking into account the start and end times [18].

4.5. Cogeneration with Non-Electric Applications

Due to SMRs' ability to provide CO_2-free energy, applications in the district heating field are mainly being discussed. Although smaller than traditional nuclear power plants (NPPs), SMR plant designs can provide increased safety through passive systems, reduce costs, and increase quality through factory-based manufacturing and other advantages [19]. The primary side circuits of the SMR can be operated at rated capacity, and only a fraction of the heat can be converted for other purposes, such as district heating, desalination, or hydrogen production [20].

The requirements that the cogeneration SMR must have in order to perform a load-following operation are as follows: First, the distance between the SMR and the heat utilization plant should be reasonable. This is because heat is not easier to transmit than electricity. Second, the steam extraction amount and control speed suitable for SMR to perform load-following operations should be technically acceptable from the perspective of coproduction plants and be economically feasible. Third, the temperature of the steam must be adequate to produce a coproduct.

The higher the temperature, the more types of cogeneration facilities are available, so this is a key parameter. Most light water reactor (LWR) coolant outlet temperatures are around 300 °C [21]. Future high-temperature reactors can operate at higher temperatures. For example, the coolant outlet temperature of a sodium-cooled high-speed reactor is approximately 500~550 °C compared to 850~950 °C for a high-temperature gas reactor (HTGR) [22,23].

5. Other Considerations for Load Following

5.1. Regulatory Requirements

(1) Safety Regulations

The reactor will continue to generate significant heat from the decay of fission products that persist on a logarithmic timescale, even when the chain reaction is completely stopped. The principle of providing "defense in depth" against scenarios where the NPP is unable to provide long-term core decay heat removal shall be provided [24]. US nuclear regulatory commission (US NRC) safety and licensing criteria related to electric power are contained in general design criteria (GDC) 17 [25]. The design criteria of preferred power supply (PPS) and its interface with the class 1E power system, switchyard, transmission system, and alternate ac (AAC) source are described in IEEE Std 765, 'IEEE Standard for Preferred Power Supply (PPS) for Nuclear Power Generating Stations' [26]. IAEA safety standard series No. SSG-34, 'Design of Electrical Power Systems for Nuclear Power Plants' [27], provides the safety guide on the necessary characteristics of electrical power systems for nuclear power plants and of the process for developing these systems.

If the safety-related systems actuate by passive means and their continued operation relies on natural cooling principles, a safety-related electrical system is not required. For this reason, NuScale requested in the license document that it be excluded from GDC 17 [28].

5.2. Technical Considerations

(1) Physical aspects of power regulation

In terms of load following by fuel rod control, the following factors affect the maneuverability. By the moderator effect and Doppler effect, if the temperature of the primary coolant is increased, reactivity is decreased. When the reactor power increases, the power distribution is pushed to a lower part of the fuel [11]. If power variation is made by control rods, they deform the axial distribution of power and ^{135}Xe. Thus, it is an additional challenge for the load following with large magnitudes of power variations. At the end of the fuel cycle, the margins for the maneuverability decrease because the boron concentration is almost zero and the control rods are in the upper position [11]. The use of the control rod alone for power control has negative consequences, such as flux distribution disturbance, component material fatigue, mechanical wear, and adverse impacts on the burn-up balance in the core [29].

(2) Influence of the load following on the lifetime of components

Operating the NPP in load-following mode introduces technical disadvantages, as the plant components are exposed to numerous thermal stress cycles. This results in faster aging and requires a more sophisticated system for reactor monitoring and control [29]. Load cycling results in variations in the coolant temperature and thus in the temperatures of different components. Repeated temperature changes can create cyclic changes in the mechanical load of a part of the equipment and cause local structural damage (fatigue) to these elements. As a result, the maximal number of load-following operation cycles during the whole operational lifetime of the plant should be considered in the equipment qualification of the safety-related components [11].

5.3. Interaction of Grid Characteristics with Nuclear Power Plants

(1) Effects of Grid Frequency Change on NPP

Changes in frequency affect NPP operation through the speed governor of the turbine generator and through the speed change in the pump that delivers the flow to the reactor and the secondary coolant circuit. If the frequency drops, the turbine/generator examines the load based on the governor droop setting and frequency deviation. The mismatch between the reactor output and the produced electric power requires intervention from the control system. As the frequency rises, the turbine speed governor closes the throttle valve on the turbine to reduce power. When the reactor output has not changed, the reactor output is greater than the power drawn by the turbine. This mismatch causes transient overtemperature and overpressure in pressurized water reactors. Modern turbines for the grids of developed countries can only operate for a few minutes at a frequency below their rated frequency. These adverse operations have a cumulative effect and are only allowed for a certain total period over the lifetime of the turbines [30].

(2) Effect of Grid Voltage Change on NPP

A rapid voltage drop is mainly caused by electric fault on a transmission line. The magnitude of the voltage dip depends upon the distance from the fault, the type of the fault, and upon the performance of the automatic voltage regulator (AVR) of the generators connected to the grid [31]. During sharp voltage dip conditions, all motors connected to the auxiliary power system of the NPP will be retarded. The magnitude of the retardation is determined by the voltage dip and its duration, the characteristics of the motor, and the mass moment of inertia of the motor pump assembly [30].

If the grid has a light load and the NPP remains connected with long lines at the remote end, the grid voltage may be higher than generator voltage. If high grid voltage is continued for a long period, then the generator connected to the grid may be unstable because the generator must consume a large value of reactive power (Mvar). On the other hand, if the grid voltage remains low, the large motors of the NPP cannot start or can be retarded.

6. Discussion and Conclusions

Calls for a reduction in CO_2 emissions triggered by the climate change crisis pose a significant challenge for most countries around the world. Therefore, all the countries have to reduce the fossil fuels that have been used to generate electricity and increase the solar and wind energies, which are called clean energy. The normative IEA net zero emission scenario (NZE) by 2050 shows a narrow but achievable path for the global energy sector to achieve net zero CO_2 emissions by 2050 [32]. Developed countries reach net zero emissions before others [33]. This trend poses significant challenges in operating power grids. As the penetration of variable renewable energy (VRE) within the grid increases, many factors require greater grid flexibility to accommodate changes in generation. In particular, PV power does not generate power at night, so it has natural difficulties associated with its diurnal cycle [8]. Nuclear power generation is the most effective alternative among the ways to supply clean energy and at the same time complement the intermittent renewable energy. Due to this reason, the European Parliament decided to include nuclear and gas in the European green tax system, which was designed to promote the energy transition.

The way to supplement intermittent renewable energy is to operate nuclear power plants in a load-following mode. Large-capacity nuclear power plants are also capable of load following, but an SMR fleet is more appropriate, considering various technical and economic aspects. Therefore, this paper introduced a typical SMR currently in operation or development and elaborated various other problems in the load-following operation of the SMR. In addition, this paper specified the various requirements for an SMR load-following operation. Finally, the interaction between the SMR and the main power grid during the load-following operation of the SMR were described. In terms of the load following of SMRs, the main characteristics and contributions of this paper are that it deals with the reactor power control method, the turbine generator output control method, and the coordinated power control of the reactor and turbine generator. It is noticed that the interaction between the SMR and the main power grid is also important for the load following of SMRs.

Author Contributions: C.-k.C. wrote the paper and H.C.O. reviewed the references and conducted editorial reviews. All authors have read and agreed to the published version of the manuscript.

Funding: This research received no external funding.

Institutional Review Board Statement: Not applicable.

Informed Consent Statement: Not applicable.

Data Availability Statement: Not applicable.

Acknowledgments: This research was supported by the 2022 Research Fund of the KEPCO International Nuclear Graduate School (KINGS), Ulsan, Republic of Korea.

Conflicts of Interest: The authors declare no conflict of interest.

Abbreviations

AAC	Alternative AC
AGC	Automatic Generation Control
AGR	Advanced Gas-cooled Reactor
AVR	Automatic Voltage Regulator
EUR	European Utilities Requirements
FOAK	First of A Kind
FP	Full Power
GCR	Gas-Cooled Reactor
GDC	General Design Criteria
GF	Governor-Free
GFL	Grid-Following
GFM	Grid-Forming
HTGR	High-Temperature Gas-Cooled Reactor
HWR	Heavy Water Reactor
LCFR	Lead-Cooled Fast Reactor
LNG	Liquefied Natural Gas
LWR	Light Water Reactor
MSR	Molten Salt Reactor
NRC	Nuclear Regulatory Commission
NPP	Nuclear Power Plant
PPS	Preferred Power Supply
RES	Renewable Energy System

RTP	Rated Thermal Power
SCFR	Sodium-Cooled Fast Reactor
SMR	Small Modular Reactor
TSO	Transmission System Operator
VRE	Variable Renewable Energy

References

1. IAEA. *Advances in Small Modular Reactor Technology Developments;* A Booklet Supplement to the IAEA Advanced Reactors Information System (ARIS); IAEA: Vienna, Austria, 2020; p. 354.
2. Kuznetsov, A.L.V. *Current Status, Technical Feasibility and Economics of Small Modular Reactors;* OECD: Paris, France, 2011.
3. International Atomic Energy Agency. *Technology Roadmap for Small Modular Reactor Deployment;* NR-T-1.18; International Atomic Energy Agency: Vienna, Austria, 2021.
4. Zohuri, B. *Small Modular Reactors as Renewable Energy Sources;* Springer: Cham, Switzerland, 2019.
5. IAEA. *IAEA-TECDOC-1972 Benefits and Challenges of Small Modular Fast Reactors;* IAEA: Vienna, Austria, 2021.
6. IRENA. *Renewable Capacity Statistiques De Capacité Estadísticas De Capacidad;* IRENA: Abu Dhabi, United Arab Emirates, 2022.
7. U.S. Energy Information Administration. *Administration. International Electricity Data..* Available online: https://www.eia.gov/international/data/world (accessed on 12 July 2022).
8. Kroposki, B.; Johnson, B.; Zhang, Y.; Gevorgian, V.; Denholm, P.; Hodge, B.M.; Hannegan, B. Achieving a 100% Renewable Grid: Operating Electric Power Systems with Extremely High Levels of Variable Renewable Energy. *IEEE Power Energy Mag.* **2017**, *15*, 61–73. [CrossRef]
9. Pierre, I.; Lorubio, G. *Flexible Generation: Backing Up Renewables;* Union of the Electricity Industry-EURELECTRIC Depot legal: D/2011/12.105/47; RESAP: Paris, France, 2011.
10. Bose, D.; Hazra, A.; Mukhopadhyay, S.; Gupta, A. A Co-ordinated Control Methodology for Rapid Load-Following Operation of a Pressurized Water Reactor Based Small Modular Reactor. *Nucl. Eng. Des.* **2020**, *367*, 110748. [CrossRef]
11. Nuclear Energy Agency. *Technical and Economic Aspects of Load Following with Nuclear Power Plants;* NEA, OECD: Paris, France, 2011; pp. 1–51.
12. IAEA (International Atomic Energy Agency). *Non-Baseload Operation in Nuclear Power Plants: Load Following and Frequency Control Modes of Flexible Operation;* IAEA Nuclear Energy Series; IAEA: Vienna, Austria, 2018; pp. 1–190.
13. Ackermann, T.; Prevost, T.; Vittal, V.; Roscoe, A.J.; Matevosyan, J.; Miller, N. Paving the Way A Future without Inertial Is Closer Than You Think. *IEEE Power Energy Mag.* **2017**, *15*, 65–67. [CrossRef]
14. Pattabiraman, D.; Lasseter, R.H.; Jahns, T.M. Comparison of Grid Following and Grid Forming Control for a High Inverter Penetration Power System. In Proceedings of the 2018 IEEE Power & Energy Society General Meeting (PESGM), Portland, OR, USA, 5–10 August 2018; pp. 1–5.
15. Bruynooghe, C.; Eriksson, A.; Fulli, G. *Load-Following Operating Mode at Nuclear Power Plants (NPPs) and Incidence on Operation and Maintenance (O&M) Costs. Compatibility with Wind Power Variability;* European Commission: Luxembourg, 2010.
16. Peakman, A.; Merk, B.; Hesketh, K. The potential of pressurised water reactors to provide flexible response in future electricity grids. *Energies* **2020**, *13*, 941. [CrossRef]
17. Ingersoll, D.T.; Colbert, C.; Houghton, Z.; Snuggerud, R.; Gaston, J.W.; Empey, M. Can Nuclear Power and Renewables be Friends? In Proceedings of the ICAPP 2015, Nice, France, 3–6 May 2015; p. 9.
18. Ludwig, H.; Salnikova, T.; Waas, U. Load-following Capability of German Nuclear Power Plants (NPPs). *ATW Int. J. Nucl. Power* **2010**, *55*, 4–5.
19. World Nuclear Association. Small Nuclear Power Reactors. Available online: https://www.world-nuclear.org/information-library/nuclear-fuel-cycle/nuclear-power-reactors/small-nuclear-power-reactors.aspx (accessed on 5 May 2022).
20. Locatelli, G.; Boarin, S.; Fiordaliso, A.; Ricotti, M.E. Load following of Small Modular Reactors (SMR) by cogeneration of hydrogen: A techno-economic analysis. *Energy* **2018**, *148*, 494–505. [CrossRef]
21. IAEA Advanced Reactors Information System. Characteristics of Advanced Reactors. Available online: https://aris.iaea.org/sites/operating.html (accessed on 7 July 2022).
22. Locatelli, G.; Mancini, M.; Todeschini, N. Generation IV nuclear reactors: Current status and future prospects. *Energy Policy* **2013**, *61*, 1503–1520. [CrossRef]
23. Kunitomi, K.; Yan, X.; Nishihara, T.; Sakaba, N.; Mouri, T. JAEA'S VHTR for Hydrogen and Electricity Cogeneration: GTHTR300C. *Nucl. Eng. Technol.* **2007**, *39*, 9–20. [CrossRef]
24. Bickel, J.H. Grid Stability and Safety Issues Associated with Nuclear Power Plants. *Evergr. Saf. Reliab. Technol.* **2001**, *14*, 1–9.
25. USNRC. Appendix A to Part 50—General Design Criteria for Nuclear Power Plants. US Nuclear Regulatory Commission. *Med. Dosim.* **2020**, *13*, 87–93.
26. *IEEE Std 765-1983;* IEEE Standard for Preferred Power Supply (PPS) for NUclear Power Generating Stations. IEEE: New York, NY, USA, 1995. [CrossRef]
27. IAEA. *Design of Electrical Power Systems for Nuclear Power Plants;* Specific Safety Guide, No. SSG-34; IAEA: Vienna, Austria, 2012; pp. 1–144.

28. NuScale Power. *Nuscale SMR Technology: An Ideal Solution for Repurposing U.S. Coal Plant Infrastructure and Revitalizing Communities*; NuScale Power: Corvallis, OR, USA, 2021.
29. Wenisch, A.; Becker, O. *NPP Output Flexibility Expectations in the Light of Reality*; Österreichisches Ökologie-Institut: Vienna, Austria, 2010.
30. IAEA. *Interaction of Grid Characteristics with Design and Performance of Nuclear Power Plants. A Guidebook*; Technical Reports Series No. 224; International Atomic Energy Agency: Vienna, Austria, 1983.
31. Chowdhury, A.H.; Rabby, M.K.M. A study on low grid voltage problem near Rooppur nuclear power plant. In Proceedings of the 8th International Conference on Electrical and Computer Engineering, ICECE, Dhaka, Bangladesh, 20–22 December 2014; pp. 289–292.
32. World Bank. *Tracking SDG7: The Energy Progress Report 2018*; World Bank: Washington, DC, USA, 2018; p. 193.
33. OECD/IEA. *World Energy Model Documentation*; OECD: Paris, France, 2021.

energies

MDPI

Article

Enhancing LOCA Breach Size Diagnosis with Fundamental Deep Learning Models and Optimized Dataset Construction

Xingyu Xiao [1,†], Ben Qi [1,†], Jingang Liang [1,*], Jiejuan Tong [1], Qing Deng [2] and Peng Chen [3]

[1] Institute of Nuclear and New Energy Technology, Tsinghua University, Beijing 100084, China; xxy23@mails.tsinghua.edu.cn (X.X.); qib22@mails.tsinghua.edu.cn (B.Q.); tongjj@tsinghua.edu.cn (J.T.)
[2] Research Institute of Macro-Safety Science, University of Science and Technology Beijing, Beijing 100083, China; dengqing@ustb.edu.cn
[3] School of Computer Science and Technology, University of Chinese Academy of Sciences, Beijing 100049, China; chenpeng23@mails.ucas.ac.cn
* Correspondence: jingang@tsinghua.edu.cn
† These authors contributed equally to this work.

Abstract: In nuclear power plants, the loss-of-coolant accident (LOCA) stands out as the most prevalent and consequential incident. Accurate breach size diagnosis is crucial for the mitigation of LOCAs, and identifying the cause of an accident can prevent catastrophic consequences. Traditional methods mostly focus on combining model algorithms and utilize intricate composite model neural network architectures. However, it is crucial to investigate whether greater complexity necessarily leads to better performance. In addition, the consideration of the impact of dataset construction and data preprocessing on model performance is also needed for model building. This paper proposes a framework named DeepLOCA-Lattice to experiment with different preprocessing approaches to fundamental deep learning models for a comprehensive analysis of the diagnosis of LOCA breach size. The DeepLOCA-Lattice involves data preprocessing via the lattice algorithm and equal-interval partitioning and deep-learning-based models, including the multi-layer perceptron (MLP), recurrent neural networks (RNNs), convolutional neural networks (CNNs), and the transformer model in LOCA breach size diagnosis. After conducting rigorous ablation experiments, we have discovered that even rudimentary foundational models can achieve accuracy rates that exceed 90%. This is a significant improvement when compared to the previous models, which yield an accuracy rate of lower than 50%. The results interestingly demonstrate the superior performance and efficacy of the fundamental deep learning model, with an effective dataset construction approach. It elucidates the presence of a complex interplay among diagnostic scales, sliding window size, and sliding stride. Furthermore, our investigation reveals that the model attains its highest accuracy within the discussed range when utilizing a smaller sliding stride size and a longer sliding window length. This study could furnish valuable insights for constructing models for LOCA breach size estimation.

Keywords: deep learning; breach size estimation; lattice algorithm; fault diagnosis; loss-of-coolant accident (LOCA)

Citation: Xiao, X.; Qi, B.; Liang, J.; Tong, J.; Deng, Q.; Peng C. Enhancing LOCA Breach Size Diagnosis with Fundamental Deep Learning Models and Optimized Dataset Construction. *Energies* **2024**, *17*, 159. https://doi.org/10.3390/en17010159

Academic Editor: Roman Vadimovich Davydov

Received: 19 October 2023
Revised: 14 December 2023
Accepted: 16 December 2023
Published: 27 December 2023

1. Introduction

Nuclear power plants (NPPs) consist of multiple intricate, nonlinear, and dynamic systems. The availability of large amounts of information from operators, due to advances in digital technology [1], has made it challenging to swiftly diagnose fault information. Furthermore, research has established human error as the primary cause of accidents in NPPs [2], in particular, the intrinsic human factors of uncertainty [3] and the impact of human–computer interface design [4]. To ensure safe operations, minimize the impact of human factors during accidents, and mitigate economic losses, automating the precise and accurate recognition of fault information in NPPs is crucial.

A loss-of-coolant accident (LOCA) is a prevalent type of mishap in nuclear power plants, posing significant potential hazards. As shown in Figure 1, the LOCA originates from a leakage in the reactor system, which causes the coolant to escape and leaves the reactor vulnerable to high temperatures and subsequent damage [5]. The severity of the LOCA varies depending on the size of the breach, which requires customizing emergency measures accordingly. Consequently, estimating the breach size of the LOCA is a crucial step in accident response planning, as it provides essential information for determining appropriate emergency response actions. Neural network models have emerged as an effective and popular strategy for describing accident behavior [6]. Therefore, in this study, our aim is to thoroughly investigate the performance metrics of fundamental neural networks within the domain of deep learning. The names and abbreviations of the primary parameters in this article are shown in Table 1.

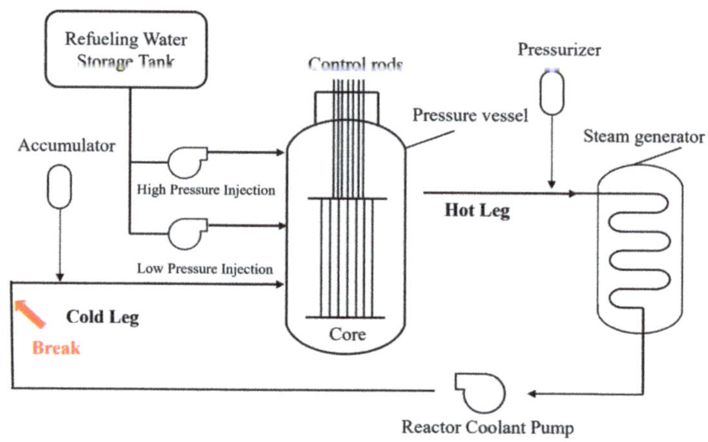

Figure 1. A schematic representation of a LOCA in a nuclear reactor.

Table 1. Notation and abbreviations.

Index	Parameter Name	Parameter Abbreviation
1	Loss-of-coolant accident	LOCA
2	Simplified multi-layer perceptron	Simplified MLP
3	Multi-layer perceptron	MLP
4	Long short-term memory	LSTM
5	Gated recurrent unit	GRU
6	Convolutional neural network	CNN
7	Deep learning	DL
8	Nuclear power plants	NPPs
9	Deep learning framework for LOCA breach size diagnosis based on the lattice algorithm	DeepLOCA-Lattice
10	Pressurized water reactor	PWR

The rest of the paper is organized as follows: Section 2 introduces the related work. Section 3 contains relevant concepts, and Section 4 details the DeepLOCA-Lattice adopted for LOCA breach size diagnosis. Section 5 presents scenario deduction and sensitivity analysis, and Section 6 provides the discussion and conclusions.

2. Related Work

With the progress of deep learning (DL), several hybrid models composed of multiple deep learning models for diagnosis have been successfully applied in the nuclear fault diagnosis field [7]. As shown in Table 2, She et al. combined CNN, LSTM, and convolutional LSTM (ConvLSTM) for the diagnosis and prediction of LOCAs, and this hybrid model has been proven to be functional, accurate, and divisible [8]. Meanwhile, Choi et al. used a cascaded fuzzy neural network (CFNN) to estimate the size of the LOCA breach [9]. Furthermore, Wang et al. proposed a PKT algorithm to extract more general fault information for NPP fault diagnosis and construct the coarse-to-fine knowledge structure intelligently [10], while Mandal et al. applied a deep belief network (DBN) to classify fault data in an NPP [11]. Yao et al. optimized convolutional neural networks with small-batch-size processing for assembly in the NPP diagnostic system [12], and Wang et al. presented a highly accurate and adaptable fault diagnosis technique based on the convolutional gated recurrent unit (CGRU) and improved particle swarm optimization (EPSO) [13]. Saghafi et al. defined a type of recurrent neural network model known as a nonlinear auto-regressive model with exogenous input (NARX) for LOCA breach size diagnosis [14]. However, there exists a significant complexity within the models used in prior research. The exploration of simpler, fundamental models for the accurate diagnosis of loss-of-coolant accident (LOCA) faults has not yet been undertaken.

In various domains, there is work leveraging deep learning for the purpose of assimilating data from monitoring points to forecast critical parameters. Xu et al. [15] presented, in the form of the transfer learning framework based on transformer (TL-Transformer), an accurate prediction of flooding in data-sparse basins. El-Shafeiy et al. [16] introduced and applied a pioneering technology, multivariate multiple convolutional networks with long short-term memory (MCN-LSTM), to real-time water quality monitoring. Liapis et al. [17] used deep learning in conjunction with models that extract emotion-related information from text to predict financial time series. Islam et al. [18] presented a combined architecture of a convolutional neural network (CNN) and recurrent neural network (RNN) to diagnose COVID-19 patients from chest X-rays.

Table 2. Overview of existing work.

Author	Main Work
She et al. [8]	Integrated CNN, LSTM, and ConvLSTM for diagnosing and predicting LOCAs. Demonstrated functionality, accuracy, and divisibility.
Choi et al. [9]	Employed a cascaded fuzzy neural network (CFNN) for estimating LOCA breach sizes.
Wang et al. [10]	Proposed a PKT algorithm for extracting more generalized fault information in NPP fault diagnosis and intelligent construction of a coarse-to-fine knowledge structure.
Mandal et al. [11]	Utilized a deep belief network (DBN) for classifying fault data in NPPs.
Yao et al. [12]	Optimized CNNs with small-batch-size processing for assembly in NPP diagnostic systems.
Wang et al. [13]	Developed a highly accurate and adaptable fault diagnosis technique using CGRU and improved particle swarm optimization (EPSO).
Saghafi et al. [14]	Defined a nonlinear auto-regressive model with exogenous input (NARX) for diagnosing LOCA breach sizes.

In addressing the critical challenge of LOCAs in nuclear power plants, our research marks a significant departure from traditional, complex neural network architectures towards a more streamlined methodology. Previous studies, such as those by She et al. [8] and Choi et al. [9], have predominantly employed intricate, composite models. While these models are effective, they often require substantial computational resources and a deep understanding of varied neural network configurations.

Our study, in contrast, explores the untapped capabilities of fundamental neural network models, illustrating that simplicity can coexist with high accuracy. We have methodically developed and applied a simplified neural network model that, despite its reduced complexity, achieves an accuracy rate over 90%. This not only questions the prevailing dependence on complex models in LOCA fault diagnosis but also offers a more accessible and efficient alternative.

This research fills a vital gap in the literature, demonstrating that a well-designed, simpler model can match or even outperform complex systems. This is of paramount importance in the high-stakes environment of nuclear power plant operations, where rapid and reliable diagnostics are crucial. Simplifying the model structure also increases the interpretability and user-friendliness, making it more viable for practical applications, especially in settings with limited resources and expertise.

Additionally, previous research has focused primarily on combining model algorithms, with little consideration given to the impact of dataset construction and data preprocessing on model performance. To avoid the issues mentioned above, we propose a framework named DeepLOCA-Lattice for integrated analysis of LOCA breach size diagnosis. This framework has the potential to expand the scope of fault severity diagnosis for other types of accidents in NPPs. Furthermore, note that there are few studies on the high-precision diagnosis of LOCA breach size. Only the cascaded fuzzy neural networks proposed by Choi et al. [9] in 2016 and the NARX neural network suggested by Saghafi et al. [14] in 2019 are known to the authors.

3. Methodology

3.1. Structure of Proposed DeepLOCA-Lattice Framework

As illustrated in the flow diagram in Figure 2, the proposed DeepLOCA-Lattice framework consists of four main stages: data collection and preprocessing, expanding the diagnostic scale, model building, and estimation of the size of LOCA breaches. In the data preprocessing stage, the lattice algorithm is used to segment the original data into smaller windows, allowing the capture of temporal dependencies between data samples and the extraction of meaningful features for classification. After data preprocessing, four classes are defined based on the size of the LOCA breach using an equal-interval partitioning approach for expanding the diagnostic scale. For stage III, we experimented with five different deep learning models for the estimation of the LOCA breach size and the detection of key factors for the estimation model, namely, the simplified MLP, MLP, LSTM, GRU, CNN, and transformer models. Lastly, in a nuclear power plant during normal operation, if a LOCA incident occurs, the parameters from various sensors are input into our model, enabling the specific LOCA breach sizes to be estimated.

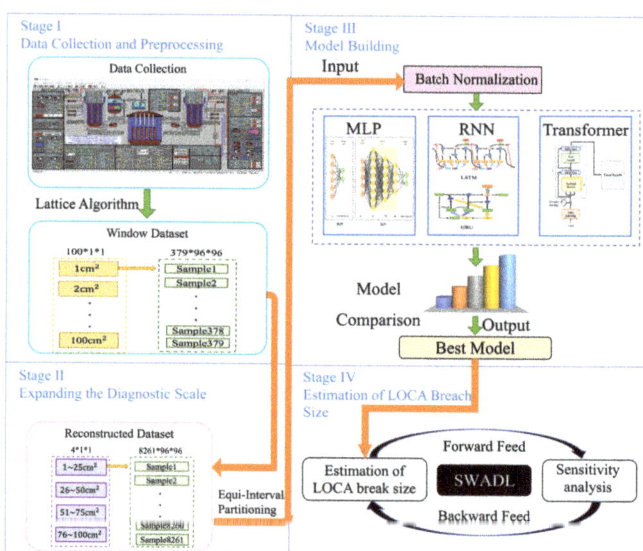

Figure 2. Structure of proposed DeepLOCA-Lattice framework. The proposed DeepLOCA-Lattice framework consists of four main stages: stage I: data collection and preprocessing, stage II: expanding the diagnostic scale, stage III: model building, and stage IV: estimation of the LOCA breach size.

3.2. Lattice Algorithm

The lattice algorithm is also known as the sliding window method. The concept of the algorithm was proposed by Rubinger in 1974 for the first time [19]. Since it has the advantages of being simplified, easy, operable, practical, and accurate to implement, it has been successfully applied in many fields, such as speech recognition [20], data processing [21], and anomaly detection [22]. Figure 3 shows a simplified diagram of the lattice algorithm. s and l are the step and window sizes, respectively. The window pane of length l slides s to the next window.

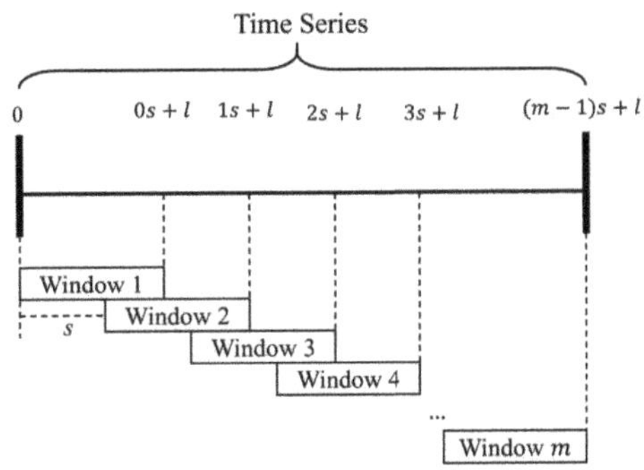

Figure 3. A diagram of the lattice algorithm.

3.3. DL Architectures

With the advent of big data, deep learning technology has become an important research hotspot as a perfect performing solution and has been widely applied in image processing [23], natural language processing [24], speech recognition, large language modeling, online advertising, and so on [25]. DL architectures can be divided into supervised learning and unsupervised learning [26]. In supervised learning, the training data are composed of labeled examples, whereas unsupervised learning relies on unlabeled data. In this study, the available data are labeled.

Deep learning was proposed by [25] in 2015. The training process of deep neural networks can be dissected into a series of core steps, each contributing indispensably to the network's overall efficacy. The first step in building a successful neural network is to collect, clean, and label the data. The second step is network construction. In this study, we chose the MLP [27], RNNs, CNN, and transformer [28] as fundamental DL architectures to experiment with. The GRU model [29] and LSTM [30] are two popular types of RNN architectures that have been widely used in various applications. Subsequently, the training data were input into the network, and the output was calculated using forward propagation. Afterward, the loss function was calculated to measure the difference between the predicted output and the true output. Then, the gradient of each weight and bias was calculated using the backpropagation algorithm based on the value of the loss function. In the next step, an optimization algorithm was used to update the weights and biases of the network based on the gradient direction. Then, the model was validated and adjusted. Lastly, the trained network was tested using the testing set to assess its performance and accuracy.

4. DeepLOCA-Lattice Construction

4.1. Data Description and Preprocessing

The reactor type selected for this research was the PWR, which is the most popular in the world. The dataset used in this study was obtained from our previous work [31] and was generated from the PCTRAN v1.0 software. PCTRAN, a reactor transient and accident simulation software, can be operated on a personal computer. Leveraging a diverse array of mathematical models, PCTRAN facilitates the simulation of nuclear reactor cores and the reactor coolant system during transients and accidents. Its effectiveness has been validated through Fukushima nuclear accident simulations and two operating conditions in our previous work [31], which presents a first-of-its-kind open dataset created using PCTRAN.

PCTRAN is capable of generating an NPP accident dataset based on the type of accident, the severity of the accident, and the simulation time [31]. The generated data are informative, containing 97 parameters such as time stamp, temperature, flow speed, and power. For the LOCA, the dataset comprises various sizes of fault data ranging from 1% to 100% of a 100 cm^2 breach in intervals of 1% of 100 cm^2. In our previous work, the simulation time was set to longer than 3600 s. The time-step value for the simulation is automatically generated by the PCTRAN software, with a constant interval of 10 s. Following the implementation of equal-interval processing, the sample count for each category experienced a substantial increase from 379, as observed in the original class division, to 8261. In this paper, the structure of the generated data for the LOCA (i.e., 100% of 100 cm^2) is depicted in Table 3. Each parameter within the dataset carries distinct physical interpretations. For instance, in Table 3, TAVG corresponds to the temp RCS average, while RRCO signifies the ratio core flow.

In this study, data preprocessing involved two aspects: data standardization and the utilization of the sliding window algorithm to construct the dataset. The PCTRAN simulation provides samples of full-process accident data, while real-world samples are often taken from a specific period. Thus, after employing the PCTRAN simulation data, to align with practical scenarios, we applied the lattice algorithm to partition the dataset along the temporal dimension for data preprocessing, which is more suitable for the requirements of NPPs. The lattice algorithm was used to segment the original data into smaller windows.

Table 3. The time series of the status parameters with a 1% of 100 cm^2 break of LOCA.

Time	P	TAVG	THA	...	RRCO	WFLB
0	155.5000	310.0000	327.8240	...	1.0000	0
10	155.4682	309.9801	327.8055	...	1.0000	0
20	155.4674	309.9777	327.8105	...	1.0000	0
30	155.4719	309.9802	327.8112	...	1.0000	0
40	155.4711	309.9779	327.8114	...	1.0000	0
...	...	...	...	...	...	...
...	...	...	...	...	...	...
...	...	...	...	...	...	...
4710	157.0756	292.0552	292.2995	...	1.0621	0
4720	157.0776	292.0528	292.2970	...	1.0621	0
4730	157.1084	292.0511	292.2944	...	1.0621	0
4740	157.0882	292.0486	292.2919	...	1.0621	0

4.2. Equal-Interval Partitioning and Data Reconstruction

As described before, the number of breach sizes in the original data is 100. This study provides a method to estimate the fault breach size. However, if the number of categories is 100, a complex, easily over-fitted, time-consuming neural network model will be built [32]. Conversely, a lower number of intervals can more effectively showcase the characteristics and differences of the dataset while mitigating the risk of excessive detail and confusion. Additionally, fewer intervals may enhance the accuracy of subsequent calculations. Thus, this paper proposes dividing the degree of the fault into equally spaced intervals for optimal results.

Given that four is a commonly employed partitioning number [33], it strikes an optimal balance between simplicity and complexity while accurately reflecting data distribution. Therefore, this study divided the PCTRAN simulation data into four intervals with certain representativeness for data description and analysis. Our data were separated into four categories for the subsequent estimation task evenly, where sizes ranging from 1 to 25 cm^2, 26 to 50 cm^2, 51 to 75 cm^2, and 75 to 100 cm^2 were each assigned one level.

4.3. Estimation of LOCA Breach Size

To estimate the breach size of the LOCA, DeepLOCA-Lattice designed and constructed a range of popular fundamental deep learning models for testing, including the MLP, RNNs, CNN, and transformer. To assess the adaptability of the simplified neural network model, we deliberately crafted a single-hidden-layer MLP without activation functions. In this article, it is referred to as the simplified MLP.

4.3.1. Multi-Layer Perceptron (MLP)

The MLP, also known as the artificial neural network (ANN), was first proposed by Rosenblatt in 1957 [27]. Subsequently, JL McClell and DE Rumelhart proposed the back propagation (BP) algorithm based on this model, which made the neural network algorithm applicable to nonlinear problems [34]. Ackley et al. introduced the concept of hidden layers and proposed the Boltzmann machine, which led to the idea of a multi-layer perceptron consisting of input, hidden, and output layers [35].

Since the MLP model serves as the fundamental model in deep learning networks, this paper presents the design and construction of both a simplified MLP model and an MLP model. The simplified MLP has emerged as a refined iteration of the traditional MLP architecture, characterized by a solitary hidden layer and a deliberate absence of activation functions. The motivation for developing a simplified model is to explore the adequacy of a straightforward approach in addressing the fault diagnosis of LOCAs. The specific architecture of the simplified MLP model is shown in Figure 4.

The MLP model comprises an input layer, three hidden layers, and an output layer. The input layer has a total of 96 neurons. The number of neurons in each hidden layer is set to 512, 256, and 64, sequentially, while the number of neurons in the output layer remains as 4.

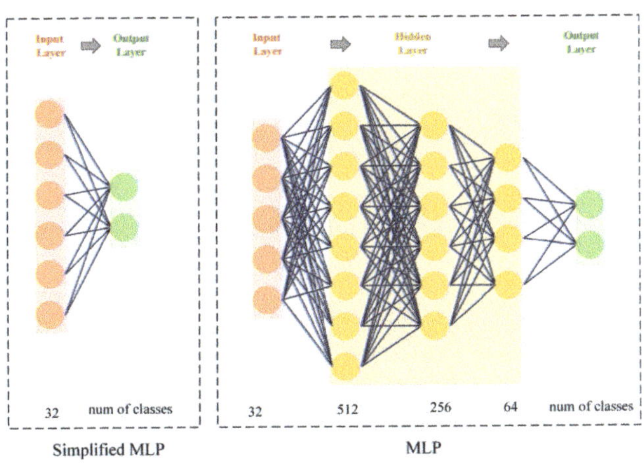

Figure 4. Simplified MLP and MLP diagrams of DeepLOCA-Lattice framework.

4.3.2. Recurrent Neural Networks (RNNs)

In the processing of temporal data using traditional neural networks, the correlation between temporal data is often ignored. To address this issue, RNNs are proposed and have been widely applied in various fields such as natural language processing (NLP), machine translation, speech recognition, and fault diagnosis [36]. However, because of the problems of gradient explosion and vanishing gradients, RNNs are unable to handle long-term sequences effectively. As a result, researchers introduced improved RNN algorithms, namely, long short-term memory networks (LSTMs) and gated recurrent units (GRUs), which have successfully overcome these issues.

In this study, the dataset consists of time-series data. Therefore, we utilized the RNN model to uncover patterns in predicting and diagnosing the size of the LOCA breach. The aim of this study was to explore the effectiveness of enhanced RNN models, which are LSTM and GRU models, in the task of estimating the LOCA breach size. Figure 5 illustrates the LSTM and GRU network architectures designed in this study. The constructed RNN network architecture includes bidirectional computations, dropout regularization, and batch normalization operations. Specifically, this paper defines two layers of a recurrent neural network with 300 neurons per layer. For example, when using the dataset with four categories and a window pane of 96 adopted in DeepLOCA-Lattice and setting the batch size to 32, the input data shape fed into the RNN network is 32 × 96 × 96 and the output data shape is 32 × 96 × 300. The shape of the hidden state data for the final time step of each layer of the network is 32 × 300. In this study, the hidden state of the final time step of the second layer, i.e., h_N^2, was selected as input for the next stage to predict the classification task. Finally, h_N^2 is passed through a batch normalization operation and a linear classifier to output the predicted classification results.

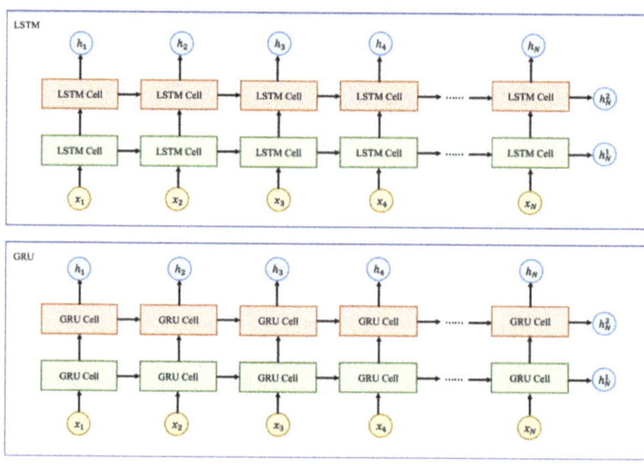

Figure 5. LSTM and GRU of DeepLOCA-Lattice framework.

4.3.3. Convolutional Neural Networks (CNNs)

Convolutional neural networks (CNNs), often referred to as ConvNets or convolutional networks, represent a class of deep learning models specially designed for tasks involving grid-structured data, such as images [37], video frames [38], and spatially organized data. As shown in Figure 6, our CNN architecture incorporates convolutional and fully connected layers with appropriate activation functions for feature extraction and classification. It consists of two convolutional layers with ReLU activation. After the convolutional layers, there are two fully connected layers, with ReLU activation in the first and linear transformation in the second. The final layer produces the network output logit without an activation function.

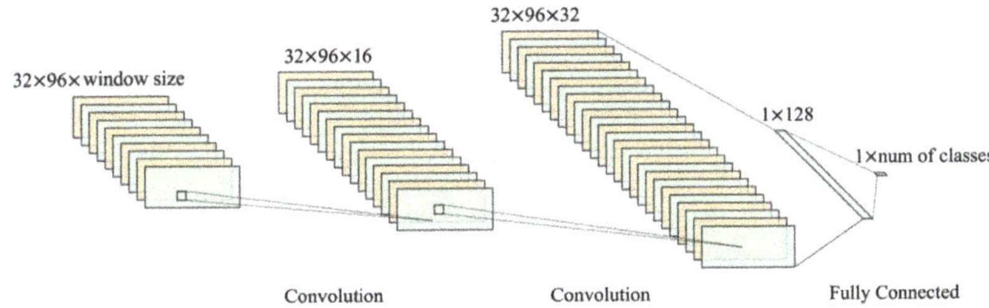

Figure 6. CNN of DeepLOCA-Lattice framework.

4.3.4. Transformer Model

The transformer model [28], consisting of an encoder and decoder, offers the key advantage of parallel computation. Our DeepLOCA-Lattice-based implementation incorporates a positional encoding layer, two transformer encoder layers, and a linear decoder, as shown in Figure 7. The positional encoding function generates position embedding vectors using the sin–cos formula, and the network weights are initialized before applying transformations to the input data. This process converts the input tensor size from 32 × 96 × 96 to 96 × 32 × 96 while normalizing it. The positional encoding vector is then added to the input data, resulting in a new input vector. The input is then encoded using the transformer encoder based on the set-masked input tensor mask that conceals information from certain positions in the input tensor. Finally, the output of the encoder is decoded linearly to obtain the prediction results.

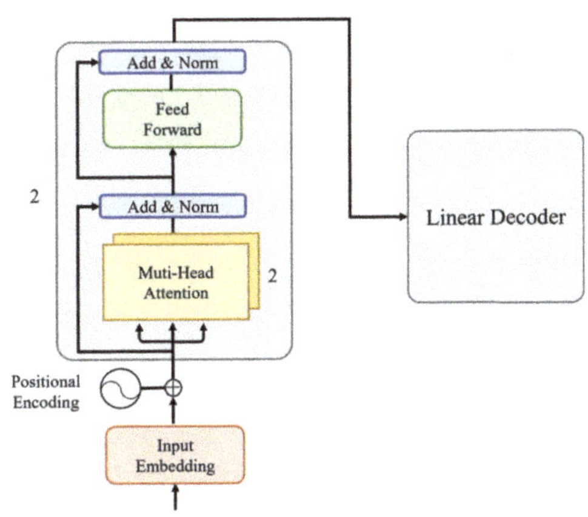

Figure 7. Transformer of DeepLOCA-Lattice framework.

5. Results and Discussion

The subsequent sections will present scenario predictions and sensitivity analysis of data construction in DeepLOCA-Lattice. Furthermore, we also compared the precision of the models proposed by Choi et al. [9] and Saghafi et al. [14] with the fundamental and simple model in our framework. Choi et al. designed a cascaded fuzzy neural network (CFNN) model, which is a computational framework designed for modeling complex systems using fuzzy logic. It is a T-S FIS model with a specific structure and utilizes genetic algorithm optimization to train and fine-tune the model's parameters. Saghafi et al. defined a type of recurrent neural network model known as a nonlinear auto-regressive model with exogenous input (NARX). The NARX has an input layer; three hidden fully connected (FC) layers with dimensions of 10, 20, and 10 neurons; and an output layer. Our experiments revealed a particularly intriguing phenomenon. And it is worth noting that our study design incorporates a robust validation strategy, consisting of distinct training, validation, and testing datasets, to ensure an unbiased evaluation of model performance.

5.1. Scenario Deduction

LOCAs are a significant safety concern in NPPs. The size of the breach determines the severity of the accident, and our research provides immediate information on the breach size to help operators minimize consequential damage caused by the LOCA. These findings are expected to contribute significantly to improving the safety standards of NPPs.

In this study, we have employed both grid search and random search techniques for hyperparameter tuning to ensure optimal performance of our deep learning models, as detailed in the strategies implemented. Table 4 presents the hyperparameters of the simplified MLP, MLP, LSTM, GRU, CNN, and transformer models used in DeepLOCA-Lattice. To ensure model convergence, it is crucial to set an appropriate learning rate. A high learning rate can result in gradient explosion, while a low learning rate can lead to overfitting and trap the model in a 'local optimum'. After conducting multiple experiments, we set the initial learning rate to 0.0001 in this study. The preliminary experiment indicated that the model converges in around 100 iterations. Therefore, we observed convergence progress for 250 iterations. A batch size of 32 was used to optimize the hardware computing resources. Additionally, the dropout rate for the simplified MLP, MLP, and CNN was set to 0. Our optimization process for the LSTM, the GRU model, and transformer models revealed their susceptibility to overfitting. To mitigate this issue and enhance the performance of the model, we introduced a random dropout rate of 0.1 for the three models.

Table 4. Neural network hyperparameter optimization.

Hyperparameter	Number
Learning rate	0.0001
Number of iterations	250
Batch size	32

This study analyzed a lattice algorithm with a window size of 96 and a sliding stride of 1. Figure 8 presents a comparative visualization of different neural network architectures in terms of their computational complexity and performance. The x-axis represents the floating-point operations per second (FLOPs), a measure of computational complexity, indicating the number of floating-point operations executed per second by the model. The y-axis displays the logarithmically scaled number of parameters (params) in each model, which is a common proxy for model size or capacity. Each dot represents a distinct neural network architecture, including simplified MLP, MLP, LSTM, GRU, CNN, and transformer. And the size of each dot correlates with the accuracy percentage of the respective model, as denoted by the color bar on the right side of the graph.

Models positioned towards the bottom left corner, such as the transformer, indicate lower computational complexity and a smaller number of parameters, while those towards the top right indicate higher computational complexity and a higher number of parameters. The color gradient from blue to red illustrates a scale from lower to higher accuracy, suggesting that models with more FLOPs and params tend to achieve higher accuracy, with the GRU model marked by the largest and darkest dot reflecting the highest accuracy among the compared models. Our results show that all models achieved high accuracy rates, exceeding 90%. Among these models, the GRU model achieved the highest accuracy rate of 99.90%, followed by the transformer model with an accuracy of 99.69%. The MLP, LSTM, simplified MLP, and CNN achieved accuracies of 98.64%, 96.87%, 93.41%, and 92.28%, respectively.

To visualize the neural network's classification process, a random sample from the 76–100 cm^2 LOCA break size area is selected. Figure 9 illustrates this using a multi-layer MLP, presented as a probabilistic heat map across the four layers. The axes represent the transformation of a one-dimensional vector into a two-dimensional space, indicating positional data rather than physical dimensions. The color intensity correlates with higher probability values.

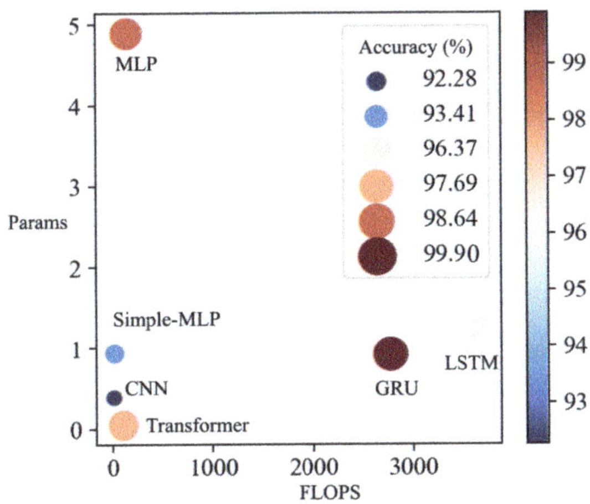

Figure 8. Results of DeepLOCA-Lattice framework.

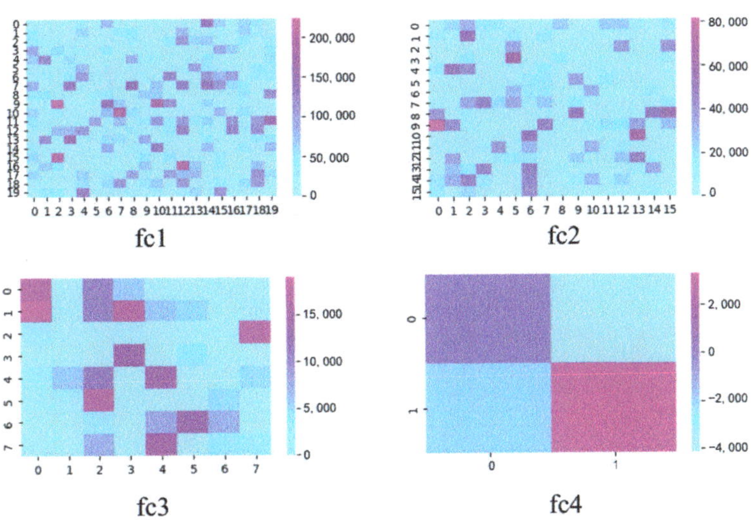

Figure 9. Heat map of LOCA break size (78 cm^2) during four-layer MLP processing. In the fc4 results, the coordinate (0,0) represents a LOCA break size area of 1–25 cm^2, (1,0) indicates 26–50 cm^2, (0,1) signifies 51–75 cm^2, and (1,1) denotes 76–100 cm^2.

The data processing in the MLP starts with the first fully connected layer (fc1), enhancing the dimensionality to 512. Subsequently, it traverses the second layer (fc2), reducing the dimensionality to 256, and then the third layer (fc3), further decreasing to 128 dimensions. Considering our four-class classification task, the final fourth layer (fc4) output is a four-dimensional vector. The map shows the highest probability value at position (1,1), accurately classifying the break size in the 76–100 cm^2 range, aligning with actual conditions.

Additionally, we have replicated the NARX of Saghafi et al. and the CFNN of Choi et al. Saghafi et al. that primarily revolve around nonlinear programming, a domain that deviates from our research emphasis on neural networks, and the result of the NARX was 36.46%. We also found that the execution time of the NARX was considerably prolonged. The NARX would require approximately 7 h on a GeForce RTX 3080 Ti GPU, whereas our model typically completes within a range of 1 to 3 h. Otherwise, the optimal accuracy of Choi's CFNN is merely 35.48%, which is less than half of the accuracy rate achieved by the simplified MLP in our proposed DeepLOCA-Lattice. Therefore, we draw an intriguing and substantial inference. That is, the complexity of the model structure does not invariably correlate with an improvement in its accuracy, which means that even a simple single model can effectively perform the fault diagnosis significantly.

As shown in Table 5, in terms of the FLOPs-to-params (FLOPs/params) ratio, the simplified MLP, MLP, and NARX exhibit proximity, both hovering around 32, indicating a balance achieved between computational efficiency and model capacity. On the contrary, LSTM and the GRU model demonstrate significantly higher FLOPs-to-params ratios compared to the other models, signifying their elevated computational complexity and potentially necessitating greater computational resources. On the other hand, the CNN exhibits a relatively high FLOPs-to-params ratio, indicating its outstanding computational efficiency. Similarly, the transformer boasts a comparably high FLOPs-to-params ratio, suggesting an equilibrium attained between computational efficiency and model capacity. As for CFNN, it is incapable of calculating FLOPs. The estimation of FLOPs (floating-point operations) necessitates an enumeration of the floating-point arithmetic operations per computational act. Within a CFNN, the operations are restricted to matrix manipulations and fuzzy logic computations. Fuzzy logic operations diverge from the standard linear algebraic operations and do not subscribe to a fixed FLOP count. These findings suggest that networks of the same type display consistency between their number of parameters and floating-point numbers relative to each other. These observations provide insight into the resource requirements and performance aspects of various deep learning models used in LOCA breach size diagnosis.

Table 5. Model metrics comparison.

Model	FLOPs	Params (MB)	FLOPs/Params	Accuracy (%)
Simplified MLP	30.6708	0.9401	32.65	93.41
MLP	157.2045	4.8935	32.11	98.64
LSTM	3702.402	1.2202	3031.23	96.37
GRU	2778.9588	0.9202	3019.45	99.90
CNN	18.9153	0.3929	48.18	92.28
Transformer	120.3241	0.0568	2119.29	97.69
NARX	0.0461	0.0015	31.05	36.46
CFNN	/	0.0066	/	35.48

The superior performance of the GRU and transformer models can be attributed to their stronger modeling and memory capabilities, allowing them to handle more complex data relationships and sequences. Surprisingly, the relatively simplified MLP and MLP models also performed well in this fault diagnosis task, suggesting the presence of smaller nonlinear relationships among different LOCA breach sizes. As for RNNs, LSTM is more prone to gradient vanishing and exploding problems due to the introduction of three gate units (input gate, output gate, and forget gate) to filter and process the input data, making training difficult and increasing the likelihood of encountering gradient issues. In contrast, the GRU model only uses two gate units (update gate and reset gate), reducing computational complexity, improving training efficiency, and minimizing the risk of gradient vanishing and exploding. Moreover, the GRU model is better suited for processing small batches of data compared to LSTM. These findings have significant implications for selecting appropriate deep learning models for breach size diagnosis in LOCAs.

Notably, the accuracy of the model is ascertained not from the training or test sets but rather from a separate validation set. This method ensures a more unbiased evaluation of the model's performance on unseen data, thus providing a more realistic gauge of its generalization capabilities. Furthermore, in subsequent sections, we delve deeper by conducting sensitivity analysis, altering various sample parameters within the data. Such an analysis is instrumental in understanding how different factors influence the model's performance and in identifying potential areas of improvement or vulnerabilities in the modeling process.

5.2. Sensitivity Analysis

In this study, we conducted an in-depth discussion and analysis of the impact of key data construction parameters on model performance. These parameters included various window sizes, sliding stride sizes, and equal-interval partitions. The goal is to identify critical parameters in the dataset that significantly affect model performance. Our findings provide insight into the optimal configurations of the dataset for the DL models used in the diagnosis of LOCA breach size and have important implications for enhancing safety standards in NPPs.

5.2.1. Window Size and Sliding Stride

This paper presents a comprehensive analysis that compares different window sizes and sliding strides using the simplified MLP, MLP, LSTM, GRU, CNN, and transformer models. The objective is to identify the optimal window parameters for computing performance. Specifically, we evaluated window sizes of 10 and 96 with stride values of 1 and 5, as well as data without lattice operations. Table 6 provides detailed parameter selections for each dataset.

Table 6. Various window sizes and stride values.

Index	Window	Stride
1	10	1
2	10	5
3	96	1
4	96	5
5	none	

Our analysis reveals that when using a window size of 96 and a stride of 1, all models demonstrate optimal performance in terms of accuracy. This is because this configuration enables a more comprehensive and complete representation of sequence data by leveraging temporal information to a greater extent. As a result, this configuration improves both the accuracy and generalizability of the models, making it ideal for LOCA breach size diagnosis tasks in NPPs.

Figures 10 and 11 provide a comprehensive analysis of the model accuracy across different sliding stride and window sizes. In these figures, 'c' denotes the number of categories, 'c = 4' represents the model accuracy after reconstructing category divisions, and 'c = 100' corresponds to the model accuracy without any category division.

As illustrated in Figure 10, in addition to the CNN, when the sliding stride is held constant, there exists a positive correlation between the model accuracy and window size. Irrespective of whether the number of categories is 4 or 100, increasing the window size improves the accuracy of all models under the stride = 5 and stride = 1 conditions. Although increasing the window size results in a reduction in the number of samples, it concurrently amplifies the continuity of information within each sample. Consequently, for the diagnosis of LOCA breach size in NPPs, stronger continuity in individual sample sequences and higher information entropy significantly contribute to the model's enhanced reasoning capabilities. However, for the CNN model, in the case where the stride is 5 and the number of categories is 4, the model accuracy decreases as the window size increases. This also indicates that larger window sizes do not necessarily lead to higher model accuracy, and this is also dependent on the model characteristics and the number of categories.

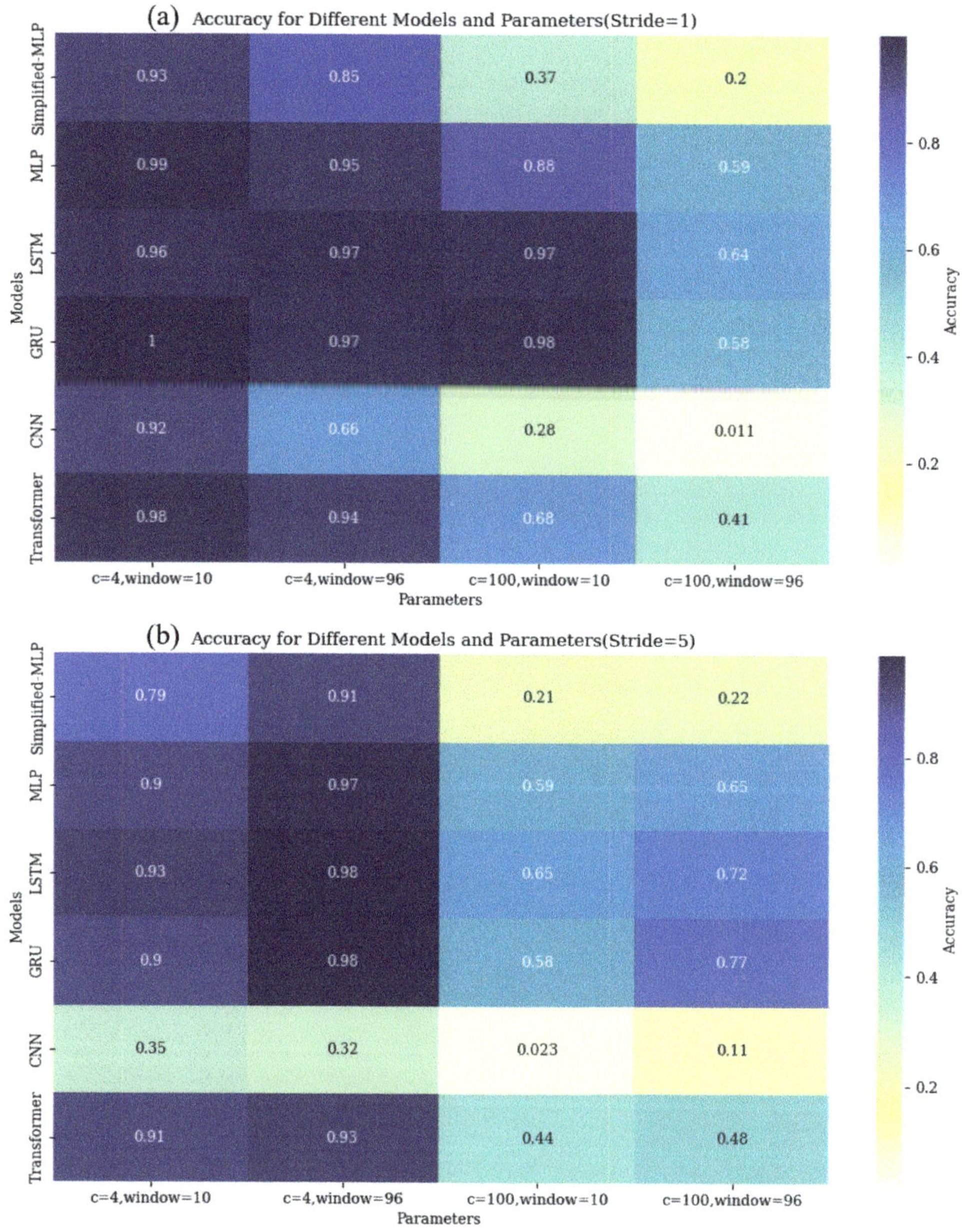

Figure 10. The accuracy of different models for data of various categories and window sizes: (**a**) The distribution of model accuracy for different categories and different window sizes when the sliding stride is fixed at 1. (**b**) The distribution of model accuracy for different categories and different window sizes when the sliding stride is fixed at 5.

When considering a fixed window size, as depicted in Figure 11, the model accuracy demonstrates a negative correlation with the sliding stride for a data category of 4. Decreasing the sliding stride size leads to an increase in the number of samples, thus contributing to improved model reasoning. However, for a data category of 100, with a sliding window size of 10, reducing the sliding stride size from 5 to 1 results in a slight decrease in model accuracy. Conversely, when the sliding window size is set to 96, decreasing the stride from 5 to 1 significantly enhances the model's accuracy. These findings imply that in scenarios where sample information is limited, simply increasing the number of samples may not be as effective as enriching the information content within individual samples.

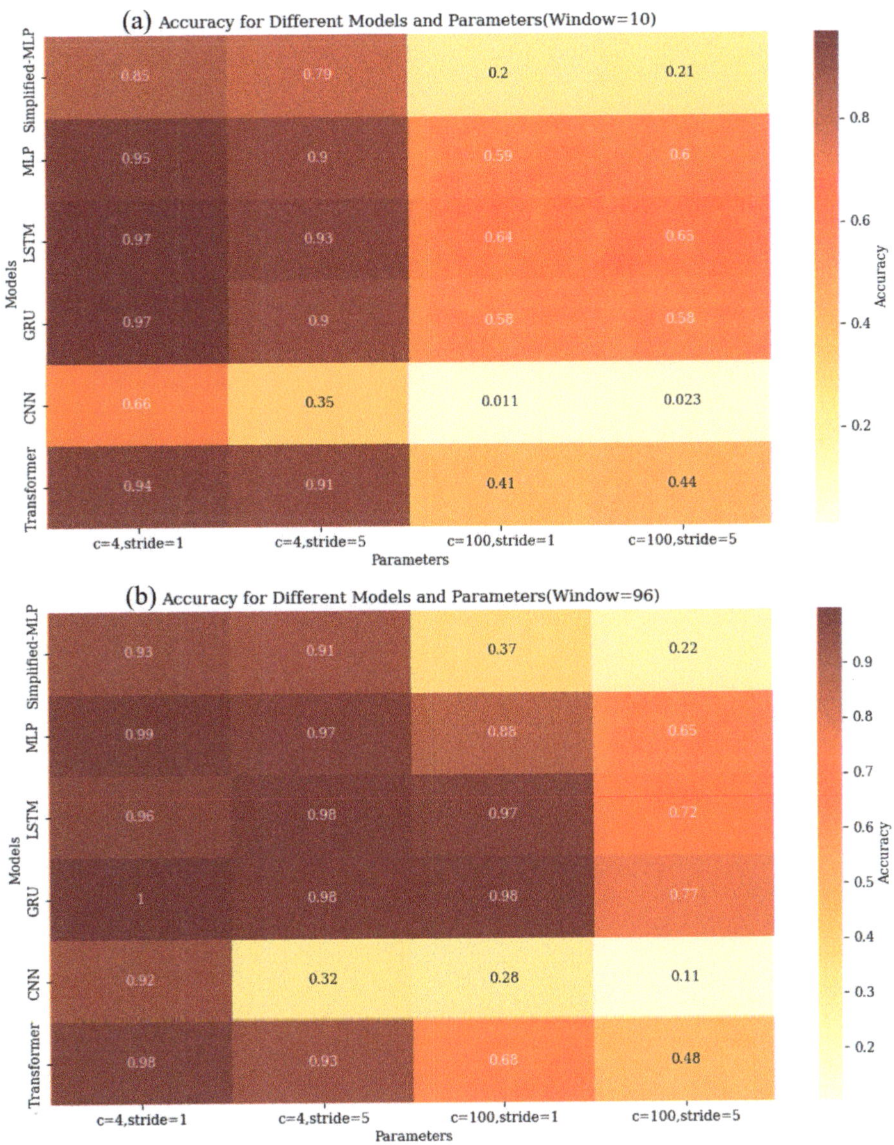

Figure 11. The accuracy of different models for data of various categories and sliding strides: (**a**) The distribution of model accuracy for different categories and different window strides when the window size is fixed at 10. (**b**) The distribution of model accuracy for different categories and different sliding strides when the window size is fixed at 96.

It is well known that model precision demonstrates a positive correlation with window size and a negative correlation with sliding stride size [39]. However, in the context of diagnosing the size of the LOCA breach in NPPs, there exists a nuanced relationship among the number of window sizes, the number of categories, and the sliding stride through our experiments. The relationship between them is not simply a straightforward positive or negative correlation.

5.2.2. Diagnostic Scales

In this study, we investigated the impact of size-equivalent partitioning on the entire model framework by applying both graded and non-graded partitioning methods to scale the data. The data are divided into two scales, which are the graded 4-category scale and the non-graded 100-category scale.

Table 7 presents the average improvement factors for the simplified MLP, MLP, LSTM, GRU, CNN, and transformer models on different diagnostic scales. The analysis reveals that the diagnosis accuracy of the graded scale is higher than the non-graded. This indicates that there is an enhancement in model accuracy after size-equivalent partitioning. The results demonstrate that the CNN model shows the highest improvement factor, reaching 19.92, followed by the simplified MLP, which achieves 3.63. The third position is held by the transformer model, with an improvement factor of 1.94. Lastly, the MLP, LSTM, and GRU models are quite similar, all hovering around 1.4 in terms of the improvement factor.

Table 7. The accuracy distribution of different models across varying numbers of categories.

Model	Number of Categories	Stride = 5 Window = 10	Stride = 5 Window = 96	Stride = 1 Window = 96	Stride = 1 Window = 10	Average Factor
Simplified MLP	4	79.08%	91.23%	93.41%	84.59%	3.63
	100	21.33%	22.09%	36.90%	20.38%	
MLP	4	90.46%	96.63%	98.64%	95.17%	1.46
	100	59.70%	64.92%	87.67%	58.64%	
LSTM	4	92.81%	98.48%	96.37%	96.95%	1.33
	100	64.83%	72.01%	96.70%	64.34%	
GRU	4	89.90%	98.48%	99.90%	96.63%	1.38
	100	58.21%	76.56%	97.86%	57.98%	
CNN	4	34.55%	31.98%	92.28%	66.04%	19.92
	100	2.31%	10.56%	28.22%	1.13%	
Transformer	4	90.97%	92.55%	97.69%	93.79%	1.94
	100	43.66%	47.55%	67.66%	40.59%	

The significant improvement in the accuracy of the CNN is attributable to the advantageous structure of the convolutional and pooling layers, facilitating feature extraction. The structure of the simplified MLP may lead to lower computation, especially when dealing with data with simpler structures, hence achieving a higher improvement factor. The transformer and MLP models follow closely. Their complexity and larger number of parameters may impose limitations on their improvement factor. Finally, the LSTM and the GRU model exhibit similar improvement factors. Their memory capabilities enable them to effectively capture long-term dependencies in time-series data. As a result, the potential for accuracy improvement is smaller for the LSTM and GRU models.

In the contemporary landscape of statistical analysis within the realm of computational sciences, the Friedman test has emerged as a robust non-parametric alternative to the one-way ANOVA with repeated measures. It is particularly adept at discerning significant discrepancies across multiple treatment conditions when the dependent variable being measured is ordinal. The test is predicated on the ranks of data rather than their raw values, thus it is intrinsically immune to the parametric assumptions of normality and homogeneity of variances that often constrain the applicability of parametric tests.

The Friedman test is most efficacious when employed to evaluate the performance across a suite of computational models or algorithms that have been subjected to a series of experimental conditions. As shown in Table 8, upon the application of the Friedman test, comprising the performance results of multiple models, the derived test statistic stands at 33.5663. This statistic, which follows a chi-squared distribution under the null hypothesis, is a measure of the overall divergence among the model ranks. Accompanying this statistic, the p-value obtained is 3.1048×10^{-6}. This is significantly below the conventional alpha threshold of 0.05, unequivocally indicating that we can reject the null hypothesis with a high degree of confidence.

Table 8. The results of the Friedman test.

Friedman Test Statistic	p-Value
33.5663	3.1048×10^{-6}

In conclusion, the results of the Friedman test are indicative of statistically significant differences in the performance of the evaluated models. Furthermore, we use the post hoc Nemenyi test to assess the differences between groups. As shown in Figure 12, there is a matrix of p-values indicating the statistical significance of differences between pairs of models. The colors range from dark to light, with darker shades representing lower p-values, and thus, more significant differences between model performances. The color bar on the right-hand side indicates the scale of significance levels. It can be interpreted from the heatmap that the comparison between the GRU and CNN models results in a very low p-value (0.001), shown by the dark color, suggesting a statistically significant difference in their performance rankings. Conversely, the comparison between the LSTM and GRU models, with a p-value of 0.9, is represented by a much lighter color, indicating no significant difference between these two models' rankings.

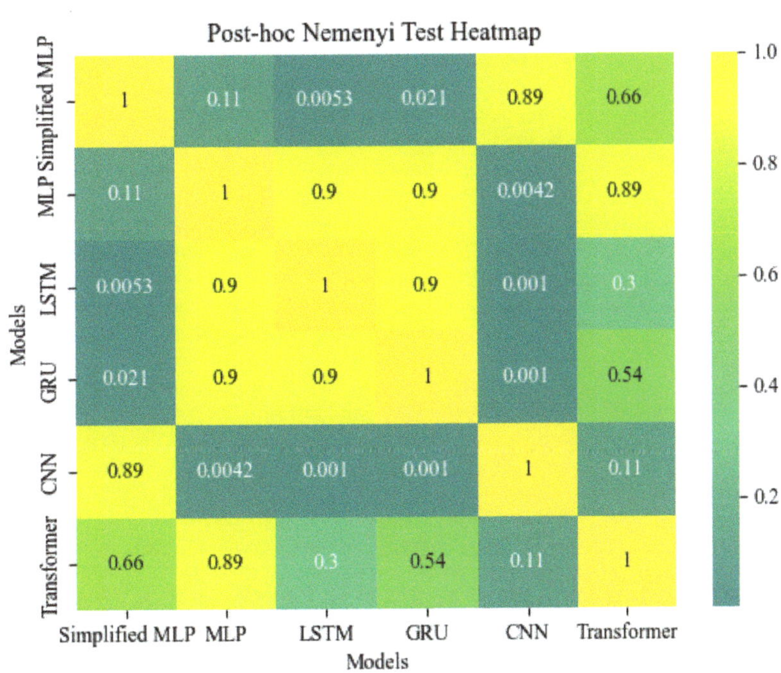

Figure 12. Friedman test results highlighting significant differences in model performance.

6. Conclusions

In NPPs, the early and accurate diagnosis of LOCA breach size is crucial for effective emergency response. This study proposes an effective LOCA breach size diagnosis framework (DeepLOCA-Lattice) to experiment with different data construction approaches and fundamental deep learning models. The main conclusions are as follows:

- The complexity of a model does not necessarily equate to its performance. In this study, even the simplest deep learning models can achieve accuracy rates that exceed 90% in LOCA breach size diagnoses, while the accuracy of the complex CFNN and NARX models is less than 40%. On the other hand, the high accuracy of 90% also underscores the idealized nature of the PCTRAN simulated data, emphasizing the necessity of considering the disparity between simulated and real data in genuine research endeavors.
- The findings reveal the existence of an intricate relationship among diagnostic scales, sliding window size, and sliding stride. It is not the case that larger sliding windows and smaller stride lengths consistently yield higher model accuracy. Specific outcomes are also influenced by factors such as the number of categories and the precise architecture of the model. For instance, as discussed in Section 5.2.1, in the scenario where the number of categories is four and the stride is 5, increasing the window size results in a decrease in the model accuracy. In contrast, with a window size of 10 and 100 categories, reducing the stride leads to an increase in the accuracy of the model.
- Our analysis reveals that when using a window size of 96 and a stride of 1, all models demonstrated optimal performance in terms of accuracy. This can serve as a reference for the construction of datasets for subsequent LOCA breach size estimation models. Researchers can attempt to use smaller window sizes with larger stride sizes for LOCA breach size diagnosis.

Our proposed DeepLOCA-Lattice framework exhibits versatile applicability across multiple domains. Specifically, it thrives in environments characterized by high dimensionality, such as applications producing voluminous data akin to the sensors in NPPs. Its robust design excels in scenarios necessitating critical fault detection, where timely and precise diagnoses are paramount to averting catastrophic outcomes or substantial economic repercussions. Additionally, the technology within our architecture renders it adept at handling temporal data sequences, making it particularly apt for time-series data or sequential datasets. Furthermore, the framework is designed for conducting in-depth ablation studies, catering to domains where dissecting the influence of individual components or parameters is imperative for further optimization and refinement.

These results have significant implications for the improvement of safety standards in NPPs and contribute to the development of more advanced and reliable fault diagnosis methods in nuclear energy systems. While the DeepLOCA-Lattice framework exhibits commendable performance with simulated datasets, its applicability to real-world scenarios from nuclear power plants (NPPs) remains a subject of inquiry. The inherent variability and noise prevalent in data sourced from NPPs could substantially influence the model's accuracy and reliability, signaling a limitation in the study due to the absence of extensive real-world validation and testing. Given that our data are derived from simulation models, future researchers might consider introducing noise or leveraging data trend variations in order to align the diagnostic process more closely with real-world scenarios.

Author Contributions: Writing—original draft preparation and formal analysis, X.X.; investigation, methodology, and visualization, X.X. and P.C.; data curation and conceptualization, B.Q.; software, X.X., B.Q., and P.C.; validation, X.X., B.Q., J.L., Q.D. and J.T.; writing—review and editing, project administration, supervision, and resources, J.L., Q.D. and J.T.; funding acquisition, J.L. All authors have read and agreed to the published version of the manuscript.

Funding: The research was supported by the Innovation Funds of CNNC–Tsinghua Joint Center for Nuclear Energy R&D (Project No. 20202009032) and a grant from the National Natural Science Foundation of China (Grant No. T2192933).

Data Availability Statement: The data that support the findings of this study are available upon request from the corresponding author.

Conflicts of Interest: The authors declare no conflict of interest.

References

1. Qi, B.; Zhang, L.; Liang, J.; Tong, J. Combinatorial techniques for fault diagnosis in nuclear power plants based on Bayesian neural network and simplified Bayesian network-artificial neural network. *Front. Energy Res.* **2022**, *10*, 920194. [CrossRef]
2. Lee, S.; Kim, J.; Arigi, A.M.; Kim, J. Identification of Contributing Factors to Organizational Resilience in the Emergency Response Organization for Nuclear Power Plants. *Energies* **2022**, *15*, 7732. [CrossRef]
3. Lin, Y.; Zhang, W.J. Towards a novel interface design framework: Function–behavior–state paradigm. *Int. J. Hum.–Comput. Stud.* **2004**, *61*, 259–297. [CrossRef]
4. Lin, Y.; Zhang, W.J. A function-behavior-state approach to designing human–machine interface for nuclear power plant operators. *IEEE Trans. Nucl. Sci.* **2005**, *52*, 430–439. [CrossRef]
5. Yamanouchi, A. Effect of core spray cooling in transient state after loss of coolant accident. *J. Nucl. Sci. Technol.* **1968**, *5*, 547–558. [CrossRef]
6. Zhang, C.; Chen, P.; Jiang, F.; Xie, J.; Yu, T. Fault Diagnosis of Nuclear Power Plant Based on Sparrow Search Algorithm Optimized CNN-LSTM Neural Network. *Energies* **2023**, *16*, 2934. [CrossRef]
7. Qi, B.; Liang, J.; Tong, J. Fault Diagnosis Techniques for Nuclear Power Plants: A Review from the Artificial Intelligence Perspective. *Energies* **2023**, *16*, 1850. [CrossRef]
8. She, J.; Shi, T.; Xue, S.; Zhu, Y.; Lu, S.; Sun, P.; Cao, H. Diagnosis and prediction for loss of coolant accidents in nuclear power plants using deep learning methods. *Front. Energy Res.* **2021**, *9*, 665262. [CrossRef]
9. Choi, G.P.; Yoo, K.H.; Back, J.H.; Na, M.G. Estimation of LOCA breach Size Using Cascaded Fuzzy Neural Networks. *Nucl. Eng. Technol.* **2017**, *49*, 495–503. [CrossRef]
10. Wang, Y.; Liu, R.; Lin, D.; Chen, D.; Li, P.; Hu, Q.; Chen, C.L.P. Coarse-to-fine: Progressive knowledge transfer-based multitask convolutional neural network for intelligent large-scale fault diagnosis. *IEEE Trans. Neural Netw. Learn. Syst.* **2021**, *34*, 761–774. [CrossRef]
11. Mandal, S.; Santhi, B.; Sridhar, S.; Vinolia, K.; Swaminathan, P. Nuclear power plant thermocouple sensor-fault detection and classification using deep learning and generalized likelihood ratio test. *IEEE Trans. Nucl. Sci.* **2017**, *64*, 1526–1534. [CrossRef]
12. Yao, Y.; Wang, J.; Long, P.; Xie, M.; Wang, J. Small-batch-size convolutional neural network based fault diagnosis system for nuclear energy production safety with big-data environment. *Int. J. Energy Res.* **2020**, *44*, 5841–5855. [CrossRef]
13. Wang, H.; Peng, M.; Ayodeji, A.; Xia, H.; Wang, X.; Li, Z. Advanced fault diagnosis method for nuclear power plant based on convolutional gated recurrent network and enhanced particle swarm optimization. *Ann. Nucl. Energy* **2021**, *151*, 107934. [CrossRef]
14. Saghafi, M.; Ghofrani, M. Real-time estimation of break sizes during LOCA in nuclear power plants using NARX neural network. *Nucl. Eng. Technol.* **2019**, *51*, 702–708. [CrossRef]
15. Xu, Y.; Lin, K.; Hu, C.; Wang, S.; Wu, Q.; Zhang, L.; Ran, G. Deep transfer learning based on transformer for flood forecasting in data-sparse basins. *J. Hydrol.* **2023**, *625*, 129956. [CrossRef]
16. El-Shafeiy, E.; Alsabaan, M.; Ibrahem, M.; Elwahsh, H. Real-Time Anomaly Detection for Water Quality Sensor Monitoring Based on Multivariate Deep Learning Technique. *Sensors* **2023**, *23*, 8613. [CrossRef]
17. Liapis, C.M.; Kotsiantis, S. Temporal Convolutional Networks and BERT-Based Multi-Label Emotion Analysis for Financial Forecasting. *Information* **2023**, *14*, 596. [CrossRef]
18. Islam, M.M.; Islam, M.Z.; Asraf, A.; Al-Rakhami, M.S.; Ding, W.P.; Sodhro, A.H. Diagnosis of COVID-19 from X-rays using combined CNN-RNN architecture with transfer learning. *Benchcouncil Trans. Benchmarks Stand. Eval.* **2022**, *2*, 100088. [CrossRef]
19. Zaki, M.J.; Hsiao, C.J. Efficient algorithms for mining closed itemsets and their lattice structure. *IEEE Trans. Knowl. Data Eng.* **2005**, *17*, 462–478. [CrossRef]
20. Yang, Z.; Yang, H.; Chang, C.-C.; Huang, Y.; Chang, C.-C. Real-time steganalysis for streaming media based on multi-channel convolutional sliding windows. *Knowl.-Based Syst.* **2022**, *237*, 107561. [CrossRef]
21. Yaroslavsky, L.P.; Egiazarian, K.O.; Astola, J.T. Transform domain image restoration methods: Review, comparison, and interpretation. *Nonlinear Image Process. Pattern Anal. XII* **2001**, *4304*, 155–169.
22. Chang, C.-I.; Wang, Y.; Chen, S.-Y. Anomaly detection using causal sliding windows. *IEEE J. Sel. Top. Appl. Earth Obs. Remote Sens.* **2015**, *8*, 3260–3270. [CrossRef]
23. Rubinger, B. Performance of a sliding window detector in a high interference air traffic environment. In Proceedings of the IEEE Conference on Computer Vision and Pattern Recognition, Las Vegas, NV, USA, 27–30 June 2016.
24. Kłosowski, P. Deep learning for natural language processing and language modelling. In Proceedings of the 2018 Signal Processing: Algorithms, Architectures, Arrangements, and Applications (SPA), Poznań, Poland, 19–21 September 2018.
25. LeCun, Y.; Bengio, Y.; Hinton, G. Deep learning. *Nature* **2015**, *521*, 436–444. [CrossRef] [PubMed]
26. Baker, B.; Gupta, O.; Naik, N.; Raskar, R. Designing neural network architectures using reinforcement learning. *arXiv* **2016**, arXiv:1611.02167.

27. Rosenblatt, F. The perceptron: A probabilistic model for information storage and organization in the brain. *Psychol. Rev.* **1958**, *65*, 386. [CrossRef]
28. Vaswani, A.; Shazeer, N.; Parmar, N.; Uszkoreit, J.; Jones, L.; Gomez, A.N.; Kaiser, Ł.; Polosukhin, I. Attention is all you need. *Adv. Neural Inf. Process. Syst.* **2017**, *30*, 1–11.
29. Chung, J.; Gulcehre, C.; Cho, K.; Bengio, Y. Empirical evaluation of gated recurrent neural networks on sequence modeling. *arXiv* **2014**, arXiv:1412.3555.
30. Hochreiter, S.; Schmidhuber, J. Long short-term memory. *Neural Comput.* **1997**, *9*, 1735–1780. [CrossRef]
31. Qi, B.; Xiao, X.; Liang, J.; Po, L.; Zhang, L.; Tong, J. An open time-series simulated dataset covering various accidents for nuclear power plants. *Sci. Data* **2022**, *9*, 766. [CrossRef]
32. Schmidhuber, J. Deep learning in neural networks: An overview. *Neural Netw.* **2015**, *61*, 85–117. [CrossRef]
33. Ostrand, T.J.; Balcer, M.J. The category-partition method for specifying and generating fuctional tests. *Commun. ACM* **1988**, *31*, 676–686. [CrossRef]
34. McClell, J.L.; Rumelhart, D.E.; PDP Research Group. *Parallel Distributed Processing, Volume 2: Explorations in the Microstructure of Cognition: Psychological and Biological Models*; MIT Press: Cambridge, MA, USA, 1987.
35. Ackley, D.H.; Hinton, G.E.; Sejnowski, T.J. A learning algorithm for Boltzmann machines. *Cogn. Sci.* **1985**, *9*, 147–169.
36. Bao, Y.; Wang, B.; Guo, P.; Wang, J. Chemical process fault diagnosis based on a combined deep learning method. *Can. J. Chem. Eng.* **2022**, *100*, 54–66. [CrossRef]
37. Arena, P.; Basile, A.; Bucolo, M.; Fortuna, L. Image processing for medical diagnosis using CNN. *Nucl. Instruments Methods Phys. Res. Sect. Accel. Spectrometers Detect. Assoc. Equip.* **2003**, *497*, 174–178. [CrossRef]
38. Saponara, S.; Elhanashi, A.; Gagliardi, A. Real-time video fire/smoke detection based on CNN in antifire surveillance systems. *J. Real-Time Image Process.* **2021**, *18*, 889–900. [CrossRef]
39. Beane, S.R.; Bedaque, P.F.; Parreno, A.; Savage, M.J. Two nucleons on a lattice. *Phys. Lett. B* **2004**, *585*, 106–114. [CrossRef]

energies

Article

iMAGINE—Visions, Missions, and Steps for Successfully Delivering the Nuclear System of the 21st Century

Bruno Merk [1,*], Dzianis Litskevich [1], Anna Detkina [1], Omid Noori-kalkhoran [1], Lakshay Jain [1], Elfriede Derrer-Merk [1], Daliya Aflyatunova [2] and Greg Cartland-Glover [3]

1 School of Engineering, University of Liverpool, Liverpool L69 3GH, UK
2 School of Physical Science, University of Liverpool, Liverpool L69 3GH, UK
3 Scientific Computing Department, Daresbury Laboratory, Science and Technology Facilities Council, UK Research and Innovation, Warrington WA4 4AD, UK
* Correspondence: b.merk@liverpool.ac.uk

Abstract: Nuclear technologies have the potential to play a major role in the transition to a global net-zero society. Their primary advantage is the capability to deliver controllable 24/7 energy on demand. However, as a prerequisite for successful worldwide application, significant innovation will be required to create the nuclear systems of the 21st century, the need of the hour. The pros (low harmful emissions, high reliability, low operational expenses, and high energy density) and cons (environmental damage, fuel waste disposal concerns, limited uranium reserves, and long construction time-frame) of nuclear are discussed and analysed at different levels—the societal and public recognition and concerns (accidents, weapons, mining, and waste) as well as the scientific/engineering and economic level—to assure a demand-driven development. Based on the analysis of the different challenges, a vision for the nuclear system of the 21st century is synthesised consisting of three pillars—**unlimited nuclear energy**, **zero waste nuclear**, and **accident-free nuclear**. These three combined visions are then transformed into dedicated and verifiable missions that are discussed, in detail, regarding challenges and opportunities. In the following, a stepwise approach to the development of such a highly innovative nuclear system is described. Essential steps to assure active risk reduction and the delivery of quick progress are derived as answers to the critique on the currently observed extensive construction time and cost overruns on new nuclear plants. The 4-step process consisting of basic studies, experimental zero power reactor, small-scale demonstrator, and industrial demonstrator is described. The four steps, including sub-steps, deliver the pathway to a successful implementation of such a ground-breaking new nuclear system. The potential sub-steps are discussed with the view not only of the scientific development challenges but also as an approach to reduce the regulatory challenges of a novel nuclear technology.

Keywords: nuclear; nuclear energy; nuclear reactors; nuclear waste management; iMAGINE; strategic development; vision development; mission development

check for updates

Citation: Merk, B.; Litskevich, D.; Detkina, A.; Noori-kalkhoran, O.; Jain, L.; Derrer-Merk, E.; Aflyatunova, D.; Cartland-Glover, G. iMAGINE—Visions, Missions, and Steps for Successfully Delivering the Nuclear System of the 21st Century. *Energies* **2023**, *16*, 3120. https://doi.org/10.3390/en16073120

Academic Editor: Roman Vadimovich Davydov

Received: 10 February 2023
Revised: 10 March 2023
Accepted: 28 March 2023
Published: 29 March 2023

1. Introduction and Background

In October 2021, the UK committed to the use of advanced nuclear technologies as a significant share of the decarbonisation of the economy and delivering on its future net-zero obligations, which is ably highlighted in the following statement: "A clean, reliable power system is the foundation of a productive net zero economy as we electrify other sectors—so we will fully decarbonise our power system by 2035, subject to security of supply. Our power system will consist of abundant, cheap British renewables, cutting edge new nuclear power stations, ... " [1]. This is a strong, positive message since the change to net-zero, with the elimination of hydrocarbons, will have a tremendous influence on the whole energy system due to the reduction of freely storable energy resources (such as storage-based hydro and hydrocarbon-based systems), which can be turned into secondary energy on demand [2], and defines the challenge for a future energy system. Nuclear energy production will give us the opportunity to fill this gap in a sufficient and sustainable long-term way, but only if we are able to close the fuel cycle and use fertile materials, such as U-238, as additional fuel resources [3].

However, at least acceptance, and ideally a clear positive recognition of nuclear, is one of the key factors for the future success of nuclear energy technologies. It is a prerequisite in order to achieve the development goals—by delivering the required contribution to energy production and positively influencing worldwide development. Problems in public perception and recognition have, for example, in Germany, led to the phase-out of all nuclear power plants even if they could have played an essential role in the "Energiewende" [4].

Historical accidents at nuclear power plants, such as in Three Miles Island (TMI), USA, Chernobyl, erstwhile USSR, or Fukushima, Japan, have increased the public's risk perception and reduced the acceptance of nuclear plants significantly [5,6]. These accidents were associated with reduced trust in nuclear power and an increase in environmental damage recognition and attitudes towards risk avoidance. These accidents represent some vulnerabilities experienced by society due to the operation of nuclear power plants and related consequences of accidents, for example, radiation exposure and its inherent perceived horror, rumours about adverse impact on individual's health and environment, and lack of trustworthiness due conflicting risk communication [7]. Bromet [7] found that people affected by such accidents had lower self-reported health, known as a strong predictor of people's risk of morbidity, mortality and social outcome, and suffered unexplained medical issues, such as anxiety. These results might be explained by the well-known discrepancy of individually perceived risk and the actual measurable risk [8,9]. The risk perception of nuclear power was historically impacted by the lack of transparency in reporting about the accidents, which also led to distrust and hostile attitudes towards governments and the scientific community.

Other concerns are related to the unsolved nuclear waste disposal problem and its perceived health threat in society. Till now, people do not approve of any plans to dispose of nuclear waste neither near their homes or further away [10]. A study from Finland by Vilhunen et al. [11] talked about the "intragenerational and intergenerational injustices" (p. 1) from community experiences when becoming a host for the final disposal of nuclear waste. Furthermore, radioactive waste is perceived by society as dangerous for "health, safety and environment" [12] (p. 69); [13]. The ignorance of societal concerns regarding nuclear waste by nuclear scientists contributed to increased negative attitudes in society against any final disposal decisions [10].

Furthermore, the public's perception that uranium mining is dangerous for individuals' health and the environment is based on the early stages of uncontrolled mining for military and monetary purposes [14]. The danger of uranium mining concerns the "health and safety of miners and mine sites; health and safety of people in the immediate vicinity who might be affected by the spread of radioactivity from the tailings or tailing ponds; and global health and environmental effects of increasing background radiation and water contamination" [15] (p. 470). Increasingly, research is carried out exploring the impact of uranium mining on the environment [16]. The study by Dewar et al. [15] states that uranium mining has a detrimental effect on the environment due to contamination with dust, radon gas and water-borne toxins and impacts people's risk perception. This negative risk perception might be caused by the historical and current release of ionizing radiation and limited interest in caring for the safety and protection of humans and the environment whilst mining uranium. However, the safety of the people and environment during uranium mining should have the highest priority, and the concerns of society should be taken seriously.

Finally, the experience of the use of nuclear weapons in Hiroshima and Nagasaki (1945) has proven that massive consequences occur when nuclear weapons are detonated. The risk of using nuclear weapons and, in consequence, the anxiety regarding nuclear warfare has risen again and is fostered by the war in Ukraine [17,18]. Research suggests that some people suffer from the anxiety of nuclear weapons—sometimes also called "Nucleomituphobia"—which is not unreal and represents a real danger for society [18,19]. People's concerns reach much beyond the use of nuclear weapons, with the mere existence and fallout of radiation during testing causing severe distress in society. It is known that exposure to large doses of radioactive substances has detrimental consequences on humans and the environment, such as death shortly after or cancer from longer term [20], and the widespread use of nuclear weapons would "lead to a cooling of the atmosphere, shorter growing seasons, food shortages, and a global famine" [21].

Thus, the societal challenges seen in public, given in Figure 1, can be summarised into the following points:

- Fear of accidents such as TMI, Fukushima and Chernobyl and their potential consequences
- Anxiety due to the nuclear waste problem—there is no final disposal. Thus we pass a problem on to the future generations
- Fear of environmental damage and CO_2 production due to the mining of uranium
- Fear of proliferation of nuclear weapons and the materials required for their manufacture through the use of civil nuclear technologies

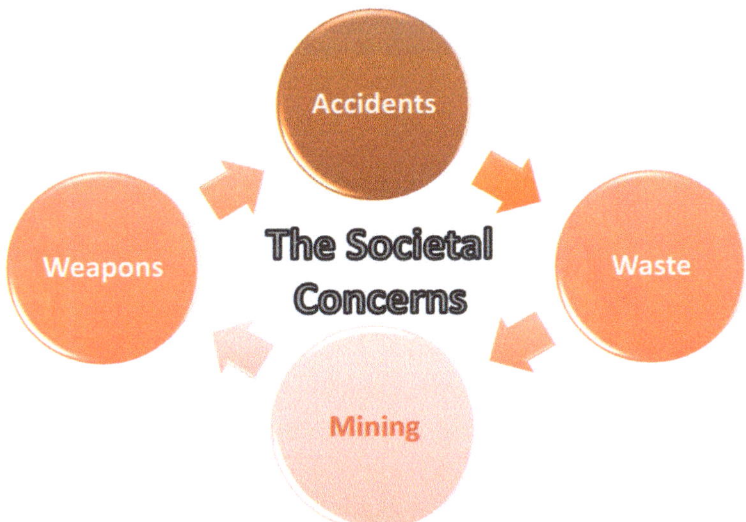

Figure 1. The societal concerns influencing the perception of nuclear energy at a glance.

Over the last decades, little has been done to address these societal challenges and fears and promote higher trust in nuclear power. However, a study by the Department for Business, Energy and Industrial Strategy [22] found that workshops and training sessions helped to increase people's positive views of nuclear power. Based on this experience, the vision of iMAGINE aims to consider these challenges and contribute towards lower risk perception and reduced risk for society and the environment.

Besides this public perception, there are independent, scientific/technical and business-oriented evaluations, for example, one recently published in NS Energy [23] highlighting the pros of nuclear (see Figure 2), proven through operational experience and physical/chemical boundary conditions. Key points are the low harmful emissions "Electricity produced from a nuclear power plant emits fewer greenhouse gas emissions compared to those released by coal power plants and other traditional power generation sources" [23], high reliability "When compared to renewable sources of energy such as solar and wind, power generation from nuclear power plants is more reliable. . . . Nuclear power units can produce power continuously for several months without any interruption" [23], low operational expenses "Although building nuclear power plants requires huge initial investment, the costs associated with operating them are low. The fuel costs of nuclear power plants are also low and the electricity produced from them is relatively inexpensive" [23], and high energy density "Nuclear energy sources have a higher density than fossil fuels and release massive amounts of energy" [23].

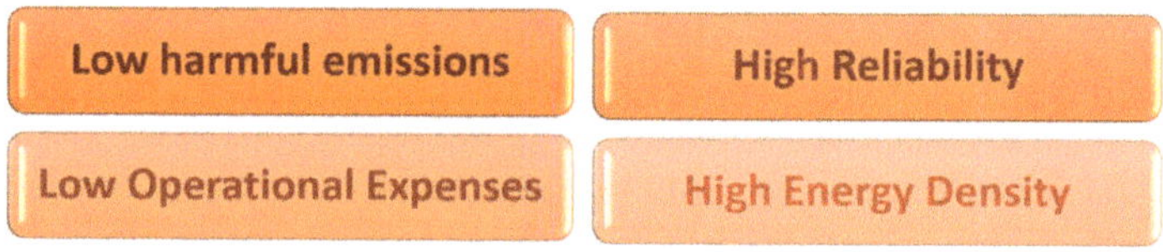

Figure 2. Pros of nuclear technologies as given in the NS Energy publication.

The already proven pros are contrasted with the cons (see Figure 3) based on scientific and economic analysis for a long-term and widespread sustainable operation of nuclear technologies. Interestingly, two points coincide with the public perception of nuclear, environmental damage and waste concern, while the two other points are long-term sustainability and economic attractiveness. The core points are on environmental damage "One of the major negatives of nuclear energy is the impact of uranium on the environment. While transportation of nuclear fuel to a power plant can cause pollution, the process involved in mining and refining uranium is also a concern" [23], on the fuel waste concerns "The vast amount of nuclear waste created by power plants can lead to high radiation and raise temperature levels. . . . The cost of managing nuclear waste is also high" [23], on the limitation of uranium reserves "Similar to fossil fuels, uranium reserves are limited and are found in few nations, while the processes carried out to mine and refine uranium involve huge costs . . . . As large quantities of waste are created during the refinement of uranium, any mishandling of the processes can affect the environment and pose health risks to human beings" [23], and on the long construction time-frame "The construction of nuclear power plants usually takes several years to complete as they require large infrastructure . . . . Massive investments are also required to build a nuclear power plant, as the associated costs of installing radiation containment systems are high" [23].

Figure 3. Cons of nuclear technologies as given in the NS Energy publication.

Limited uranium reserves are presently not seen as a problem for the current reactor operation. This is reflected in the investment in light water reactors without any discussion on fuel availability (using only 5% of the energy content of the fuel) and the decision for direct final disposal (discarding a potential massive energy source underground). Both approaches may be considered acceptable considering the current share of nuclear in the global energy mix [24], but they will not be a sustainable long-term solution if nuclear energy is envisioned to contribute substantially to the worldwide net-zero strategy. In order to avoid a massive increase in the fuel demand and waste generation, relying solely on existing technologies would require a massive increase in the energy content harnessed from nuclear fuel than the current maximum 10% delivered by new reactors.

The long construction time is another problem that is often discussed as one of the factors limiting the growth in the contribution of nuclear to electricity production, but it is, in addition, a problem of the financing of nuclear reactors due to the high share of upfront investment [25]. It has to be seen as one of the big problems in attracting investors since delays and related cost overruns do not allow a robust determination of the investment risk and the potential payback of the investment, which, in the end, makes the projects more and more costly [26]. The delays are often highlighted with respect to the current nuclear projects, such as Vogtle and VC Summer, USA, Olkiluoto, Finland, and Flamanville, France which face further schedule delays. However, a more detailed analysis using IAEA PRIS data [27], see Table 1, indicates that the problem had already appeared for other reactors with construction or project start/re-start after the Three Mile Island (TMI) accident, see Watts Bar, USA, Civaux, Golfech and Chooz-B, France, compared to the last nuclear power plants developed before the TMI accident, see Emsland, Germany (even if the physical construction started in 1982), or Chinon B 1 to 4, France. A conclusion could be that the increased complexity and the sharpened regulatory demands after the analysis of the TMI accident could be one reason. Another reason could be: "Did we lose the experience and the qualified people due to the massively reduced building activity after TMI?". Indeed, this seems to be the case as highlighted by the statement—"As the western nuclear industry flounders, Russia's Rosatom is building nuclear power plants (NPPs) on time and under budget around the world ... " [28]—since other major players are still able to deliver on time and budget. This has to be seen as a challenge, especially when considering that the currently delivered VVER reactors are "claimed" to fulfil comparable safety standards as western products and clearly points to the lack of capabilities and capacities. Both had declined substantially in the decade after the TMI accident due to a lack of orders in the western world.

Table 1. Construction time of various nuclear power plant projects initiated before and after the TMI accident [27].

Country	Nuclear Power Plant Project	Construction/Completion Time (in Years)	Project Start
Germany	Emsland	6	Before TMI accident
France	Chinon B 1 to 4	5–6	
France	Civaux 1 and 2	9 and 8	After TMI accident
	Golfech 1 and 2	8 and 9	
	Chooz-B 1 and 2	12 and 11	
USA	Watts Bar 1 and 2	23 and 12 (+9)	

Based on the aforementioned discussion, we propose a vision for a nuclear system for the 21st century. The aim is to go well beyond the conceptual framework of the Generation IV international forum, not only working on reactor development but thinking about a comprehensive nuclear system incorporating the complete fuel cycle from cradle to grave. This vision will then be refined into a set of useful, tangible and achievable missions based on the approach of Fredmund Malik [29], followed by the approach proposed for the successful delivery of such a new challenge through a consequent stepwise paradigm, thus the implementation.

2. Vision for a 21st-Century Nuclear System

The demand analysis, as given above, indicates three partly interlaced areas:

- Fuel usage, the related environmental damage and the uranium reserves
- The system-inherent accumulation of nuclear waste and the related final disposal challenge
- Safe operation, fear of accidents and fear of nuclear weapon distribution

The first two themes are related to the efficiency of fuel utilisation since efficient usage of fuel will stretch the uranium reserves, reduce the environmental damage due to mining and also reduce the amount of waste that has to be disposed of. The third point coincides with "prevention of abnormal operation and failures" as level one of defence in depth strategy and the subsequent higher levels, see [30]. The last broader concern about nuclear technologies—long construction time-frame—falls under the topic of implementation and will be covered later in the section on delivery.

The core challenge for the development of the vision is now to get these demands reflected in an "as far reaching dream", as proposed by Malik as the point of origin for the mission development. "It [the mission] often follows from a very broad and far-reaching idea which could be called a vision or a dream. That dream, however, has to be transformed into a viable mission: this is the only way to distinguish useful from useless visions" [29]. In the beginning, only a singular vision—unlimited energy, or, more controversially, the Perpetuum Mobile—had been developed as the working basis for iMAGINE [31]. This has been expanded into a ternary vision now to reflect the full demand, see Figure 4.

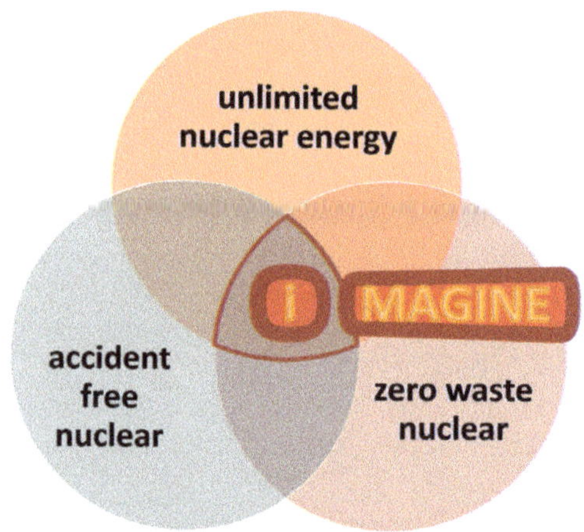

Figure 4. The ternary vision as the basis for the further development of the iMAGINE project.

In general, the vision for developing a new, comprehensive nuclear system, instead of just a reactor, is rather complex and should be very far-reaching. Thus, it seems appropriate to split it into three different core visions—**unlimited nuclear energy**, **zero waste nuclear** and **accident-free nuclear**. All three visions seem to be far-reaching enough to give guidance for the development on a very high level and all three visions are dreams, since it is clear that unlimited nuclear energy cannot be fully achieved due to the limited character of natural resources, whether it be uranium or uranium and thorium. The same can be said about zero waste nuclear since nuclear fission produces such a wide range of fission products—with some producing a high level of radiation and some producing a certain level of radiation for a very long time—that it seems unreasonable to claim that all materials can be re-used. Similar to the first two cases, accident-free nuclear cannot be absolute since engineered systems cannot be designed to be completely accident free, and the system's inherent probability for unexpected behaviour/failure increases with the number of systems being employed.

The next step in strategic development is now to translate these visions into viable missions.

3. Missions for iMAGINE

The following missions have been defined as a part of the strategic philosophy of iMAGINE, based on the visions highlighted above as guidance for the developers to find solutions to the given challenges.

The vision, **unlimited nuclear energy**, is obviously closely related to closed fuel cycle operation since the latter is already well recognized [32] as the gateway to improved uranium utilization. However, only limited progress has been made up to now in the successful implementation of closed fuel cycle operation in the nuclear industry. Even if it can potentially allow the release of a factor of 100 more energy out of the already mined nuclear material, such as spent fuel and tailings, compared to today's light water reactor technology. The mission aims to create a significant amount of energy without mining new resources, see Figure 5.

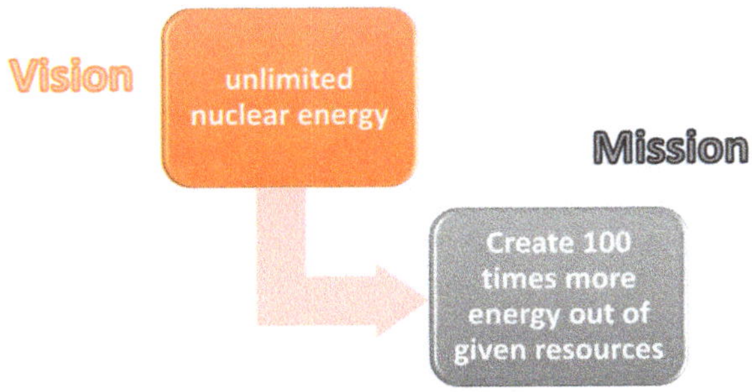

Figure 5. Translating the vision of unlimited nuclear energy into a viable mission.

The aim here is to make the already mined resources available through advanced technology development without creating proliferation issues while massively reducing the complexity of the fuel cycle compared to the one with external reprocessing proposed for solid-fuelled reactors, see Figure 6. The mission, in addition to the massively improved resource utilisation, delivers a significant improvement in resource security for all countries that have operated nuclear power plants in the past since stockpiles of spent fuel and tailings will be already available. At the same time, it also enables other nations the option to start the iMAGINE system with enriched uranium and subsequently feed it with the tailings accumulated during the enrichment process. The mission should be accomplished through the development of the closed fuel cycle in an integrated system. Rather, this disregards the complex split fuel cycle consisting of fuel production, reactor operation, fuel cooling and reprocessing in multiple cycles to ease future industrial implementation along with reduced investment into the whole nuclear system.

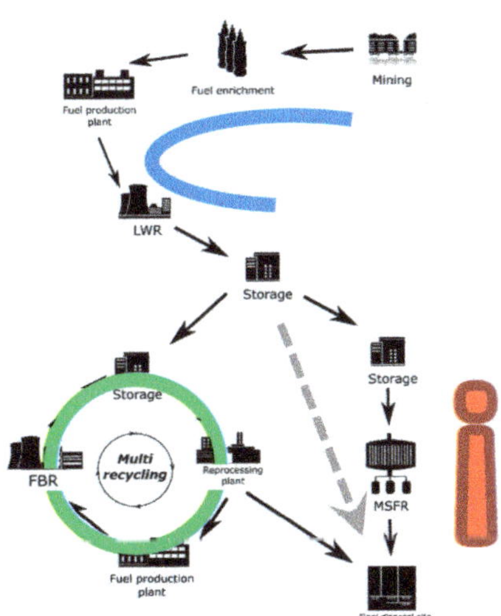

Figure 6. Fuel cycle options: open fuel cycle, closed fuel cycle and the envisaged implementation of iMAGINE.

The vision, **zero waste nuclear**, is closely related to improving fuel usage, but it should not be forgotten that nuclear waste—not having disposal solutions or a sustainable strategy implemented for the nuclear waste—in addition, is one of the major impediments of more widespread societal acceptance of nuclear energy. Improved fuel usage will ideally help to avoid the disposal of valuable material into the waste stream, as currently happens with U-238 in the spent fuel of LWRs, while the number of fission products created per unit of energy could be seen as a natural constant of nuclear energy conversion. Thus, integrated closed fuel cycle operation is one of the aims reflecting the demand of reducing the waste amount per unit of energy produced by releasing almost all energy from the material that has already been mined; this is the part that links to the mission of unlimited energy. The objective is to reduce the waste per unit of energy to 1% or lower, compared to LWR open fuel cycle operation, see Figure 7.

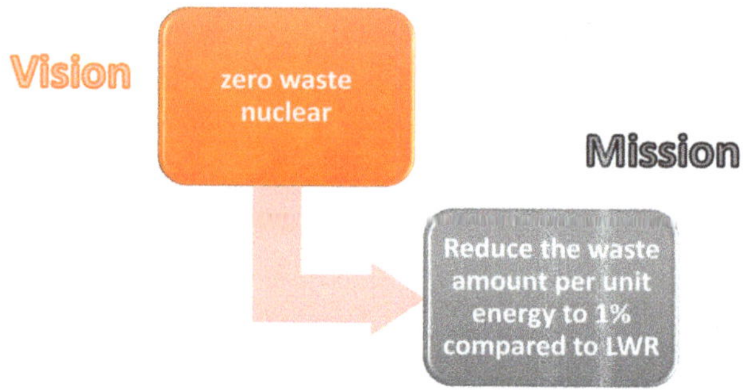

Figure 7. Translating the vision of zero waste nuclear into a viable mission.

This can be achieved partly through the subsequent use of almost all fissile and fertile material, as well as by developing reasonable strategies for the required fission product removal. However, this approach should ideally be accompanied by a recycling strategy—can we create sustainable use for some of the discarded material, as these are often required for the development of processes in other technologies [33]—thinking about a cascade of potential uses with reduced quality before final disposal of the material, as given in Figure 8. All these approaches will help reduce the amount of material to be disposed of. Even when the material has to be disposed of, it will allow the finding of better solutions due to the massively reduced amounts to be handled. This approach is currently not followed in nuclear energy production, especially not when applying the open fuel cycle accompanied by direct final disposal of spent fuel. The core idea of the cascading down approach will be identifying strategies for the use of fission products separated from the reactor instead of just declaring all fission products as waste.

The vison, **accident-free nuclear**, surely has economic as well as societal components and, notably, applies to the completely integrated nuclear system in the case of iMAGINE. The economic components point to the availability/reliability of the facility, the cost of preventing accidents and their effect on the outside world. These points are even reflected in the GEN-IV objectives "Generation IV nuclear energy systems operations will excel in safety and reliability. . . . will eliminate the need for offsite emergency response" [34]. The societal component seems to be based on the fear of large-scale accidents with a massive release of radioactive materials and the loss of territory due to radioactive contamination, such as what happened in the case of the Chernobyl accident through the distribution of radioactive materials due to graphite fire. Thus, this vision is transformed into strategically reducing the driving forces for potential accidents (reducing the potential for release and spread of contamination) as well as limiting the consequences of accidents in the facility. The key points are relying on a low-pressure primary system and ideally developing a low-pressure energy conversion system that could deliver a higher efficiency as the potential link to energy. Other important factors are eliminating accident initiators, such as avoiding excess reactivity, and reducing the potential radiological source term of the system, see Figure 9.

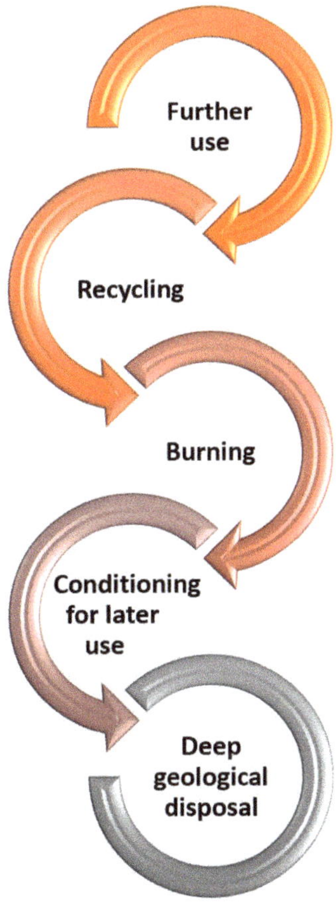

Figure 8. The cascade of potential re-use of materials before these materials should be considered waste needing disposal.

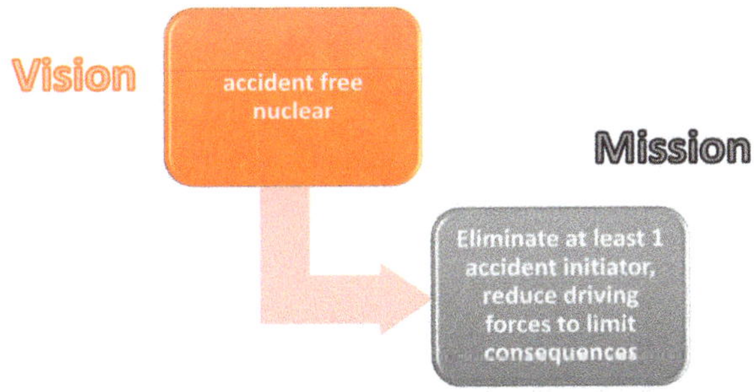

Figure 9. Translating the vision of accident-free nuclear into a viable mission.

Another objective is limiting the potential of proliferation and other high-risk incidents in the integrated nuclear system. The most prominent ones besides the risk of proliferation are the risk of misuse and theft of fissile material and the risk of unintended release of radioactive materials.

4. The Technology

The described vision aims much higher than the approaches of GEN-IV [34] since it requires an integrated fuel cycle system to deliver a closed fuel cycle approach within the energy production system instead of just a reactor technology. The proposal is already supported by a significant body of research. The development is based on integrative thinking of the complete nuclear system for energy production [31] as well as for waste management [35] instead of focussing on reactor development while locating it, in the best case, in a complex, partly already existing, fuel cycle to allow the recycling of fissile material.

iMAGINE is based on molten salt fast reactor technology with a highly integrated fuel cycle operating with a continuous salt clean-up system based on the approach of reverse reprocessing based on a demand list, systematic optimization of the chemical approaches [36], inter-disciplinary studies [37] and operational analysis [38]. The system operates on a tertiary chlorine salt system Nacl-UCl_3-UCl_4 to allow a high heavy metal content in the core to support self-sustained breeding in a eutectic with reasonably low operational temperatures [39]. It is designed for operating on SNF or depleted uranium, making waste and tailings an energy resource to avoid the demand for the mining of new fuel materials [3]. The reactor is mainly controlled through very strong inherent feedback effects [40] and does not require excess reactivity due to the opportunity for online feeding. It is supported by online reverse reprocessing to improve recycling and conditioning options for fission products and to eliminate the demand for the separation of fissile material [33]. The design basis delivering homogeneous breeding and reverse reprocessing helps to avoid the separation of fissile material and thus will reduce proliferation issues. The stepwise approach for the development has already been studied to deliver the understanding of dimensions [41], control [42] and optimization [43] of a zero power experiment, as well as a general discussion of the role of such an experimental facility for the development of new technology [44]. The potential opportunities of this new approach to waste management and the final disposal situation of nuclear have already been discussed in [33].

The technology that comes closest could be BREST, developed in the Proryv project [45], demonstrating partly comparable approaches such as a fully integrated fuel cycle [46] and advanced reactor design with strong limitations on excess reactivity [47], but due to the use of solid fuel, the demand for the separation of fissile material and thus upcoming proliferation issues, cannot be avoided in the current development.

5. Implementation

Providing the vision and developing the missions provide a strong foundation for the development of iMAGINE as a nuclear solution for the 21st century, see Figure 10. However, the whole approach could still be seen as a dream without concrete plans for its implementation and delivery. This will also encompass the point of extensive construction times, the only point not tackled as a part of the vision and mission development.

First of all, we need a good reason for the investment in the implementation of new technology. To make this argument, it is important to see the opportunities of the new technology as described in the vision and missions. It is also necessary to understand the risks of new development along with risk mitigation measures for potential investment at different levels. The long construction time is only one of these aspects, and the focus should be on identifying and reducing the broader technological risks. However, it is important to note that potential reasons for long construction times might be totally non-technical in nature and, instead, be rooted in a lack of political and/or societal support, which results in the withdrawal of required political will and/or in demonstrations leading, in the worst case, to civil disobedience. Nonetheless, from a technical point of view, some steps have to be delivered, and the aim has to be to develop a system that is simpler and quicker to build. Typical points are the use of a low-power system, a reduced number of highly complex safety and mitigation systems as well as the consequent use of inherent safety and stabilization processes already in the early stages of the design.

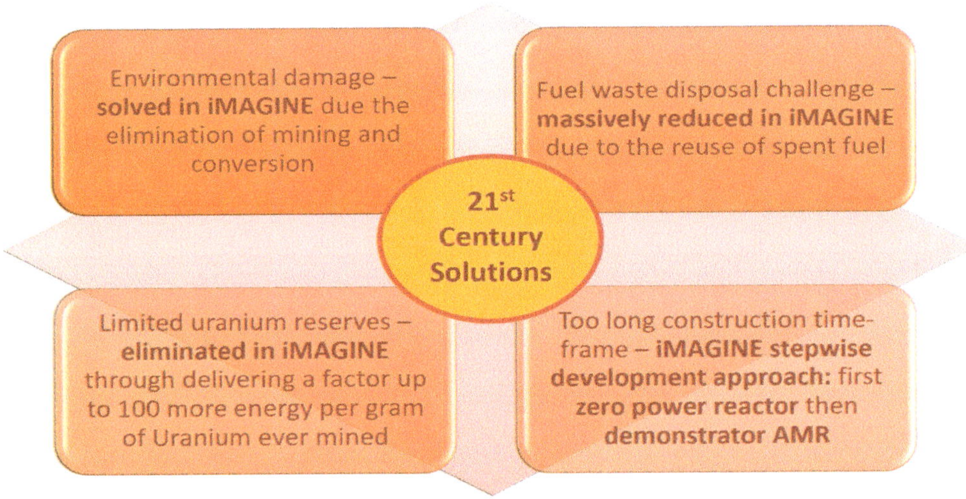

Figure 10. Overview of the solutions offered by iMAGINE to resolve the challenges of the 21st century in the view of the public as well as in the scientific and business community.

The general multi-dimensional risk reduction strategy in iMAGINE is as follows:

- Financial
  - A stepwise plan to mitigate the development risk by creating an approach to deliver quick feedback, early recovery from problems during the development phase and, in addition, the capabilities and capacities required for the successful implementation of a new reactor system [44].
  - Operational safety risk reduction due to a low-pressure system with significantly reduced accident risks and initiators and early safety demonstrations through experiments to enable lowering of insurance and off-site response requirements.
  - Consequent use of inherent safety approaches to reduce the reliance on complex, redundant technical solutions.
- Political/Societal
  - Mitigation of energy and resource security risk through the utilisation of materials that are already stored within the country's borders and transforming the waste disposal problem into reservoirs of huge energy resources and wealth.
  - Reduction of the nuclear waste storage challenge by achieving a new level of waste recycling and, ideally, harnessing additional accessible material resources as well as improving the chances of finding a final disposal site.
  - Decreasing the instability risks in national electrical grids by delivering reliable and controllable, 24/7 net-zero energy production based on existing resources.
  - Limiting the risk of proliferation, misuse and theft of nuclear materials by eliminating the enrichment process and the separation of fissile material in the fuel cycle.
  - Eliminating, by far, the largest environmental damage by avoiding mining and conversion and even reducing the very long-term release risk from final disposal.
- Building trust in society whilst considering health and safety concerns.

It is not only important to talk about risk reduction itself but also about effective risk communication, a point that was raised in public recognition. In most cases, the problem lies in not being able to effectively and transparently communicate with the general public about nuclear facilities, including their advantages, the existing or non-existing risks and mitigation measures. A future approach should be based on working with communities and listening to the concern of the people affected; we could call this a participatory approach.

6. Delivery

Acting on the long construction time is an essential part of success, and the fundamental philosophy of iMAGINE is returning to the development pathway used back in the 1950s, when nuclear really was a new technology, by applying a gradual stepwise approach to develop this highly innovative nuclear system. Such a paradigm shift is essential to enable the fast creation of operational experience, drive active risk mitigation and deliver quick progress [3,48]. An up-to-date four-step process has been developed on a historic basis, consisting of basic studies followed by zero power and other demonstration experiments, a small-scale demonstrator and an industrial demonstrator, see Figure 11. A comparable process is followed by Rosatom for the development of their molten salt reactor programme: national programme, research reference facility, research reactor and large-scale reactor, as published in 2019 [49]. This is in contrast to many of the recently proposed solutions for innovative reactors delivered by the private industry.

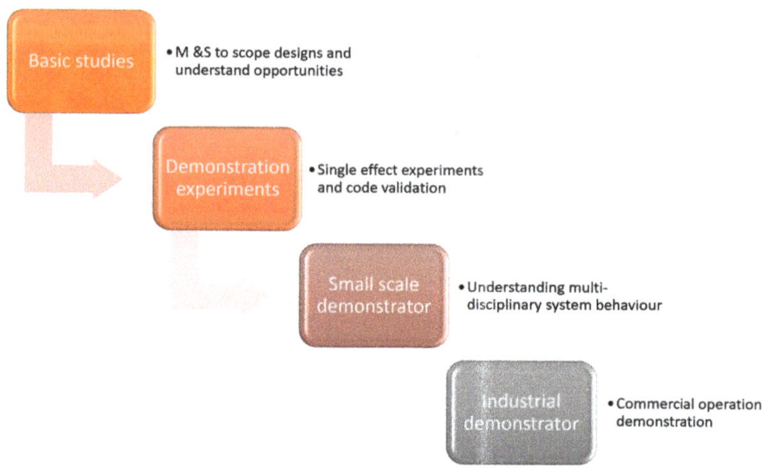

Figure 11. The four-step process proposed for the development of a breakthrough reactor system.

However, these four steps are only the beginning and will have to be filled with an additional set of small, intermediate steps within each main step, keeping in mind that the current regulations have been developed for light water reactors and completely new demands will arise for a system such as iMAGINE, see Figure 12. This challenge will have to be treated collaboratively between the developers and the regulators, similar to the situation when nuclear technologies were nascent and completely new. The key challenge for success will be for the developer to start a journey together with the regulator by defining the detailed steps in a mutually convenient shape for both partners as well as larger society. The process should be based on assuring timely feedback and stepwise learning in successive, partly overlapping projects. The aim must be to deliver an innovative key-step approach to assure rapid and sustainable progress, which is essential to make nuclear ready for a significant contribution to the net-zero goals in 2050. For this, a concrete fundament for discovering a highly innovative breakthrough technology has to be delivered by following a step-by-step process to open a game-changing opportunity. However, the key to success will not only be to get the regulator engaged early but also other future stakeholders and the broader public. The stepwise approach has to be delivered here, too, geographically from the wider to the narrower engagement while taking care of the sensibilities of the local host communities as soon as a site selection process has been started to receive positive and broad support from the host community while demonstrating the sensibility for the local concerns.

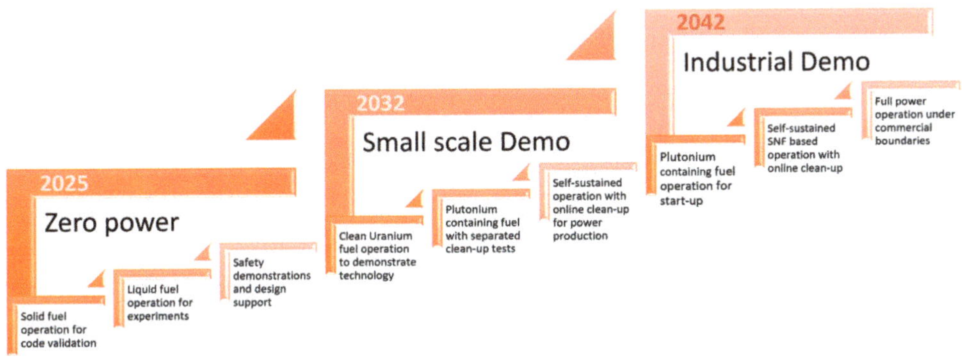

Figure 12. Proposed approach for a step-by-step approach for the development of iMAGINE with indicative dates.

The first steps that most probably have to be delivered in the framework of a national program are:

- A zero-power experimental facility for fast and inexpensive learning and delivery, as the first step into a new reactor technology, the related fuel production and regulation, as proven in the past [48]
- A small demonstrator AMR, operating ideally within 10 years for an estimated budget of £1Bn.

Interestingly this approach for the development and delivery of really new, innovative reactor systems through the initiation of national programs coincides again with the historic experience described by the Electric Power Research Institute (EPRI) in [49]. In addition, the new Russian programme on developing a molten salt reactor as a tier two burner for waste management follows a comparable stepwise approach [49] with the research programme recently investigated in an international project [50].

7. Conclusions

In order to ensure that nuclear technologies can attain their massive potential in enabling a global net-zero future, a highly strategic approach for the development of a set of demand-driven visions has been applied. The research for the demand is not limited to only a techno-economic analysis of the pros and cons of nuclear but is also based on the analysis of public perception and the fears articulated by the affected people. The proposed strategic, demand-driven approach should support the successful worldwide application of nuclear technologies by delivering significant progress compared to existing solutions with the aim of creating and delivering an innovative nuclear system of the 21st century, the need of the hour.

To create the basis for a truly demand-driven development, the pros and cons of nuclear are discussed and analysed on different levels—the societal and public recognition as well as a techno-economic level. Based on these analyses, a three-fold vision is delivered containing the three pillars **unlimited nuclear energy**, **zero waste nuclear**, and **accident-free nuclear**. After defining the visions, they are translated into explicit and verifiable missions, given as follows. A detailed discussion of these missions with respect to the evaluation of different approaches and support for future development is presented.

- **Releasing a factor of 100 more energy** out of the already mined nuclear material.
- **Reducing the waste per energy to 1% or lower**, compared to LWR open fuel cycle operation.
- **Reducing the driving forces for potential accidents** as well as **limiting the consequences of accidents.**

This is followed by the description of a stepwise approach for the development of such a highly innovative nuclear system to assure active risk reduction and the delivery of quick progress in response to the critique on the currently observed extensively long construction time associated with new nuclear plants. The four-step process—basic studies, experimental zero power reactor, small-scale demonstrator and industrial demonstrator—as the pathway to a successful implementation of a ground-breaking new nuclear system is presented.

The four-step process has been further refined with multiple intermediate sub-steps and risk mitigation at each stage. The process is rounded up with the proposal to work in close collaboration with the regulator to assure fast development and delivery of a highly innovative and holistic nuclear energy technology. However, the key to success will not only be to get the regulator engaged early but also other future stakeholders and the broader public.

The stepwise approach has to be delivered here, too, geographically from the wider to the narrower engagement while taking care of the concerns of the local host communities as soon as a site selection process has been started to receive positive and broad support from the host community whilst caring for the societal needs and public value.

Author Contributions: Conceptualization, B.M.; Methodology, B.M.; Validation, L.J., E.D.-M. and G.C.-G.; Investigation, B.M. and E.D.-M.; Writing—original draft, B.M.; Writing—review & editing, D.L., A.D., O.N.-k., L.J., E.D.-M., D.A. and G.C.-G.; Visualization, A.D.; Supervision, B.M. All authors have read and agreed to the published version of the manuscript.

Funding: This research was funded by Royal Academy of Engineering grant number CiET 2021/161.

Data Availability Statement: No new data were created or analyzed in this study. Data sharing is not applicable to this article.

Conflicts of Interest: The authors declare no conflict of interest.

References

1. Net Zero Strategy: Build Back Greener, HM Government. October 2021. Available online: https://assets.publishing.service.gov.uk/government/uploads/system/uploads/attachment_data/file/1033990/net-zero-strategy-beis.pdf. (accessed on 19 December 2022).
2. Saurugg, H. Energy Supply after the Phase out—No Nuclear Energy at all? In Proceedings of the JK'22, Leipzig, Germany, 21–22 June 2022.
3. Merk, B.; Litskevich, D.; Peakman, A.; Bankhead, M. IMAGINE—A Disruptive Change to Nuclear or How Can We Make More Out of the Existing Spent Nuclear Fuel and What Has to be Done to Make it Possible in the UK? *Atw-Int. J. Nucl. Power* **2019**, *64*, 353–359.
4. Appunn, K. The History Behind Germany's Nuclear Phase-Out, Clean Energy Wire. Available online: https://www.cleanenergywire.org/factsheets/history-behind-germanys-nuclear-phase-out (accessed on 9 January 2023).
5. Huang, L.; Zhou, Y.; Han, Y.; Hammitt, J.K.; Bi, J.; Liu, Y. Effect of the Fukushima nuclear accident on the risk perception of residents near a nuclear power plant in China. *Proc. Natl. Acad. Sci. USA* **2013**, *110*, 19742–19747. [CrossRef]
6. Prati, G.; Zani, B. The Effect of the Fukushima Nuclear Accident on Risk Perception, Antinuclear Behavioral Intentions, Attitude, Trust, Environmental Beliefs, and Values. *Environ. Behav.* **2013**, *45*, 782–798. [CrossRef]
7. Bromet, E.J. Emotional consequences of nuclear power plant disasters. *Health Phys.* **2014**, *106*, 206–210. [CrossRef] [PubMed]
8. Sjöberg, L. Explaining Individual Risk Perception: The Case of Nuclear Waste. *Risk Manag.* **2004**, *6*, 51–64. Available online: http://www.jstor.org/stable/3867934 (accessed on 9 February 2023). [CrossRef]
9. Paul, S. Perceptions of Risk: Reflections on the Psychometric Paradigm. In *Theories of Risk*; Golding, D., Krimsky, S., Eds.; Praeger: New York, NY, USA, 1990. Available online: https://core.ac.uk/download/pdf/84755706.pdf/ (accessed on 6 February 2023).
10. Ramana, M.V. Technical and social problems of nuclear waste. *WIREs Energy Environ.* **2018**, *7*, e289. [CrossRef]
11. Vilhunen, T.; Kojo, M.; Litmanen, T.; Taebi, B. Perceptions of justice influencing community acceptance of spent nuclear fuel disposal. A case study in two Finnish nuclear communities. *J. Risk Res.* **2022**, *25*, 1023–1046. [CrossRef]
12. National Research Council. Disposition of High-Level Waste and Spent Nuclear Fuel: The Continuing Societal and Technical Challenges. In *Chapter: 5 Societal Issues in Radioactive Waste Management, National Academies of Sciences, Engineering, and Medicine*; The National Academies Press: Washington, DC, USA, 2001. [CrossRef]
13. Kim, Y. The Radiation Problem and Its Solution from a Health Communication Perspective. *J. Korean Med. Sci.* **2016**, *31*, S88–S98. [CrossRef]

14. NEA. Perceptions and Realities in Modern Uranium Mining Extended Summary NEA No. 7063 Nuclear Energy Agency Organisation for Economic Co-Operation and Development. 2014. Available online: https://www.oecd-nea.org/upload/docs/application/pdf/2019-12/7063-mehium-es.pdf (accessed on 6 February 2023).
15. Dewar, D.; Harvey, L.; Vakil, C. Uranium mining and health. *Can. Fam. Physician* **2013**, *59*, 469–471.
16. Poinssot, C.; Bourg, S.; Ouvrier, N.; Combernoux, N.; Rostaing, C.; Vargas-Gonzalez, M.; Bruno, J. Assessment of the environmental footprint of nuclear energy systems. Comparison between closed and open fuel cycles. *Energy* **2014**, *69*, 199–211. [CrossRef]
17. International Campaign to Abolish Nuclear Weapons, Dealing with Nuclear War Anxiety. 2023. Available online: https://www.icanw.org/dealing_with_nuclear_anxiety (accessed on 6 February 2023).
18. Alex, O. Nuclear Anxiety Is Nothing New. Here's How to Handle It. 2022. Available online: https://www.discovermagazine.com/mind/nuclear-anxiety-is-nothing-new-heres-how-to-handle-it (accessed on 6 February 2023).
19. Smith, T.W. A Report: Nuclear Anxiety. *Public Opin. Q.* **1988**, *52*, 557–575. Available online: http://www.jstor.org/stable/2749262 (accessed on 6 February 2023). [CrossRef]
20. International Committee of the Red Cross. Humanitarian Impacts and Risks of Use of Nuclear Weapons. 2020. Available online: https://www.icrc.org/en/document/humanitarian-impacts-and-risks-use-nuclear-weapons?amp#_edn8 (accessed on 6 February 2023).
21. Alan, R.; Brian Toon, O.; Xia, L.; Stenke, A.; Helfand, I. Global Famine after a Regional Nuclear War: Overview of Recent Research. Presentation to the Vienna Conference on the Humanitarian Impact of Nuclear Weapon. 2014. Available online: https://www.bmeia.gv.at/fileadmin/user_upload/Zentrale/Aussenpolitik/Abruestung/HINW14/Presentations/HINW14_S1_Presentation_Michael_Mills.pdf (accessed on 6 February 2023).
22. Goterfelt, F.; Martinez Miranda, N.; McCool, S.; Nashef, L. Public Dialogue on Advanced Nuclear Technologies (ANT). 2021. Available online: https://www.gov.uk/government/publications/public-dialogue-on-advanced-nuclear-technologies-ants (accessed on 6 February 2023).
23. NS Energy Profiling the Top Nuclear Power Pros and Cons, NS Energy. 2021. Available online: https://www.nsenergybusiness.com/features/newstop-nuclear-power-pros-and-cons-5760814/ (accessed on 9 January 2023).
24. World Nuclear Association. The Nuclear Fuel Report: Expanded Summary—Global Scenarios for Demand and Supply Availability 2021–2040, World Nuclear Association, April 2022 Report No. 2022/001. Available online: https://world-nuclear.org/getmedia/9a2f9405-1135-407a-85c8-480e2365bee7/nuclear-fuel-report-2021-expanded-summary.pdf.aspx (accessed on 9 January 2023).
25. World Nuclear Association. Financing Nuclear Energy, World Nuclear Association. 2020. Available online: https://world-nuclear.org/information-library/economic-aspects/financing-nuclear-energy.aspx#:~:text=A%20nuclear%20power%20plant%20project%20is%20characterised%20by,key%20determinant%20of%20the%20cost%20of%20electricity%20generated (accessed on 9 January 2023).
26. Dunning, H. Construction Delays Make New Nuclear Power Plants Costlier Than Ever, Imperial College News. Available online: https://www.imperial.ac.uk/news/186487/construction-delays-make-nuclear-power-plants/ (accessed on 9 January 2023).
27. IAEA | Power Reactor Information System (PRIS). Available online: https://pris.iaea.org/PRIS/About.aspx (accessed on 9 January 2023).
28. Gustafson, T. Russian Nuclear Power—Unsanctioned—Is Prospering Worldwide, Bne Intelli News. 2023. Available online: https://www.intellinews.com/gustafson-russian-nuclear-power-unsanctioned-is-prospering-worldwide-266160/?source=russia&_cldee=2dIo3ATr48BrNb-EG7yRuyXKkQ_B7fBQUgv5TBWM1s0poTIFAYCYGJhNA_mfW-_n&recipientid=contact-e856b0eb075fe7118105e0071b6ee571-fd9d23408b29403eb2e80feea4c60e15&utm_source=ClickDimensions&utm_medium=email&utm_campaign=Morgentelegramm%20Versand%20DE&esid=017571c3-9990-ed11-aad1-000d3aaa0e50 (accessed on 16 January 2023).
29. Malik, F. *Strategy—Navigating the Complexity of the World*; Campus Verlag: Frankfurt, Germany; New York, NY, USA, 2013.
30. International Nuclear Safety Group. Defence in Depth in Nuclear Safety, INSAG-10, IAEA. 1996. Available online: https://www-pub.iaea.org/MTCD/Publications/PDF/Pub1013e_web.pdf (accessed on 10 January 2023).
31. Merk, B.; Litskevich, D.; Whittle, K.R.; Bankhead, M.; Taylor, R.J.; Mathers, D. On a Long Term Strategy for the Success of Nuclear Power. *Energies* **2017**, *10*, 867. [CrossRef]
32. Merk, B.; Stanculescu, A.; Chellapandi, P.; Hill, R. Progress in reliability of fast reactor operation and new trends to increased inherent safety. *Appl. Energy* **2015**, *147*, 104–116. [CrossRef]
33. Merk, B.; Detkina, A.; Litskevich, D.; Patel, M.; Noori-kalkhoran, O.; Cartland-Glover, G.; Degueldre, C. A First Step towards Zero Nuclear Waste—Advanced Strategic Thinking in Light of iMAGINE. *Energies* **2022**, *15*, 7209. [CrossRef]
34. Generation IV Goals, Generation IV International Forum (GIF). Available online: https://www.gen-4.org/gif/jcms/c_9502/generation-iv-goals#:~:text=Eight%20technology%20goals%20have%20been%20defined%20for%20Generation,and%20reliability%2C%20and%20proliferation%20resistance%20and%20physical%20protection (accessed on 10 January 2023).
35. Merk, B.; Litskevich, D.; Bankhead, M.; Taylor, R.J. An innovative way of thinking nuclear waste management—Neutron physics of a reactor directly operating on SNF. *PLoS ONE* **2017**, *12*, e0180703. [CrossRef]
36. Merk, B.; Litskevich, D.; Gregg, R.; Mount, A.R. Demand driven salt clean-up in a molten salt fast reactor—Defining a priority list. *PLoS ONE* **2018**, *13*, e0192020. [CrossRef]

37. Merk, B.; Detkina, A.; Litskevich, D.; Drury, M.; Noori-kalkhoran, O.; Cartland-Glover, G.; Mount, A.R. Defining the Challenges-Identifying the Key Poisoning Elements to Be Separated in a Future Integrated Molten Salt Fast Reactor Clean-Up System for imagine. *Appl. Sci.* **2022**, *12*, 4124. [CrossRef]
38. Merk, B.; Detkina, A.; Litskevich, D.; Noori-Kalkhoran, O.; Cartland-Glover, G. A Helios-Based Dynamic Salt Clean-Up Study for imagine. *Appl. Sci.* **2022**, *12*, 8748. [CrossRef]
39. Merk, B.; Detkina, A.; Atkinson, S.; Litskevich, D.; Cartland-Glover, G. Evaluation of the Breeding Performance of a NaCl-UCl-Based Reactor System. *Energies* **2019**, *12*, 3853. [CrossRef]
40. Merk, B.; Detkina, A.; Litskevich, D.; Atkinson, S.; Cartland-Glover, G. The Interplay between Breeding and Thermal Feedback in a Molten Chlorine Fast Reactor. *Energies* **2020**, *13*, 1609. [CrossRef]
41. Merk, B.; Detkina, A.; Atkinson, S.; Litskevich, D.; Cartland-Glover, G. On the Dimensions Required for a Molten Salt Zero Power Reactor Operating on Chloride Salts. *Appl. Sci.* **2021**, *11*, 6673. [CrossRef]
42. Merk, B.; Detkina, A.; Atkinson, S.; Litskevich, D.; Cartland-Glover, G. Evaluating Reactivity Control Options for a Chloride Salt-Based Molten Salt Zero-Power Reactor. *Appl. Sci.* **2021**, *11*, 7447. [CrossRef]
43. Merk, B.; Detkina, A.; Atkinson, S.; Litskevich, D.; Cartland-Glover, G. Innovative Investigation of Reflector Options for the Control of a Chloride-Based Molten Salt Zero-Power Reactor. *Appl. Sci.* **2021**, *11*, 6795. [CrossRef]
44. Merk, B.; Litskevich, D.; Detkina, A.; Cartland-Glover, G.; Atknison, S.; Bankhead, M. A Zero-power Facility as a Multi-fold Opportunity to Support Quick Progress in Innovative Reactor Development. *Atw-Int. J. Nucl. Power* **2021**, *66*, 59–64.
45. Adamov, E.O.; Rachkov, V.I. New Technological Platform for the National Nuclear Energy Strategy Development, Russian Academy of Sciences Bulletin. *Energy* **2017**, *64*, 3–12.
46. Orlov, V.V.; Filin, A.I.; Lopatkin, A.V.; Glazov, A.G.; Sukhanov, L.P.; Volk, V.I.; Poluektov, P.P.; Ustinov, O.A.; Vorontsov, M.T.; Leontiev, V.F.; et al. The closed on-site fuel cycle of the brest reactors. *Prog. Nucl. Energy* **2005**, *47*, 171–177. [CrossRef]
47. Vadim, L.; Valeriy, R. BREST-OD-300—Demonstration of Natural Safety Technologies. *Atw Int. Z. Fuer Kernenerg.* **2021**, *66*, 41–45.
48. *Program on Technology Innovation: Government and Industry Roles in the Research, Development, Demonstration, and Deployment of Commercial Nuclear Reactors: Historical Review and Analysis*; 3002010478; EPRI: Palo Alto, CA, USA, 2017.
49. Rosatom, G.H.K. A Workshop on the Development of a Liquid-Salt Reactor Was Held at the MCC. 2019. Available online: https://sibghk.ru/news/9068-na-gkhk-proshlo-rabochee-soveshchanie-po-voprosu-sozdaniya-zhidkosolevogo-reaktora.html?_x_tr_sl=ru&_x_tr_tl=en&_x_tr_hl=en-GB (accessed on 16 January 2023). (In Russian)
50. Bosbach, D.; Modolo, G.; Tromm, W. Partitioning and Efficient Transmutation, Studie mit Fokus auf innovativen Strategien in RUSsland (PETRUS). 1501639A-B. 2023; *unpublished work*.

energies

MDPI

Article

Influence of Cooling Water Parameters on the Thermal Performance of the Secondary Circuit System of a Modular High-Temperature Gas-Cooled Reactor Nuclear Power Plant

Xin Wang [1], Gang Zhao [1], Xinhe Qu [1,*], Xiaoyong Yang [1], Jie Wang [1] and Peng Wang [2]

[1] Institute of Nuclear and New Energy Technology, Advanced Nuclear Energy Technology Cooperation Innovation Center, Key Laboratory of Advanced Reactor Engineering and Safety of Ministry of Education, Tsinghua University, Beijing 100084, China; xin-wang21@mails.tsinghua.edu.cn (X.W.); zhaogang88@mail.tsinghua.edu.cn (G.Z.); xy-yang@tsinghua.edu.cn (X.Y.); wjinet@tsinghua.edu.cn (J.W.)

[2] State Nuclear Electric Power Planning, Design & Research Institute Co., Ltd., Beijing 100095, China; wangpeng@snpdri.com

* Correspondence: qxh2018@mail.tsinghua.edu.cn

Abstract: This study quantitatively analysed the influence of cooling water parameters on the performance of a modular high-temperature gas-cooled reactor (MHTGR) nuclear power plant (NPP). The secondary circuit system and cold-end system were modelled using EBSILON software, version 16.0. The influence of cooling water inlet temperature and mass flow rate on the thermal performance of the secondary circuit system was analysed over the full power range with the goal of optimising net power. Under 100% rated condition, for each 1 °C increase in cooling water inlet temperature between 10 and 33 °C, the net power and cycle efficiency decreased by 0.67 MW and 0.14%, respectively, whereas the heat consumption rate increased by 28.72 kJ/(kW·h). The optimal cooling water mass flow rates corresponding to cooling water inlet temperatures of 16 °C and 33 °C were obtained. The optimal cooling water mass flow rate decreased nonlinearly with decreasing power levels. At a cooling water inlet temperature of 33 °C, an increase in cooling water mass flow rate from the designed value (7697.61 kg/s) to the optimal value (10,922.14 kg/s) resulted in a 1.03 MW increase in net power. These findings provide guidelines for MHTGR NPP operation optimisation and economic improvement, especially under high-temperature weather conditions.

Keywords: modular high-temperature gas-cooled reactor; condenser; cooling water; high-temperature weather; EBSILON software

Citation: Wang, X.; Zhao, G.; Qu, X.; Yang, X.; Wang, J.; Wang, P. Influence of Cooling Water Parameters on the Thermal Performance of the Secondary Circuit System of a Modular High-Temperature Gas-Cooled Reactor Nuclear Power Plant. *Energies* **2023**, *16*, 6560. https://doi.org/10.3390/en16186560

Academic Editor: Roman Vadimovich Davydov

Received: 31 July 2023
Revised: 31 August 2023
Accepted: 2 September 2023
Published: 12 September 2023

1. Introduction

The modular high-temperature gas-cooled reactor (MHTGR) belongs to the category of very high-temperature reactor (VHTR), which is one of six advanced reactor types for generation IV nuclear energy systems [1]. Owing to its inherent safety and higher core outlet temperature (700–950 °C), MHTGR is suitable for high-efficiency power generation and a variety of process heat applications [2–7]. The cold-end system [8–10] is an important section of an MHTGR nuclear power plant (NPP). This system is used to cool exhaust steam into condensate and discharge waste heat into the external environment, which determines the final parameters of the steam Rankine cycle. The cold-end system of the Chinese MHTGR NPP adopts an open-loop design method using seawater as the cooling water and the ocean as the final heat sink [2]. After passing through the condenser, the seawater is redischarged into the external environment; it thus undergoes only one cycle. Cooling water parameters such as inlet temperature and mass flow rate affect the performance of the cold-end system, which in turn affects the thermal performance of the secondary circuit system of an NPP.

With the frequent occurrence of extreme high-temperature weather in recent years, the inlet temperature of the cooling water in NPPs has received increasing attention. In

May 2022, Electricite de France (EDF) stated that excessive cooling water (river water) inlet temperatures may reduce the power generation of NPPs in France [11]. Owing to various factors, including high-temperature weather, the EDF's nuclear power generation in the third quarter of 2022 was 55.1 TWh, a decrease of 36.3% compared with that of the third quarter of 2021 [12]. The influence of the cooling water inlet temperature on the thermal performance of NPPs has been studied. Laskowski et al. [13] used EBSILON software to establish a simulation model for a boiling water reactor NPP. The simulation results showed that when the cooling water temperature rose from 10 °C to 28 °C, the plant's total output power decreased by 91.405 MW, and the total efficiency decreased by 2.773%. Attia [14] established a mathematical model for a typical NPP secondary circuit system and performed an energy analysis. The results showed that the system thermal efficiency and output power decrease by 0.152% and 0.444%, respectively, when the cooling water inlet temperature increases by 1 °C. Durmayaz et al. [15] studied the influence of the cooling water inlet temperature on the thermal performance of a conceptual pressurised water reactor NPP in combination with climate conditions in Turkey. This study offers valuable insights into the site selection of power plants.

The cooling water mass flow rate also plays a crucial role in optimising the operation of steam turbine units [16–18]. In a power plant, adjusting the blade angle or the rotational speed of the pump rotor can alter the mass flow rate of cooling water entering the condenser [19–21]. Zhang et al. [22] developed a coupling algorithm using a genetic algorithm–back propagation neural network and a heat transfer model for components of cold-end systems. They determined the optimal mass flow of cooling water for a 350 MW unit at different ambient temperatures. Laskowski et al. [23] approached the optimisation goal from the perspective of the second law of thermodynamics, aiming to minimise the system's entropy generation rate. They obtained a series of optimum cooling water mass flow rates for varying conditions of the power plant. Wu et al. [24] studied a 660 MW thermal power-generating unit under various loads. The optimal cooling water mass flow rates were calculated. The results demonstrated a nonlinear trend, in which the optimal mass flow rate gradually decreased with diminishing load. Błaszczyk et al. [25] focused on choosing the appropriate cooling water mass flow rate for marine power units.

Laskowski et al. [26] analysed the effect of different cooling water mass flow rates on the thermal performance of a 225 MW thermal power plant in Poland within the load range of 40% to 100%. This study revealed that within a certain range, an increment in the cooling water mass flow rate led to an increment in overall power generation; however, it also resulted in a rapid increase in electricity usage by the cooling water pumps. They concluded that when optimising the unit's power output, it is advisable to reduce the cooling water mass flow rate when the unit load falls below 60%.

Most of the existing literature considers thermal power plants and light water reactor NPPs as research objects to examine the influence of cooling water parameters on the thermal performance of power plants. However, there is no research on MHTGR NPPs in the literature. Among the studies that have been conducted, the selected range of cooling water inlet temperature variations is generally limited, and the adverse effects of high-temperature cooling water on plant operations have not been fully considered. Additionally, considering the involvement of MHTGR NPPs in peak load regulation after grid connection, the influence of cooling water parameters over the full power range of NPP operation has to be analysed.

In this study, an energy analysis of the secondary circuit system of an MHTGR NPP was conducted. The construction, as well as verification of the system model, was completed using EBSILON software. Quasi-steady-state simulations were performed within the range of 30% to 100% rated conditions (RC) to quantitatively analyse the impact of varying cooling water parameters on the thermal performance of both the cold-end and secondary circuit systems. The maximum net power was selected as the optimisation goal, and the optimisation function of the EBSILON software was employed to determine the optimal mass flow rate of cooling water for various power levels of the power plant. Furthermore, at 100% RC, the corresponding optimal cooling water mass flow rates for different cooling water inlet temperatures were obtained. The results provide a basis for evaluating the impact of high-temperature weather on the thermal performance of power plants. Moreover, they can serve as a reference for MHTGR NPPs to optimise their cold-end systems.

2. Model and Validation

2.1. Description of the Model

The secondary circuit system of an MHTGR NPP was used as the research object. It included two steam generators, a steam turbine, a deaerator, two parallel high-pressure (HP) regenerative heaters, three low-pressure (LP) regenerative heaters, pumps, and other components (Figure 1). The first and second circuits of the NPP use helium and water as coolants, respectively. The nuclear island of the plant is equipped with two reactor modules, which provide heat to two steam generators and heat the main feed water into superheated steam. After mixing in the main steam header, the main steam from the two steam generators is directed to the steam turbine. The steam turbine consists of an HP cylinder and two LP cylinders. The arrangement of the LP cylinders follows a dual-flow reverse configuration. Two steam extraction points (I and II) in the HP cylinder provide steam to the HP heaters and deaerator, respectively. The steam extracted from point I is divided into two parts that enter the two parallel HP heaters. The LP cylinders are equipped with three steam extraction points (III, IV, and V), which provide steam for the three LP heaters. Pumps A and B are the two main feed water pumps, pump C is the condensate pump and pumps D and E are the two cooling water pumps.

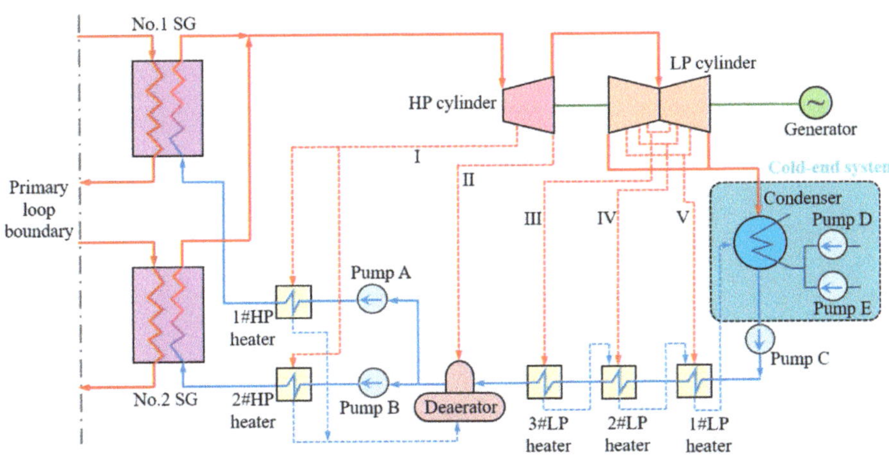

Figure 1. Schematic diagram of the secondary circuit system of the MHTGR NPP.

Table 1 presents the key parameters of the system at 100% RC. According to the design requirements, the minimum power level required for the system to maintain stable operation was 30% RC. Therefore, the power variation range selected in this study was 30–100% RC.

Table 1. Key parameters of the system at 100% RC.

Parameter	Unit	Value
Thermal power of reactor module	MW_{th}	250
Number of reactor modules	-	2
Main steam temperature	°C	566
Main steam pressure	MPa	13.24
Main feed water temperature	°C	205
Condenser back pressure	kPa	4.5

The cold-end system is an important section of NPPs, and the condenser is its core component. As indicated within the dashed box on the right side of Figure 1, the cold-end system of an MHTGR NPP mainly includes a condenser and two cooling water pumps. The condenser is a single-shell, double-pass, tube-and-shell heat exchanger arranged under the steam turbine. The exhaust steam enters from the top of the condenser and condenses into liquid water outside the coolant tubes, which then flows towards the condensate pump [2]. The local seawater, possessing a density of 1023.3 kg/m^3, is selected as the cooling water. The design value of the cooling water inlet temperature is 16 °C. Figure 2 shows a schematic diagram of the condenser. To ensure safety, the two cooling water pumps must operate simultaneously when the power plant operates.

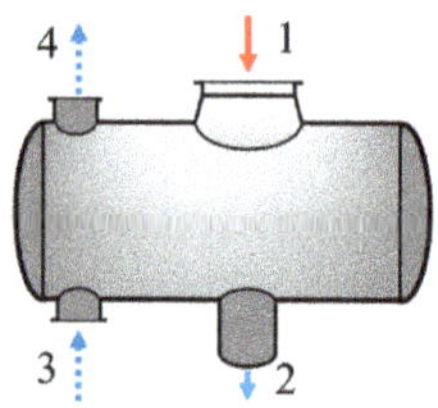

Figure 2. Schematic diagram of the condenser. 1—steam inlet; 2—condensate outlet; 3—cooling water inlet; 4—cooling water outlet.

To calculate the total heat transfer coefficient of the condenser under various operating conditions, the HEI10 (Heat Exchange Institute 10th Edition) method was selected in EBSILON software. Table 2 presents some of the condenser's design parameters, which are indispensable for the modelling process.

Table 2. Some design parameters of the condenser [2,27].

Item	Unit	Value
Tube material	-	titanium
Heat transfer surface area	m^2	12,000
Tube outer diameter	mm	28
Tube inner diameter	mm	27
Length of tubes	mm	1084
Cooling water inlet temperature	°C	16
Maximum cooling water inlet temperature	°C	33
Cleanliness factor	-	0.85
Cooling water mass flow rate	kg/s	7697.61

EBSILON software can model various thermal systems under design and off-design conditions. The simulation is based on the principles of conservation of mass, energy, and momentum. The Gauss–Seidel iteration method, which is both fast and accurate, was selected for the calculation in EBSILON. Generally, the default iteration precision is 10^{-7}, and the maximum number of iterations is 999. The basic process of EBSILON modelling is shown in Figure 3. First, individual components are selected according to engineering requirements and combined into equipment units. Second, according to the system layout, the equipment units are connected in sequence to form a thermal system framework. Then, the parameters of different pipelines, such as mass flow rate, temperature and pressure, are set, and the internal parameters of the components are set to form a complete system model. Finally, a simulation calculation is performed using the model, and the calculation data are analysed. Figure 4 shows a model of the secondary circuit thermal system of the MHTGR power plant constructed using EBSILON software. The main components are marked, which correspond to those in Figure 1. Both the red and blue lines represent pipes, with the difference being the transported fluids, namely steam and water, respectively. The numbers 1–9 correspond to the nine distinct positions within the system, matching the numbers in Figure 5. To improve the precision of the model, steam leakage from the shaft seals at all levels of the high- and low-pressure cylinders were considered during the modelling. Under 100% RC, the mass flow rate of the shaft seal steam leakage is 0.83 kg/s, accounting for approximately 0.45% of the main steam mass flow rate.

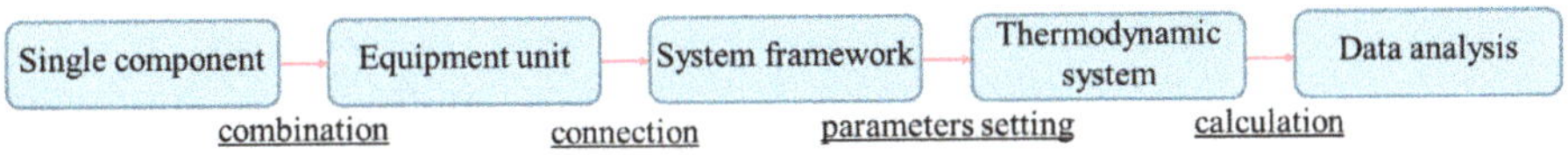

Figure 3. Modelling process in EBSILON software.

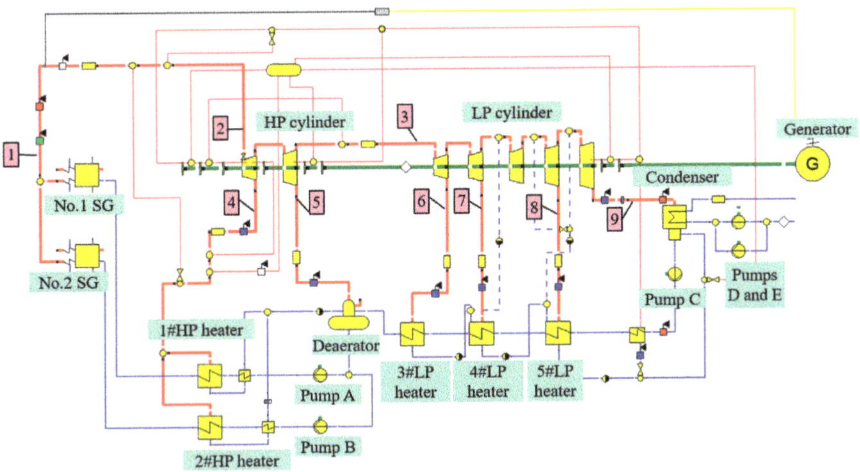

Figure 4. EBSILON model of the secondary circuit system of the MHTGR NPP.

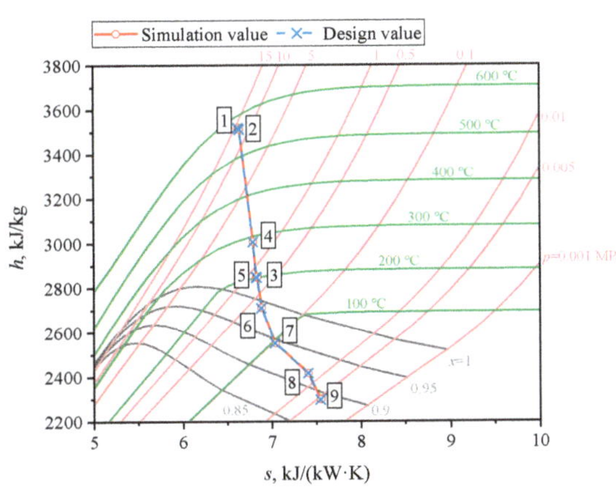

Figure 5. Enthalpy and entropy diagram of the steam expansion process in the steam turbine under 100% RC.

2.2. Model Validation

The quasi-steady-state simulations were conducted under typical operating conditions to verify the accuracy of the model. The simulation values were compared with the design values for the key parameters. The five typical operating conditions were 100%, 90%, 75%, 50% and 30% RC. In Figure 5; 2–5 represent the expansion process of steam in the HP cylinder, and 3–9 represent the expansion process of steam in the LP cylinders. The simulation values were mostly similar to the design values.

The EBSILON model employs a No. 39 component to determine the main steam mass flow rate under off-design conditions. One end of the component is connected to the generator to measure the gross power generation, whereas the other end is connected to the main steam header. The desired gross power generation can be manually set within the No. 39 component as the input under off-design conditions. The system then iteratively calculates the mass flow rate of the main steam required to achieve the target gross power generation. Table 3 lists three parameters, namely the main steam mass flow rate, main feed water temperature, and deaerator pressure, under five typical operating conditions, along with the relative errors between the simulation and design values. Table 4 displays the relative errors in the mass flow rate, pressure, and specific enthalpy of the extraction steam at the five regenerative extraction points I–V at 75% RC. For confidentiality reasons, the specific simulation and design values are not provided in Tables 3 and 4.

Table 3. Relative errors of different parameters under five typical operating conditions.

Operating Condition, % RC	Relative Error, %		
	Main Steam Mass Flow Rate, kg/s	Main Feed Water Temperature, °C	Deaerator Pressure, MPa
100	0	−0.01	0.00
90	−0.13	−0.04	−0.13
75	−0.56	−0.14	−0.38
50	−0.69	0.18	0.8
30	−2.98	−0.19	−0.57

Table 4. Relative errors of extraction parameters at extraction points under 75% RC.

Extraction Point	Relative Error, %		
	Mass Flow Rate, kg/s	Pressure, MPa	Specific Enthalpy, kJ/kg
I	−0.88	−0.54	−0.01
II	−0.37	−0.38	0.00
III	−0.69	−0.51	0.00
IV	−1.03	−0.48	−0.11
V	0.11	2.42	0.06

The results presented in Figure 5, Tables 3 and 4 demonstrate that the model performed well under five typical operating conditions: 100%, 90%, 75%, 50%, and 30% RC. The simulated values are closely aligned with the design values for the MHTGR NPP. Except for the individual parameters, the relative errors of the simulated values of the main parameters were within 1%. Therefore, the model established in this study was verified and could be used in subsequent studies.

3. Mathematical Model

Mathematical models of the condenser and pump were developed. The thermal performance evaluation indicators of the system include the net power, cycle efficiency, and heat consumption rate, each with specific definitions and formulas provided in Section 3.3. The prerequisites for the simulation experiments in this study are detailed in Section 3.4.

3.1. Condenser

In accordance with the law of energy conservation, the heat transfer rate of condenser $\dot{Q}_{CON}$ can be expressed as:

$$\dot{Q}_{CON} = \dot{m}_{CW}(h_{CW,out} - h_{CW,in}) = \dot{m}_{CON}(h_{CON,in} - h_{CON,out}). \quad (1)$$

where $\dot{m}_{CW}$ and $\dot{m}_{CON}$ are the mass flow rates of the cooling water and exhaust steam flowing into the condenser, respectively; $h_{CW,in}$ and $h_{CW,out}$ represent the specific enthalpy of the cooling water at the inlet and outlet of the condenser, respectively; $h_{CON,in}$ and

$h_{\text{CON,out}}$ are the specific enthalpy of the exhaust steam flowing into the condenser and condensate flowing out of the condenser, respectively.

The end temperature difference (δT) and logarithmic mean temperature difference ($LMTD$) are crucial indicators for assessing the condenser's heat transfer performance [28,29]. The condenser end temperature difference refers to the variance between the saturated steam temperature of the condenser and the cooling water outlet temperature. Both indicators can be expressed as follows:

$$\delta T = T_{\text{CON}} - T_{\text{CW,out}}, \tag{2}$$

$$LMTD = \frac{(T_{\text{CON}} - T_{\text{CW,in}}) - (T_{\text{CON}} - T_{\text{CW,out}})}{\ln \frac{T_{\text{CON}} - T_{\text{CW,in}}}{T_{\text{CON}} - T_{\text{CW,out}}}}, \tag{3}$$

where T_{CON} refers to the steam saturation temperature in the condenser, and $T_{\text{CW,in}}$ and $T_{\text{CW,out}}$ are the cooling water temperatures flowing into and out of the condenser, respectively.

Condenser back pressure and steam saturation temperature satisfy the following formula [30,31]. It's an empirical formula that captures solely numerical relationship.

$$p_{\text{CON}} = 0.00981 \times \left(\frac{T_{\text{CON}} + 100}{57.66} \right)^{7.46}. \tag{4}$$

3.2. Pump

The isentropic efficiency of the pump can be described as:

$$\eta_{\text{i}} = \frac{h_{\text{PU,i}} - h_{\text{PU,in}}}{h_{\text{PU,out}} - h_{\text{PU,in}}}, \tag{5}$$

where $h_{\text{PU,in}}$ and $h_{\text{PU,out}}$ represent the specific enthalpy of water at the inlet and outlet of the pump in the actual process, respectively, and $h_{\text{PU,i}}$ indicates the specific enthalpy of water at the pump outlet in the ideal isentropic process. In this study, the isentropic efficiency of the main feed water pumps and condensate pump was 0.83, and that of the cooling water pumps was set to 0.8.

3.3. Evaluation Index

In this study, net power P_{n} was defined as the difference between the gross power generation of the power plant and the power usage of the main feed waters, condensate water, and cooling water pumps. This is expressed as follows:

$$P_{\text{n}} = P_{\text{g}} - (W_{\text{PU,A}} + W_{\text{PU,B}} + W_{\text{PU,C}} + W_{\text{PU,D}} + W_{\text{PU,E}}), \tag{6}$$

where A and B represent the two main feed water pumps; C represents the condensate pump and D and E represent the two cooling water pumps.

The cycle efficiency η_{cy} is the ratio of the net power to the thermal power input into the secondary circuit system, which can be calculated as:

$$\eta_{\text{cy}} = \frac{P_{\text{n}}}{\dot{Q}_{\text{cy}}}, \tag{7}$$

where $\dot{Q}_{\text{cy}}$ indicates the thermal power input to the secondary circuit system, that is, the heat absorption power of the secondary side of the steam generator.

The heat consumption rate q refers to the heat required by the turbo-generator unit to produce 1 kW·h of electric energy [28], which is an important index for measuring the thermal performance of a power plant and can be expressed as [31]:

$$q = 3600 \times \frac{\dot{Q}_{\mathrm{cy}}}{P_{\mathrm{n}}}. \tag{8}$$

3.4. Prerequisite

Simulation experiments were carried out based on the following prerequisites.

(1) Eight operating conditions (100%, 90%, 80%, 70%, 60%, 50%, 40%, and 30% RC) were selected for research. Under the same operating conditions, the thermal power input to the secondary circuit system and main steam mass flow rate remained constant.
(2) The pipeline pressure loss at extraction point I was set to 3%, whereas that for extraction points II, III, IV, and V was assumed to be 5% based on engineering experience.
(3) Water and steam thermal properties were determined according to the IAPWS-IF97 standard [32].
(4) Heat dissipation to the external environment from other components was neglected, in addition to the main steam header and condenser. The main steam underwent a reduction in temperature and pressure from 571 °C and 13.9 MPa at the steam generator outlet to 566 °C and 13.24 MPa, respectively, at the steam turbine inlet.

4. Results and Discussion

The study was divided into two parts, encompassing a power range of 30% to 100% RC, employing a control variable approach. In the first part, the cooling water mass flow rate was maintained constant at the design value of 7697.61 kg/s, while studying the influence of changing cooling water inlet temperatures within the range of 10–33 °C. The temperature range was determined based on the combination of industrial and local climatic conditions. In the second part, a constant cooling water inlet temperature was maintained at the design value of 16 °C, and the mass flow rate of cooling water was varied from 6000 to 14,000 kg/s to examine the influence of mass flow rate. Subsequently, under 100% RC, the mass flow rate corresponding to the maximum net power is defined as the optimal cooling water mass flow rate. This optimal mass flow rate is determined for various power levels of the MHTGR NPP, with the cooling water inlet temperature set at 16 °C. Both the cooling water inlet temperature and mass flow rate were simultaneously altered, leading to the creation of a three-dimensional surface diagram illustrating the net power change. Additionally, the optimal cooling water mass flow rate is determined separately for cooling water inlet temperatures of 16 °C and 33 °C, both under 100% RC.

4.1. Influence of Cooling Water Inlet Temperature

Figures 6 and 7 show how the condenser end temperature difference and logarithmic mean temperature difference changed with respect to the cooling water inlet temperature. The data indicate that with a growth in the cooling water inlet temperature, both the condenser end and logarithmic mean temperature differences decrease. Additionally, at a given temperature, the condenser end and logarithmic mean temperature differences decreased as the power level of the NPP decreased.

Figure 8 illustrates the variation in the condenser back pressure at varying cooling water inlet temperatures. With the rise in cooling water inlet temperature, the condenser back pressure increased across the various power levels. Notably, at a cooling water inlet temperature of 33 °C, there was a significant increase in the condenser back pressure. As an example, consider the case of 100% RC, where without changing the cooling water mass flow rate, the condenser back pressure reaches 10.8 kPa at a cooling water inlet temperature of 33 °C, signifying a substantial 139.79% increase from the design parameter of 4.5 kPa at an inlet temperature of 16 °C. An elevated condenser back pressure in the Rankine cycle results in a higher exhaust steam temperature and enthalpy, leading to a decline in the cycle efficiency, as quantitatively demonstrated in Figure 9 and Table 5.

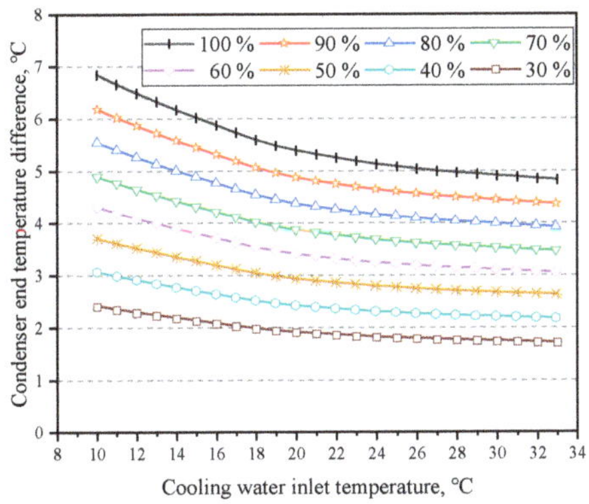

Figure 6. Variation of condenser end temperature difference with cooling water inlet temperature.

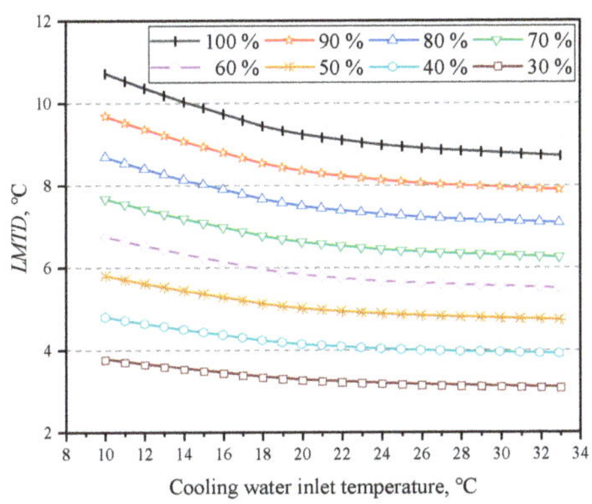

Figure 7. Variation of logarithmic mean temperature difference with cooling water inlet temperature.

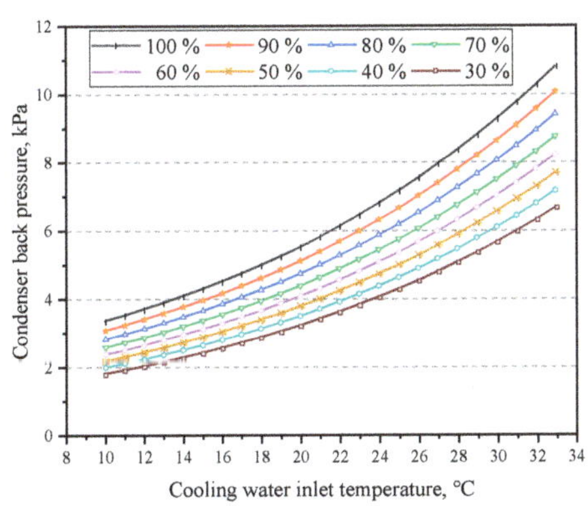

Figure 8. Variation of condenser back pressure with cooling water inlet temperature.

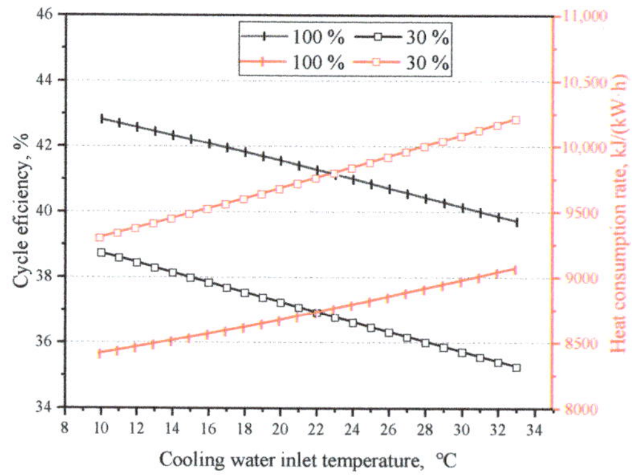

Figure 9. Cycle efficiency and heat consumption rate under 100% and 30% RC.

Table 5. Variation in cycle efficiency, heat consumption rate, and net power.

Operating Condition, % RC	Cooling Water Inlet Temperature, °C	Cycle Efficiency, %	Heat Consumption Rate, kJ/(kW·h)	Net Power, MW
100	10	0.75	−149.66	3.68
	33	−2.36	510.84	−11.64
30	10	0.90	−224.55	1.47
	33	−2.57	702.70	−4.18

As depicted in Figure 9, under 100% and 30% RC, the cycle efficiency exhibited a decline as the cooling water inlet temperature increased, whereas the heat consumption rate increased accordingly and the rate of change was approximately linear.

Figure 10 more comprehensively shows the change of unit net power with cooling water inlet temperature under 30–100% RC. Meanwhile, Table 5 displays the changes in cycle efficiency, heat consumption rate, and net power at cooling water inlet temperatures of 10 and 33 °C under 100% and 30% RC. These calculations are based on simulation results for a cooling water inlet temperature of 16 °C, the designated value for this parameter. The average value was then determined. Under 100% RC and within the temperature range of 10–33 °C, the system experiences a decrease of 0.67 MW in net power, a 0.14% reduction in cycle efficiency, and an increase of 28.72 kJ/(kW·h) in heat consumption rate for every 1 °C rise in the cooling water inlet temperature.

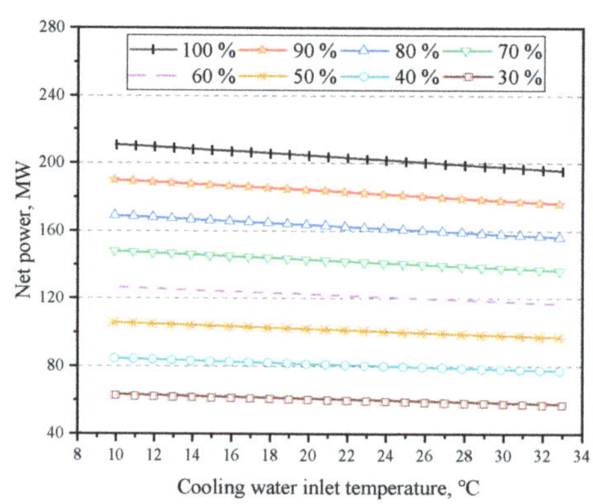

Figure 10. Variation of net power with cooling water inlet temperature.

Table 5 clearly illustrates the significant impact on unit net power when the cooling water inlet temperature reaches 33 °C without adjusting the cooling water mass flow rate.

Specifically, under 100% and 30% RC, the net power of the unit decreased by 11.64 and 4.18 MW, respectively. The reduction in net power can be attributed to the change in enthalpy drop during the steam expansion process. For instance, under 100% RC, the condenser back pressure was 4.5 and 10.8 kPa, respectively, when the cooling water inlet temperature was 16 and 33 °C. Additionally, the corresponding enthalpy drop during steam expansion work was 1219.39 and 1125.52 kJ/kg, respectively. Evidently, a higher cooling water inlet temperature resulted in a significant reduction in the enthalpy drop of the steam.

4.2. Influence of Cooling Water Mass Flow Rate

Figures 11–13 show the variations in the end temperature difference, logarithmic mean temperature difference, and back pressure of the condenser with respect to the cooling water mass flow rate. With the growth of the mass flow rate, the end temperature difference, logarithmic mean temperature difference, and back pressure of the condenser decreased under the eight operating conditions.

Although an increase in the cooling water mass flow rate results in a reduction in the condenser back pressure, which increases the gross power generation of the unit, an increase in the mass flow rate will lead to a rapid increment in the power consumption of the cooling water pumps, which will have an adverse effect on the net power of the NPP. Figure 14 shows the variation of net power with cooling water mass flow rate under 100% and 90% RC. The net power exhibits a clear trend, first increasing and then decreasing. This trend was consistently observed under the other operating conditions as well.

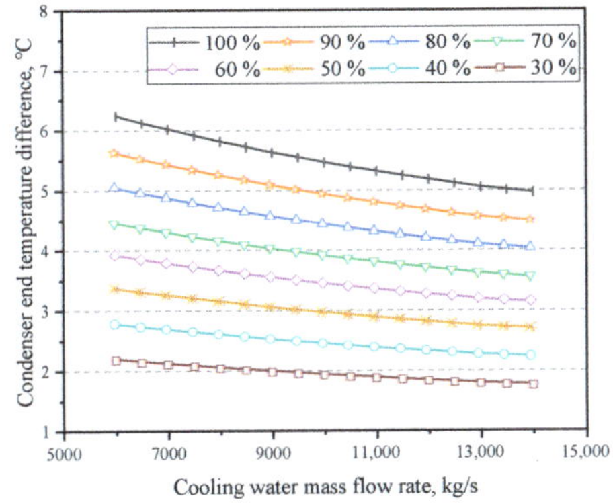

Figure 11. Variation of condenser end temperature difference with cooling water mass flow rate.

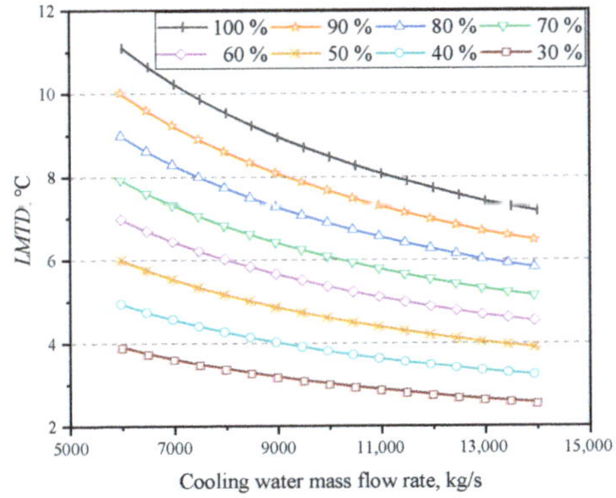

Figure 12. Variation of logarithmic mean temperature difference with cooling water mass flow rate.

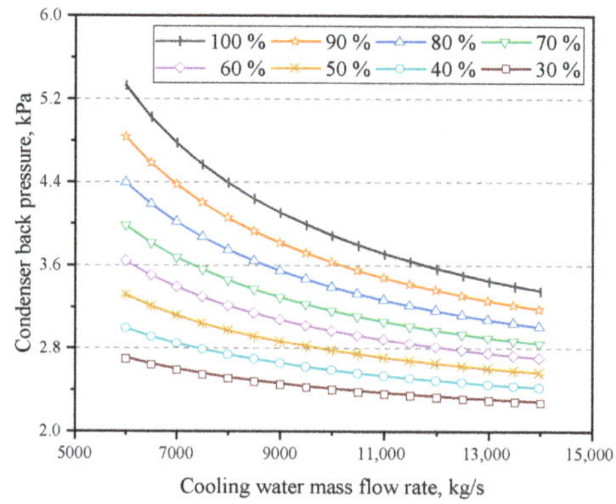

Figure 13. Variation of condenser back pressure with cooling water mass flow rate.

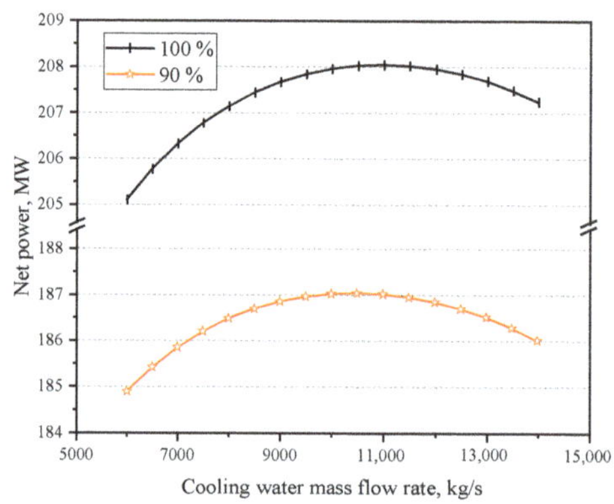

Figure 14. Variation of net power with cooling water mass flow rate under 100% and 90% RC.

Figure 15 shows the net power of the unit under eight operating conditions, where the red dotted line indicates the optimal mass flow rate of cooling water corresponding to the different power levels. As the power level decreased, the optimal cooling water mass flow rate gradually decreased, and the trend became nonlinear.

Figure 16 shows the variations in the cycle efficiency and heat consumption rate with cooling water mass flow rates under 100% and 30% RC. The net power exhibits a corresponding trend (Figure 17). When the mass flow rate increased, the cycle efficiency initially increased and then declined, whereas the heat consumption rate initially decreased and then increased. By adopting the optimal mass flow rate, the heat consumption rate was reduced by 46.11 and 15.18 kJ/(kW·h) at 100% and 30% RC, respectively. Furthermore, Figure 17 visually demonstrates the variation of the net power simultaneously with the cooling water inlet temperature and mass flow rate, aligning with the trends observed in Figures 10 and 15. At 33 °C and 6000 kg/s, the system's net power reached its minimum value of 193.57 MW.

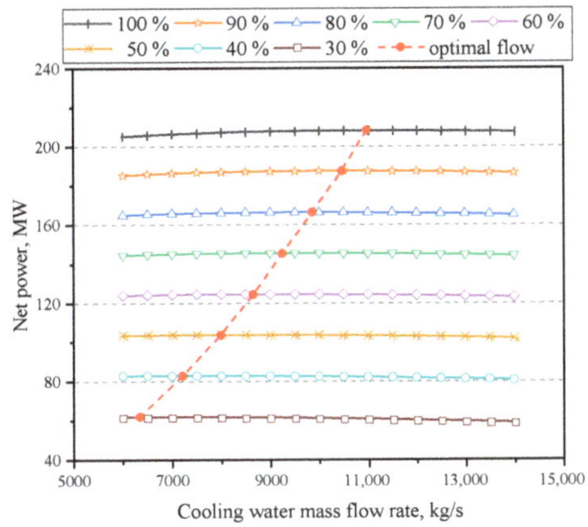

Figure 15. Net power changes with cooling water mass flow rate.

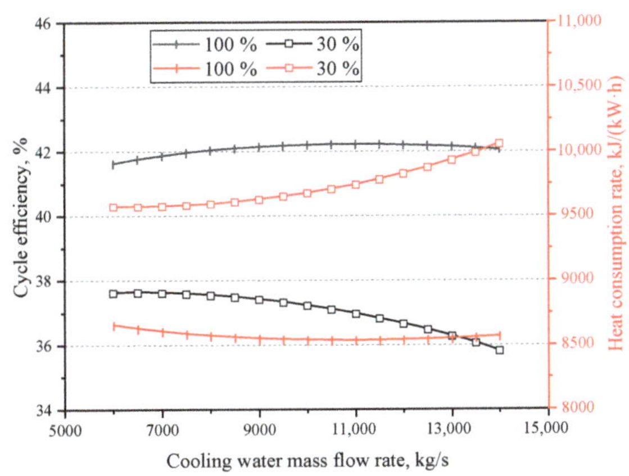

Figure 16. Cycle efficiency and heat consumption rate at 100% and 30% RC.

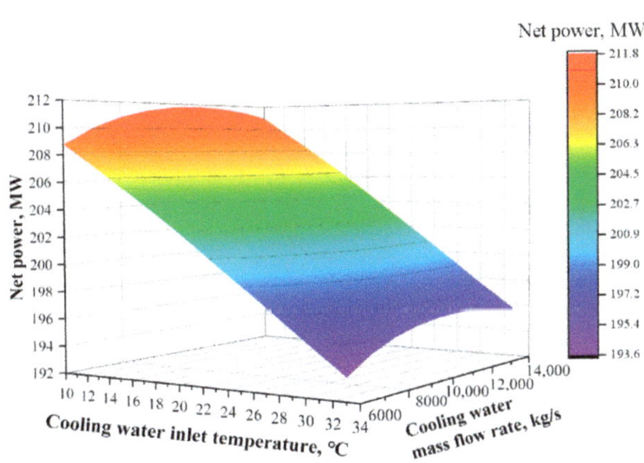

Figure 17. Influence of cooling water inlet temperature and mass flow rate on net power.

Figure 18 illustrates the net power increase of the unit under eight operating conditions when adopting the optimal cooling water mass flow rate. The green line represents the design value of the cooling water mass flow rate of 7697.61 kg/s. The deviation between the optimal and design mass flow rates first decreased and then increased as the power level decreased. At higher power levels, the net power can be significantly improved by using the optimal cooling water mass flow rate. As an illustration, when the cooling water inlet temperature was 16 °C and the cooling water mass flow rate was reduced to the optimal mass flow rate under 30% RC, the unit's net power increased by 96.62 kW. This increase translates to an additional 96.62 MWh of electricity output for every 1000 h of operation at 30% RC. Furthermore, a comparison of Figure 18a,b reveals that the optimal cooling water mass flow rates under the two temperature conditions were relatively close.

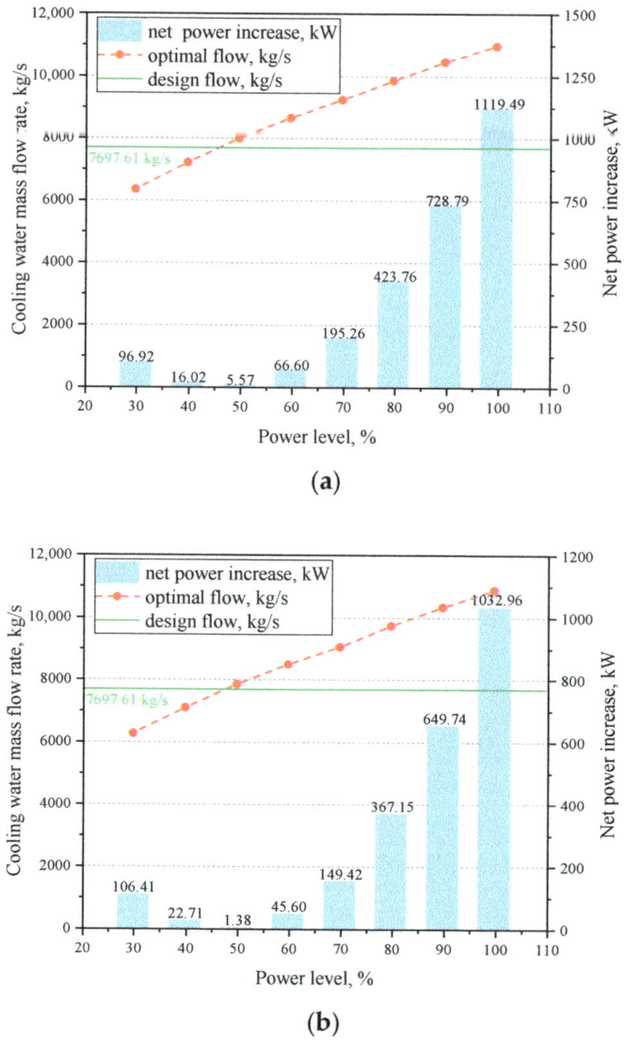

Figure 18. Net power increase under eight operating conditions. (**a**) $T_{CW,in}$ = 16 °C; (**b**) $T_{CW,in}$ = 33 °C.

5. Conclusions

The study utilised EBSILON software to establish and validate the thermal model of the secondary circuit in an MHTGR NPP. Variable operating condition simulations were performed across the full power range of stable operation of the NPP, using net power, cycle efficiency, and heat consumption rate as thermal performance evaluation indicators. A wider spectrum of cooling water inlet temperatures was selected, and the effects of both the inlet temperature and mass flow rate on the system's thermal performance were studied in detail. Taking the maximum net power as the optimisation goal, the optimal cooling water mass flow rate for various power levels of the power plant was determined when the cooling water inlet temperature was 16 °C. Additionally, the optimal cooling water mass flow rate was determined for cooling water inlet temperatures of 16 °C and 33 °C under 100% RC. The following conclusions were drawn:

(1) Excessive cooling water inlet temperature causes a substantial elevation in condenser back pressure, resulting in a decline in the thermal performance of the unit. Particularly, at a cooling water inlet temperature of 33 °C, the condenser back pressure reached 10.8 kPa, which surpasses the design parameter of 4.5 kPa by a considerable 139.79%. Additionally, under both 100% and 30% RC, the net power decreased by 11.64 and 4.18 MW, respectively, and the cycle efficiencies decreased by 2.36% and 2.57%, respectively. Moreover, the heat consumption rate increased by 510.84 and 702.70 kJ/(kW·h). Furthermore, under 100% RC, a rise in the inlet temperature from 10 °C to 33 °C led to a reduction in the net power and cycle efficiency of the system by 0.67 MW and 0.14%, respectively, with the heat consumption rate escalating by 28.72 kJ/(kW·h) for every 1 °C increment.
(2) Within a certain range, the net power of the system first increased and then decreased as the cooling water mass flow rate increased. By considering the maximum net power as the optimisation goal, the optimal cooling water mass flow rate for various operating conditions can be determined. The findings indicated a nonlinear decrease in the optimal cooling water mass flow rate as the power level decreased.
(3) At a higher power level, appropriately increasing the cooling water mass flow rate can significantly increase the net power, whereas, at a lower power level, the cooling water mass flow rate must be reduced correspondingly to increase the net power. Taking cooling water inlet temperatures of 16 °C and 33 °C, respectively, when the mass flow rate increased from the design value to the optimal value under 100% RC, the net power increased by 1119.49 and 1032.96 kW, respectively.

In summary, this study demonstrated the qualitative and quantitative relationships between the cooling water parameters and evaluation indicators, providing a basis for evaluating the impact of high-temperature weather on the thermal performance of MHTGR NPPs. The optimisation of the cold-end system is a complex process that requires a comprehensive consideration of various factors and costs. This study considers maximum net power as the optimisation goal to obtain the optimal cooling water mass flow rate under different operating conditions, which can be used as a part of the cold-end system optimisation of MHTGR NPPs and provides a reference for power plant operation optimisation and economic improvement.

Author Contributions: Conceptualization, G.Z., X.Q. and X.Y.; methodology, G.Z., X.Q. and X.Y.; software, G.Z., X.Q. and X.W.; validation, X.W.; investigation, X.Q. and X.W.; resources, X.Q. and J.W.; data curation, X.W.; writing—original draft preparation, X.W.; writing—review and editing, X.Q.; supervision, X.Q. and P.W.; project administration, X.Q. and J.W.; funding acquisition, J.W. and X.Y. All authors have read and agreed to the published version of the manuscript.

Funding: This research was funded by the National Key R&D Program of China (Grant No. 2018YFB1900500), and LingChuang Research Project of China National Nuclear Corporation.

Data Availability Statement: Due to confidentiality requirements of the partner enterprise, the data used in this study are not publicly available. For access to the data related to this research, please contact the corresponding author.

Conflicts of Interest: No conflict of interest exists in the submission of this manuscript, and the manuscript is approved by all authors for publication.

Nomenclature

h	specific enthalpy, [kJ/kg]
$LMTD$	logarithmic mean temperature difference, [°C]
$\dot{m}$	mass flow rate, [kg/s]
T	temperature, [°C]
T_{CON}	steam saturation temperature in condenser, [°C]
p	pressure, [Pa]
p_{CON}	condenser back pressure, [kPa]
s	specific entropy, [kJ/(kg·K)]
q	heat consumption rate, [kJ/(kW·h)]
P_n	net power, [MW]
P_g	gross power generation, [MW]
W	power consumption, [MW]
$\dot{Q}_{CON}$	condenser heat transfer rate, [kW]
$\dot{Q}_{cy}$	thermal power input to the secondary circuit system, [MW]
Abbreviations	
HP	high-pressure
LP	low-pressure
RC	rated condition
Greek letters	
η_{cy}	cycle efficiency of the system, [%]
η_i	isentropic efficiency of the pump, [%]
δT	condenser end temperature difference, [°C]
Subscripts	
cy	cycle
CON	condenser
CW	cooling water
i	isentropic process
in	inlet
out	outlet
PU	pump

References

1. Doe, U.S. A Technology Roadmap for Generation IV Nuclear Energy Systems. In *Nuclear Energy Research Advisory Committee and the Generation IV International Forum*; Doene: Berryville, VA, USA, 2002; pp. 9–11.
2. Kugeler, K.; Zhang, Z. *Modular High-Temperature Gas-Cooled Reactor Power Plant*; Springer: Berlin/Heidelberg, Germany, 2018. [CrossRef]
3. Wang, J.; Ding, M.; Yang, X.; Wang, J. Performance comparison and optimization of two configurations of (Very) high temperature gas-cooled reactors combined cycles. *Ann. Nucl. Energy* **2016**, *94*, 279–287. [CrossRef]
4. Qu, X.; Zhao, G.; Wang, J. Thermodynamic evaluation of hydrogen and electricity cogeneration coupled with very high temperature gas-cooled reactors. *Int. J. Hydrogen Energy* **2021**, *46*, 29065–29075. [CrossRef]
5. Ni, H.; Qu, X.; Peng, W.; Zhao, G.; Zhang, P. Study of two innovative hydrogen and electricity co-production systems based on very-high-temperature gas-cooled reactors. *Energy* **2023**, *273*, 127206. [CrossRef]
6. Haneklaus, N.; Schröders, S.; Zheng, Y.; Allelein, H.-J. Economic evaluation of flameless phosphate rock calcination with concentrated solar power and high temperature reactors. *Energy* **2017**, *140*, 1148–1157. [CrossRef]
7. Gao, Q.; Zhang, P.; Sun, Q.; Zhang, P.; Chen, S.; Peng, W. Experimental and numerical investigation of sulfuric acid decomposition for hydrogen production via iodine–sulfur cycle. *Energy Convers. Manag.* **2023**, *289*, 117167. [CrossRef]
8. Lin, X.; Song, H.; Wang, L.; Guo, Y.; Liu, Y. Cold-end integration of thermal system in a 1000 MW ultra-supercritical double reheat power plant. *Appl. Therm. Eng.* **2021**, *193*, 116982. [CrossRef]
9. Wang, C.; Liu, M.; Zhao, Y.; Qiao, Y.; Chong, D.; Yan, J. Dynamic modeling and operation optimization for the cold end system of thermal power plants during transient processes. *Energy* **2018**, *145*, 734–746. [CrossRef]
10. Yang, J.; Zhang, R. Optimization for steam turbine cold-end system of a nuclear power plant and sensitivity analysis. *J. Nucl. Eng. Radiat. Sci.* **2017**, *3*, 014501. [CrossRef]
11. Ahmad, A.; Covatariu, A.; Ramana, M.V. A stormy future? Financial impact of climate change-related disruptions on nuclear power plant owners. *Util. Policy* **2023**, *81*, 101484. [CrossRef]
12. EDF. Annual Results 2022 Presentation. Available online: https://www.edf.fr/sites/groupe/files/2023-04/annual-results-2022-presentation-2023-02-17-v3-2.pdf (accessed on 28 July 2023).

13. Laskowski, R.; Theibech, M.; Uzunow, N. The effect of cooling water temperature on the performance of a BWR nuclear power plant. *Mod. Eng.* **2021**, *3*, 5–13. Available online: https://mengineering.eu/index.php/me/article/view/85/81 (accessed on 28 July 2023).
14. Attia, S.I. The influence of condenser cooling water temperature on the thermal efficiency of a nuclear power plant. *Ann. Nucl. Energy* **2015**, *80*, 371–378. [CrossRef]
15. Durmayaz, A.; Sogut, O.S. Influence of cooling water temperature on the efficiency of a pressurized-water reactor nuclear-power plant. *Int. J. Energy Res.* **2006**, *30*, 799–810. [CrossRef]
16. Laković, M.S.; Stojiljković, M.M.; Laković, S.V.; Stefanović, V.P.; Mitrović, D.D. Impact of the cold end operating conditions on energy efficiency of the steam power plants. *Therm. Sci.* **2010**, *14*, 53–66. [CrossRef]
17. Ahmadi, G.R.; Toghraie, D. Energy and exergy analysis of Montazeri Steam Power Plant in Iran. *Renew. Sustain. Energy Rev.* **2016**, *56*, 454–463. [CrossRef]
18. Medica-Viola, V.; Pavković, B.; Mrzljak, V. Numerical model for on-condition monitoring of condenser in coal-fired power plants. *Int. J. Heat Mass Transf.* **2018**, *117*, 912–923. [CrossRef]
19. Harish, R.; Subhramanyan, E.E.; Madhavan, R.; Vidyanand, S. Theoretical model for evaluation of variable frequency drive for cooling water pumps in sea water based once through condenser cooling water systems. *Appl. Therm. Eng.* **2010**, *30*, 2051–2057. [CrossRef]
20. Xia, L.; Liu, D.; Zhou, L.; Wang, P.; Chen, Y. Optimization of a seawater once-through cooling system with variable speed pumps in fossil fuel power plants. *Int. J. Therm. Sci.* **2015**, *91*, 105–112. [CrossRef]
21. Wang, W.; Liu, J.; Zeng, D.; Lin, Z.; Cui, C. Variable-speed technology used in power plants for better plant economics and grid stability. *Energy* **2012**, *45*, 588–594. [CrossRef]
22. Zhang, W.; Ma, L.; Jia, B.; Zhang, Z.; Liu, Y.; Duan, L. Optimization of the circulating cooling water mass flow in indirect dry cooling system of thermal power unit using artificial neural network based on genetic algorithm. *Appl. Therm. Eng.* **2023**, *223*, 120040. [CrossRef]
23. Laskowski, R.; Smyk, A.; Lewandowski, J.; Rusowicz, A.; Grzebielec, A. Selecting the cooling water mass flow rate for a power plant under variable load with entropy generation rate minimization. *Energy* **2016**, *107*, 725–733. [CrossRef]
24. Wu, T.; Wei, H.; Ge, Z.; Yang, L.; Du, X. Cooling water mass flow optimization for indirect dry cooling system of thermal power unit under variable output load. *Int. J. Heat Mass Transf.* **2019**, *133*, 1–10. [CrossRef]
25. Btaszczyk, A.; Gfuch, J.; Gardzilewicz, A. Operating and economic conditions of cooling water control for marine steam turbine condensers. *Pol. Mar. Res.* **2011**, *18*, 48–54. [CrossRef]
26. Laskowski, R.; Smyk, A.; Rusowicz, A.; Grzebielec, A. Optimization of the cooling water mass flow rate under variable load of a power unit. *Appl. Therm. Eng.* **2021**, *191*, 116874. [CrossRef]
27. Dong, L.; Zhou, Z. Simulation study on the condensing system of high-temperature reactor demonstration power station HTR-PM 200 based on Vpower. *Shenyang Inst. Eng. (Nat. Sci.)* **2011**, *7*, 16–19. [CrossRef]
28. Laskowski, R. Relations for steam power plant condenser performance in off-design conditions in the function of inlet parameters and those relevant in reference conditions. *Appl. Therm. Eng.* **2016**, *103*, 528–536. [CrossRef]
29. Laskowski, R.; Smyk, A.; Rusowicz, A.; Grzebielec, A. A useful formulas to describe the performance of a steam condenser in off-design conditions. *Energy* **2020**, *204*, 117910. [CrossRef]
30. Yang, T.; Wang, W.; Zeng, D.; Liu, J.; Cui, C. Closed-loop optimization control on fan speed of air-cooled steam condenser units for energy saving and rapid load regulation. *Energy* **2017**, *135*, 394–404. [CrossRef]
31. Feng, H.; Wang, M.; Wang, N.; Xu, Y.; He, S.; Gao, M. Influence of environmental parameters on the cold-end and thermal system of coal-fired power plant based on Ebsilon simulation. *Therm. Sci. Eng. Prog.* **2022**, *32*, 101340. [CrossRef]
32. *IAPWS-IF97*; Revised Release on the IAPWS Industrial Formulation 1997 for the Thermodynamic Properties of Water and Steam. The International Association for the Properties of Water and Steam: Oakville, ON, Canada, 2007.

energies

MDPI

Review

Feasibility of Safe Operation of WWER-440-Type Nuclear Power Plants for Up to 60–70 Years

Tamás János Katona [1,*], Ágnes Biro [2] and Sándor Rátkai [3]

1 Faculty of Engineering and Information Technology, University of Pécs, 7622 Pécs, Hungary
2 VEIKI Energia+ Ltd., 2030 Érd, Hungary; janosi@hu.inter.net
3 MVM Nuclear Power Plant Paks, 7031 Paks, Hungary; ratkai@npp.hu
* Correspondence: katona.tamas@mik.pte.hu

Abstract: Most WWER-440 reactors are operated over the planned operational lifetime of 30 years. Now, the owners/operators are preparing for the second phase of extending the operation. The paper presents an overview of the most important aspects of ageing of the main components of the WWER-440-type reactors based on the operational and ageing management experiences and primary research efforts supporting the operation. The paper aims to demonstrate that the expectations for the operability of these reactors for up to 60–70 years are realistic. The scope of the review is limited to the ageing of the reactor pressure vessel, vessel internals, and steam generator, which are the lifetime-limiting components besides the containment structure. Some of the corrective actions, improvements of in-service inspections, maintenance, and ageing management are also considered in the paper. It is shown that the selection of materials and unavoidable material degradation processes do not limit the operation of the WWER-440 reactors for up to 60–70 years under the established operational environment and ageing management practice.

Keywords: reactor pressure vessel; reactor internals; steam generator; irradiation embrittlement; long-term operation; stress corrosion cracking; dissimilar material welds; ageing management

check for updates

Citation: Katona, T.J.; Biro, Á.; Rátkai, S. Feasibility of Safe Operation of WWER-440-Type Nuclear Power Plants for Up to 60–70 Years. *Energies* **2023**, *16*, 4170. https://doi.org/10.3390/en16104170

Academic Editor: Roman Vadimovich Davydov

Received: 22 April 2023
Revised: 14 May 2023
Accepted: 16 May 2023
Published: 18 May 2023

1. Introduction

Nuclear electricity generation, mainly due to geopolitical changes that dramatically affect supply chains of energy sources, is an indispensable element of the electricity system of several national economies in the European Union. In addition to developing new capacities and deploying new types, such as small and modular reactors, continuing the safe operation of existing nuclear power plants (NPPs) and extending their authorized operating time is a reasonable approach for nuclear operating countries. At many NPPs worldwide, the second period of operational time extension is already prepared.

The rationale for operating beyond the planned operating lifetime is two-fold. First, from the safety point of view, nuclear power plants shall comply with the requirements even at the very last minute of operation. This indicates that the entire functioning of plant safety systems should be ensured despite the ageing, i.e., from a safety point of view, the operation plants are working as new ones. Second, the operating plants have already reimbursed capital costs, and specific replacements, well-organized management of ageing, and maintenance require certain financial expenses to prolong the operation.

Safety enhancement programs of existing plants were rational responses to accidents at the Three Mile Island NPP (1978) and the Chernobyl NPP (1986). Since the reactors worldwide would have reached the end of their authorized operating time, preparing for the long-term operation of existing reactors was the only solution for keeping the nuclear option alive. Pioneering work has been carried out in the USA, where the NRC launched a comprehensive nuclear plant ageing research program in 1985 to identify and resolve technical safety issues related to the ageing of nuclear power plant systems, structures, and components (SSCs). In 1995, the 10 CFR Part 54 License Renewal Act was issued, which allows US nuclear power plants to extend their service life license for 40 years and receive an operating license extension for up to 20 years [1]. The technical bases for the renewal of operating licenses are summarized in [2]. In the US, the technical and scientific basis and legal framework for the second 20-year service life extension were also developed in the USA [3,4].

Similar to the US practice, many countries have established systematic ageing management programs at nuclear power plants and extended the operational lifetime over the initially defined term. At the international level, the International Atomic Energy Agency (IAEA) launched the program for Safety Aspects of Long-Term Operation (SALTO) [5] and published its first guidelines for ageing management [6]. The IAEA also established technical services for the Member States [7]. Furthermore, starting in 2010, the IAEA has been coordinating international cooperation in the frame of the International Generic Ageing Lessons Learned (IGALL) program, guiding the ageing management programs and long-term operation (LTO) activities of the Member States [8,9].

Recently, the countries in the European Union have been operating their NPPs over the originally established lifetime and preparing repeated extensions. Specific interest is devoted to the WWER-440 reactors. The WWER reactors are light-water-moderated and water-cooled pressurized water reactors. The name comes from the Russian "водо-водяной энергетический реактор", which transliterates as Vodo-Vodyanoi Energetichesky Reaktor (Water-Water Energetic Reactor, WWER, but the Russian-type acronym, VVER is also often used). The 440 indicates the nominal rated electrical power of 440 MWe. There are several subtypes of the WWER-440 reactor design. The WWER-440/213-type operates in Slovakia, Hungary, Czech Republic, Finland, Ukraine, and Russia, the WWER-440/270 in Armenia, the WWER-440/230 and WWER-440/179 in Russia [10]. The reliable power supply and the stability of the economy of these countries largely depend on the operation of the WWER-440 plants. Since the operational lifetime was initially set to 30 years, the extension for the operating time was necessary since these reactors were placed into operation between 1971 and 2000 (except for the Unit 3 of the Mochovce NPP in Slovakia, connected to the grid in 2023).

The technical and moral basis of long-term operation is the regular safety evaluation and upgrading of these plants. Practically, the safety of these reactors can no longer be judged based on their original design. After the Chernobyl accident, the safety of all WWER-440 plants has been evaluated and assessed by national and international review programs. Safety deficiencies have been identified, and extensive safety improvement programs have been implemented [11]. Moreover, the rated power has been increased above 500 MWe at several plants. These safety improvements form the technical and moral basis for the first and repeated service life extensions (e.g., in [12]). Furthermore, the post-Fukushima stress test and improvements enhanced the external hazard safety and preparedness for severe accidents.

The long-term operability of WWER-440-type reactors has been intensively investigated by the operating countries and by international organizations, such as the IAEA, as mentioned above. In the European Union In the 5th and 6th Framework Programs, the subsequent research projects called "NULIFE" and "VERLIFE" have made an essential effort regarding irradiation embrittlement of the WWER-440 reactor pressure vessel (RPV) and other components, e.g., [13]. In 2003, an "Unified Procedure for Lifetime Assessment of Components and Piping in WWER NPPs during Operation" was developed and upgraded in 2008 [14]. The final procedure, the "Guidelines for Integrity and Lifetime Assessment of Components and Piping in WWER NPPs during Operation", was developed as a joint product of the European Union and AEA project in 2011. Recently, the Electric Power Research Institute (EPRI) US coordinates the evaluation of operational experiences regarding ageing and ageing management experiences and the related research activity in the frame of the Pressurized Water Reactor Materials Reliability Program (MRP) [15].

The regulatory approval for LTO depends on national nuclear safety regulations. LTO can be approved via formal operation license renewal for the fixed term, or the prolongation of operation can be justified by periodic safety reviews every 10 years. For example, in Hungary, the regulatory authority approves and controls the long-term operation via a combination of license renewal and periodic safety reviews. In all cases, the owners of plants have some target total operational lifetime in mind. Table 1 shows the status of the operational time extension at all 35 operating WWER-440 plants. The four oldest plants, Novovoronezh-4, and Kola -1 and -2 received the first operating time extension permit for 15 years of operation from 2001 to 2004. Units 1–4 of the Paks NPP received a 20-year service extension permit between 2012 and 2017. Some WWER-440 Units (Kola-1, Kola-2, Novovoronezh-4, Armenian-2) have already been granted an operating license for ~60 years thanks to the subsequent extension of the operating time. The Finish Regulator approved the operation of the Loviisa WWER-440/213 reactors for 70 years, up to 2057. For Bohunyice 3–4 Units, it is also planned to license the 60-year operating time. At Paks NPP, Hungary, preparation for the subsequent license renewal (up to 70 years) is launched.

Table 1. The WWER-440 nuclear power plants and status of operational lifetime extension.

Plant/Type	Connected to the Grid	Years of Extension/Valid to		Approved Operation
		1st	2nd	Time
Armenian-2, WWER-440/270	5 January 1980	10/2016	20/2036	56
Bohunice-3, WWER-440/213	20 August 1984	10/2024 *	**	40
Bohunice-4, WWER-440/213	9 August 1985	10/2025 *	**	40
Dukovany-1, WWER-440/213	24 February 1985	10/2026 *	**	41
Dukovany-2, WWER-440/213	30 January 1986	10/2027 *	**	41
Dukovany-3, WWER-440/213	14 November 1986	10/2027 *	**	41
Dukovany-4, WWER-440/213	11 June 1987	10/2027 *	**	40
Kola-1, WWER-440/230	29 June 1973	15/2018	15/2033	60
Kola-2, WWER-440/230	9 December 1974	15/2019	15/2034	60
Kola-3, WWER-440/213	24 March 1981	15/2026	?	55
Kola-4, WWER-440/213	11 October 1984	25/2029	?	55
Loviisa-1, WWER-440/213	8 February 1977	20/2027	2023	70
Loviisa-2, WWER-440/213	4 November 1980	20/2030	2023	70
Mohovce-1, WWER-440/213	4 July 1998	NA	***	30
Mohovce-2, WWER-440/213	2 December 1998	NA	***	30
Mohovce-3, WWER-440/213	31 January 2023	NA	***	30
Novovoronezh-4, WWER-440/179	28 December 1972	15/2017	15/2032	60
Paks-1, WWER-440/213	28 December 1982	20/2032	**	50
Paks-2, WWER-440/213	6 September 1984	20/2034	**	50
Paks-3, WWER-440/213	28 September 1986	20/2036	**	50
Paks-4, WWER-440/213	16 August 1987	20/2037	**	50
Rovno-1, WWER-440/213	22 December 1980	10/2020	10/2030	50
Rovno-2, WWER-440/213	22 December 1981	10/2021	?	40

* Justifying the safe operation for the next 10 years by periodic safety review. ** The owner intends to prolong the operation by another 20 years. *** New units for that thedecision on the longer than 30 years operation will be made later. "No information" is indicated by the question mark "?".

The paper presents an overview of the lifetime-limiting ageing processes of the essential reactor components based on the operational and ageing management experiences of the WWER-440-type reactors to justify the operability of these reactors for the targeted period of 60–70 years.

Recently, the lifetime-limiting ageing processes identified for WWER-440 reactors are related to the material reliability of reactor pressure vessels (RPVs), especially the irradiation-assisted stress corrosion cracking (IASCC) of the core baffle-former bolts and understanding of the void swelling and its effect on material properties of WWER-440 reactor pressure vessel internals (RVIs). Understanding the susceptibility of stainless steels used in WWER-440 to IASCC and the environmental effects on the fatigue for pressure boundary components of WWERs for the targeted LTO period is essential. According to the operational experiences, at nominal water chemistry conditions, the late-life stress corrosion cracking (SCC) is not a significant limitation for the targeted term of operation of WWER primary components. At the WWER-440 plants, titanium-stabilized stainless steel is used for the steam generator heat-exchanging tubes. Although the material has an outstanding performance record, especially at WWER-440 steam generators, the outer diameter stress corrosion cracking (ODSCC) and corrosion susceptibility in the crevice environment are interesting. The ageing phenomena of the WWERs have been identified and investigated in the EPRI MRP "MRP-471-WWER Issue Management Tables: Identifies material research gaps for WWER light water reactors" [15].

In this paper, the scope of the review of the ageing processes of WWER-440 reactors is limited to the material features and ageing of the reactor pressure vessel, vessel internals, and steam generator. Although the reactor pressure vessel (RPV) is not considered limiting for the target time of 60 to 70 years, the reliability of the RPV remains the fundamental question for the justification of the long-term operation. The analyses to justify the first operational lifetime extensions are briefly presented and compared with some recent findings regarding RPV irradiation embrittlement. Based on the operational experience of the core baffle-former bolts, the ageing management plans for 50 to 70 years of operation are also presented. The questions related to ensuring the outstanding performance of the WWER-440 steam generators are also discussed. Most important corrective actions, improvements of in-service inspections, maintenance, and ageing management are also mentioned in the paper. It is shown, despite the unavoidable material degradation processes, there are no unresolvable issues that would limit the operation of the WWER-440 reactors for up to 70 years under the established operational environment and ageing management practice.

2. Materials Aspects of the Long-Term Operation of WWER-440-Type NPPs

2.1. Reactor Pressure Vessel of WWER-440/213-Type NPPs

Reactor pressure vessel (RPV) is vital for pressurized water reactors (PWRs) and WWER reactors. Their resistance against potential brittle/non-ductile failure practically determines the lifetime of PWR and WWER reactor pressure vessels. The damaging effect of neutron radiation governs this resistance.

The WWER-440 RPVs operated in Armenia, Russia, Ukraine, and Finland have been made by Soviet manufacturers, and the other RPVs were manufactured by Skoda in Czechoslovakia. Depending on the manufacturer, there are specific differences in the manufacturing procedures and chemical compositions of materials.

WWER-440 RPVs have some significant features that are different from PWR designs. First, it was assumed that they must be transportable by rails. This requirement has some crucial consequences on vessel design, such as a smaller pressure vessel diameter, which results in a smaller water gap thickness, and thus a ~5 to 10 times higher fast neutron flux ($E > 0.5$ MeV) on the RPV is relatively high, about 10^{15} m^{-2} s^{-1}. Therefore, this requires materials with high resistance against radiation embrittlement. The WWER-440 RPVs are made from low alloy steel 15Cr2MVA (12Cr2MFA at Soviet-made RPVs) without longitudinal welds, and the circumferential submerged arc welding was made using Sv-10CrMoVTi wire. The RPV is covered internally by a welded clad of two stainless steel layers. The inner layer is non-stabilized stainless steel (Sv-07Cr25Ni13) and the outer layer is niobium-stabilized stainless steel Sv-08Cr19Ni10Mn2Nb, but Sv-07Cr19Ni10Nb is in the case of the Soviet-made RPV. The phosphorus and copper contents in the welds of

WWER-440/213 range from 0.010–0.028% for P and 0.03–0.18% for Cu. For primary data of material composition of WWER-440 RPVs see, e.g., in [16–18].

The early experiences of WWER operating countries and the results and international research programs on material embrittlement for WWER-440 RPV have been summarized, e.g., in [18,19]. Before 2000, several ageing management measures were implemented for the reduction in irradiation embrittlement and its effects on the safe operation of the RPVs, as follows:

- reduction in the neutron flux on the RPV by low leakage core design and dummy assemblies;
- lessening the thermal shock by heating the water for the emergency core cooling;
- introduction of advanced surveillance programs;
- annealing of the critical circumferential RPV welds, where it was needed.

Thanks to the abovementioned measures and development of the evaluation methodology and fluence calculations, the operability of RPVs of all reactors could be justified for the first 10 to 20 years of extension above the 30 years of the original operational lifetime based on pressurized thermal shock (PTS) calculations. The critical welds of the RPVs of Loviisa-1, Rovno -1 and -2, and Kola NPPs have been annealed only.

For example, in the case of Paks NPP, the PTS calculations based on analysis of postulated embedded flaws endorsed the possibility of 50 years of operation for all four reactors. The approach applied for PTS analysis of Paks RPVs complied with the Hungarian Regulatory Guidance. According to this, the structural integrity against brittle fracture of the RPV is ensured if the actual ductile-brittle transition temperature (DBTT), T_k of its critical components is less than the maximum allowable component-specific DBTT, T_k^{allow}. The analysis compares the material's static fracture toughness K_{Ic} and stress intensity factor K_I calculated from the given loading situation. The significant steps of the analysis were as follows (see, e.g., in [20–24]):

1. Identification of the critical components of the RPV.
   Figure 1 shows the critical RPV locations subjects of fracture mechanics analysis.
   In the case of WWER-440 RPV, significant are the base metal, circumferential weld No. 5/6, weld heat affected zone, cladding in the belt line region, and other circumferential welds, including those in the nozzle region.
2. Selection of the PTS initiating events with higher frequency than 10^{-5}/year based on the probabilistic safety analysis (PSA).
3. Thermal-hydraulic analysis for each selected PTS-initiating event.
4. Neutron fluence calculations. The neutron transport calculation methodology has been validated by comparison with calculation benchmarks and measurements. The first validation test has been performed in the frame of the REDOS Project [22,23]. The experimental data stem from measurements performed on a mock-up simulating WWER-440 core and vessel wall installed in the LR-0 zero-reactor in the Nuclear Research Institute, Řež, Czech Republic. The second validation test of the calculation methodology has been performed with plant-specific data. Here, the calculated and measured reaction rates on the activation detectors besides the surveillance specimens were used to validate the calculation model. In the case of Paks NPP, the measurements performed for the 8–11 and 9–11 fuel cycles of unit 2, and the measurements performed for the 7–11 cycles of unit 4 were used. In both cases, the results of flux uncertainty estimates were less than 10% [21].
   Based on the refueling history and future core configurations, the end-of-life fluences (for 50 and 60 operating years) are calculated for the RPV wall and the surveillance position.
   In the case of Paks NPP, the end-of-life fluence for the base metal of the RPVs at Paks NPP for 60 years varies between 3.21×10^{20} cm^{-2} and 3.36×10^{20} cm^{-2} and for the critical No. 5/6 welding varies between 2.33×10^{20} cm^{-2} and 2.36×10^{20} cm^{-2}.

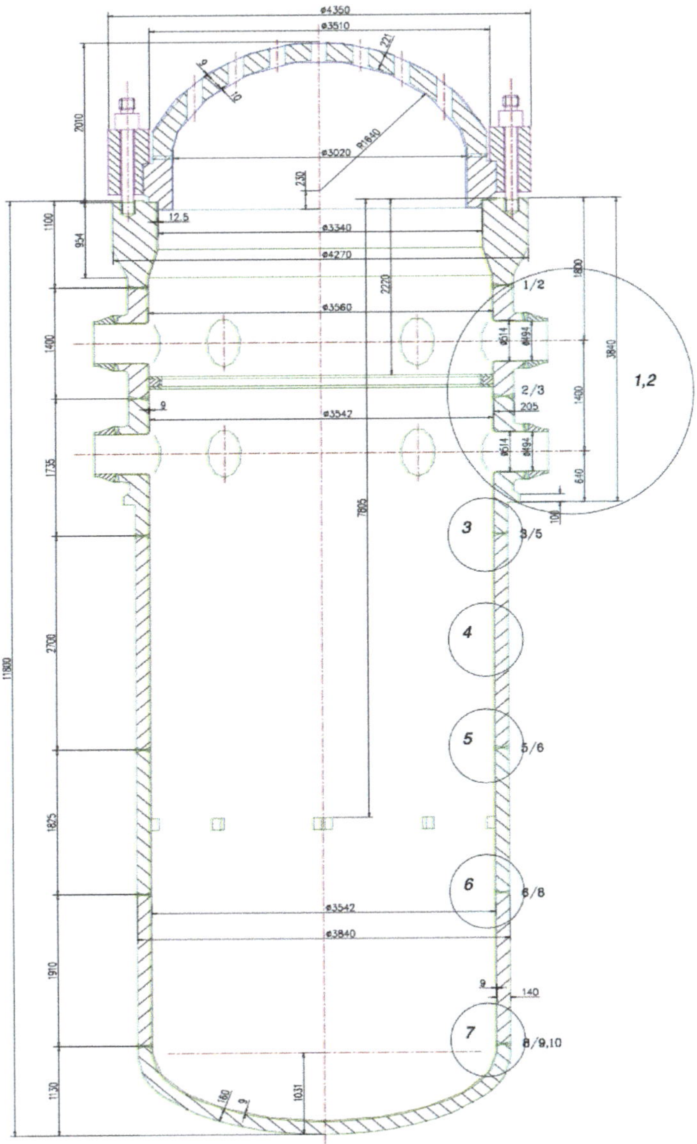

Figure 1. RPV of WWER-440/213 at Paks NPP and the locations selected for fracture mechanics analysis.

5. Temperature and stress field calculations for RPV wall. Fracture mechanics calculations assume an under-cladding crack with a depth equal to 0.1 times RPV wall thickness and with an aspect ratio of 1/3, oriented in the base metal normal to the principal stress. In the circumferential weld, it is oriented circumferentially.
The results of qualified in-service inspections justified the assumption of the embedded postulated crack. Two types of inspection were applied for the entire cladding area: (1) Ultrasonic inspection from the inner surface and (2) an additional Eddy current inspection, overlapping the first 5-mm thickness of the RPV inner wall area.
The transients with screening criteria 10^{-5}/year were analyzed using the linear elastic fracture mechanics.

The most significant transients have been analyzed by applying nonlinear fracture mechanics.
The stress intensity factor, K_I, is calculated at the crack tip and the boundary between the cladding and base metal or weld. The material's static fracture toughness, K_{Ic}, is calculated via the following reference curve:

$$K_{Ic} = min\{(26 + 36 \cdot exp[0.02(T - T_k)]); 200\}, \ [\mathrm{MPa}\sqrt{\mathrm{m}}] \quad (1)$$

where T_k is the nil-ductile temperature.
In evaluating the shift of the critical temperature (ΔT_k), the effects of irradiation, thermal ageing, and fatigue were considered.
T_k and ΔT_k were determined by evaluating the Charpy impact test results of surveillance specimens for every RPV.
The VERLIFE proposed form for the dependence of T_k versus fast neutron fluence, F_n was applied, i.e., $\Delta T_k = A_F(F_n/F_0)^n + \delta T_M$, where A_F and n are empirical constants, δT_M is equal to 10 °C for base metal and 16 °C for the weld.
For example, for reactor No. 1 at Paks NPP, the dependence of T_k versus fast neutron fluence, F_n was found for the base material as

$$T_k = \left(-34.2 + 65.747\left(F_n/10^{20}\right)^{0.511} + 10\right), \ [°\mathrm{C}], \quad (2)$$

and for the weld as

$$T_k = \left(25.0 + 62.364\left(F_n/10^{20}\right)^{0.392} + 16\right), \ [°\mathrm{C}]. \quad (3)$$

The condition of postulated defect stability is $K_I \leq K_{Ic}$ from what the allowable temperature, T_k^{allow} could be derived.

6. Qualified in-service inspections and nondestructive testing verify the integrity of the cladding [21].

Based on the analysis above, the extension for 20 years above the original operational lifetime of 30 years could be justified for the RPVs of Paks NPP.

A similar procedure as above has been applied to the other WWER-440-type reactors listed in Table 1. Among national regulations and guidelines, the references [17], and [25–29] have been followed for evaluating end-of-life RPV conditions and justification of lifetime extensions for WWER-440 RPV.

Moreover, with more than a thousand reactor years of operational experience, an extensive database of RPV material surveillances, and results of excessive research, a factual statement can be drawn on the RPV material embrittlement of the WWER-440 reactors.

Notably, statistical analysis of the data obtained during impact testing of surveillance specimens of 15 WWER-440 RPVs operating in Russia, Ukraine, Armenia, Hungary, the Czech Republic, and Slovakia was performed [30]. The raw data used for the analysis were obtained from the IAEA International Database of RPV Materials.

As a result, the empirical dependences of the shift of the nil-ductile transition temperature ΔT_k on the neutron fluence have been obtained from surveillance specimens irradiated up to a fluence of 5×10^{20} cm^{-2}. The tested specimens were grouped as per copper, nickel, and phosphorus content as "clean", "almost clean", "dirty", and "highly dirty". For example, the phosphorus copper content in "clean" base metal surveillance specimens was less than 0.012% and 0.07%, respectively.

Based on the results of the irradiation tests, the influence of P and Cu to ΔT_k was evaluated at fluences up to 5×10^{20} cm^{-2} (E > 0.5 MeV), more than double the design end-of-life WWER-440 fluence value that was 2.4×10^{20} cm^{-2} for the base metal and 1.8×10^{20} cm^{-2} for the weld.

Using the empirical correlation for T_k versus fast neutron fluence, F_n proposed in [28] and the surveillance data for RPV of reactor No. 1 at Paks NPP and extrapolating for the

70 years end-of-life fluence that is less than 5×10^{20} cm^{-2}, 70 years of safe operation can be predicted as shown in Figure 2.

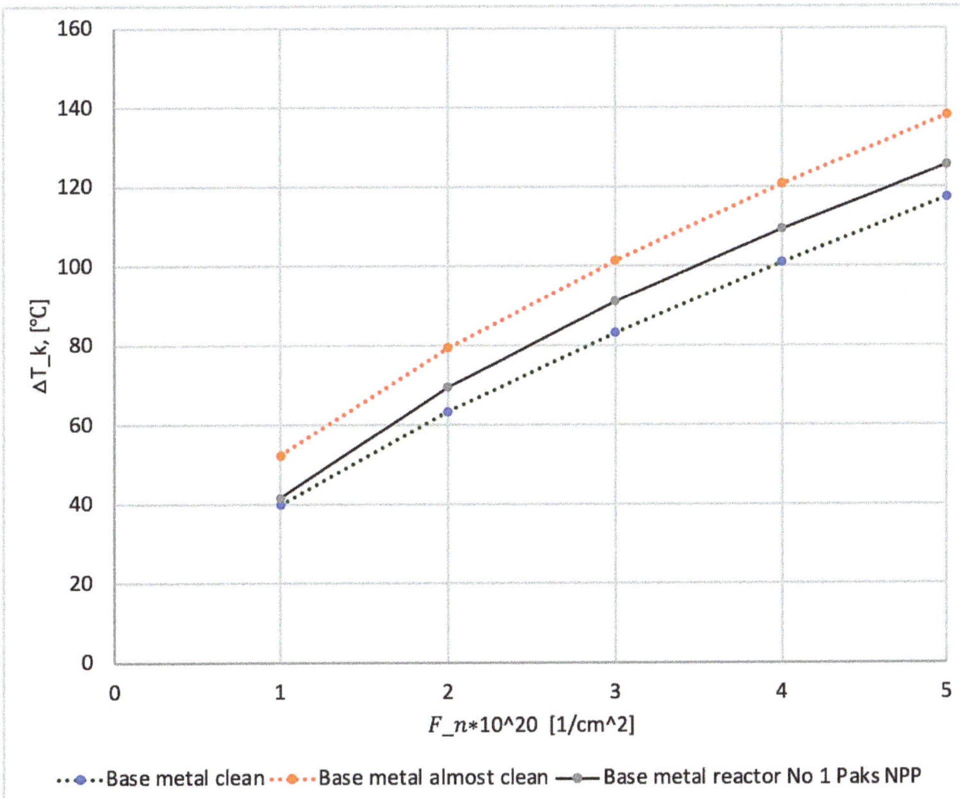

Figure 2. Nil-ductile temperature for the RPV of reactor No. 1 at Paks NPP versus fast neutron fluences compared to the prediction via the empirical formula proposed in [30].

An amendment of the regulatory limits has been proposed [30] and seems to be empirically justified. The application range of normative dependences $\Delta T_k(F_n)$ on the reactor pressure vessel can be increased from 3×10^{20} cm^{-2} to 5×10^{20} cm^{-2}.

Increasing the maximum fluence to 5×10^{20} cm^{-2} will allow the WWER-440 RPV with relatively low impurity contents (P 0.017% and Cu 0.14%) to be operated for 60–80 years without annealing the base metal and, in some cases, without annealing the irradiated welds.

2.2. Reactor Pressure Vessel Internals

The primary function of the reactor vessel internals (RVI) is to support the core and the control rod assemblies. The RVI has the additional function of directing the flow of the reactor coolant and providing shielding for the reactor pressure vessel. The RVI is subjected to neutron irradiation and exposure to the primary coolant. The core basket or barrel is part of the RVI structure consisting of the core baffle-former built from horizontal forming plates that follow the shape of the core baffles and the shape of the peripheral fuel assemblies. In the case of WWERs, the former follows the shape of the hexagonal fuel assemblies. The vertical plates/baffles surrounding the outer face peripheral fuel assemblies. The vertical plates are bolted to the horizontal former plates that are bolted to the core basket. These core baffle-former bolts hold together a structure inside the reactor vessel. This structure and the bolts are subjected to significant mechanical stress and high neutron flux.

Core baffle-former bolt degradation was first noted in the late 1980s in French PWRs. In the US, in some pressurized water reactors, the operation experiences indicate susceptibility to IASCC degradation of the baffle-former bolts [31]. Adequate in-service inspections and replacement of the damaged bolts manage the issue [32].

Figure 3 shows the core baffle-former's shape and the two bolts used for fixing the former of the WWER-440/213 reactors at Paks NPP in Hungary. In the WWER reactors, solution-annealed titanium-stabilized stainless steels (0X18H10T corresponding to Type AISI 321) have been used for RVI due to their corrosion resistance, toughness, ductility, strength, and fatigue characteristics.

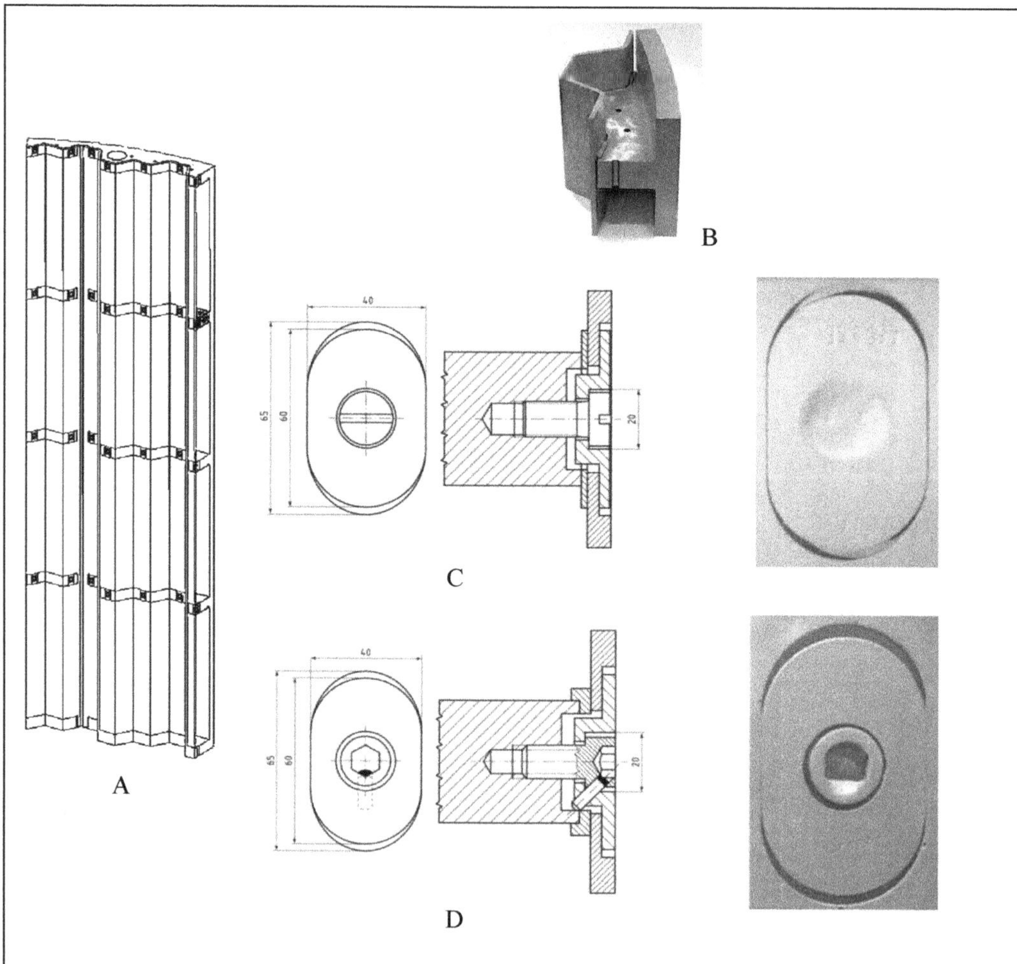

Figure 3. Core baffle-former shape (**A**), the connection of the plate to the former (**B**), and the drawing and pictures (**C**,**D**) show the two types of bolts (case Paks NPP).

There is only one case of known cracked baffle bolts in WWER-440 internals [33] found by volumetric ultrasonic examination. The damaged bolts could have cracked due to improper alignment and restricted thermal expansion. One bolt has been damaged due to irradiation-assisted intergranular stress corrosion cracking. The case has been thoroughly investigated, and the bolt has been replaced [33–36]. According to analysis [37] in the Czech Republic, the baffle-former bolts are the most susceptible subcomponent of reactor vessel internals. They are loaded by fatigue and IASCC. Both degradation mechanisms are influenced and accelerated by swelling development. Since the very limited cases of the baffle bolts were damaged, the baffle bolt IASCC has not been a high-priority issue for the first phase of the lifetime extension. Regarding swelling and baffle-former bolts ageing, the operability of the RVI structures has been justified by in-service inspections and ultrasonic testing. The performance of WWER-440 RVI materials has also been investigated and demonstrated using specimens harvested after 45 years of operation with a damaging dose of 7.9–43.0 dpa (displacement per atom) and irradiation temperature of 280–315 °C that showed negligible swelling [38].

Preparing for the extended operation above 40–50 years, the WWER operators developed in-service inspections and repair methodologies and performed ageing and stress analyses. For example, in the case of Paks NPP, Hungary, the applicable inspection and repair techniques have been reviewed. The acceptance criteria for the crack depth and cross-section for the bolts have been analyzed for different periods of testing based on the standard ASME BPVC Section III, Division 1, Subsection NG and Nonmandatory Appendix F. Despite the good operational experiences, a research effort is needed for a better understanding of the conditions contributing to void swelling and its effect on the material properties of WWER RPV internals, and for the definition of empirically justified allowable neutron doses (or dpa).

From the nuclear safety point of view, the degradation of baffle-former bolts does not directly endanger the safe operation. Since the bolts are practically replaceable, their damage does not limit the reactor's lifetime. However, the deformation of the geometry and loss of integrity of the core former structure due to the swelling and degradation of bolts should be avoided. Therefore, as part of the licensee's activity during the extended operation, especially during operations above 50 years, the utility establishes periodical inspection of the reactor vessel internals, including the volumetric inspections of the baffle-former bolts, that assure safety.

2.3. Steam Generator

The steam generators (SGs) of the type PGV-440 at WWER-440/213 are lifetime-limiting components since they are not replaceable within reasonable expenses.

The heat-exchanging tubes and the steam generator tube headers (collectors) are manufactured from titanium-stabilized stainless steel (equivalent to AISI 321). The experience regarding the ageing of WWER steam generators is summarized in the TECDOC-1577 of the International Atomic Energy Agency [39]. Based on the operational experiences, the WWER-440 steam generators demonstrated outstanding performance as it is shown in the Table 2 for the Paks NPP. Since the WWER-440 reactors are six-loop systems, there are experiences of $35 \times 6 \times 40 = 8400$ steam generator operational years available for assessing the capability to operate for 60 to 70 years (40 years is assumed as the approximate average operational time). Throughout this operational period, the steam generator's material failures could be detected at the proper time and could be repaired.

Table 2. Steam generator plugging statistics for Paks NPP, Hungary.

Unit	SG No. 1		SG No. 2		SG No. 3		SG No. 4		SG No. 5		SG No. 6	
	Tubes	%	Tubes	%	Tubes	%	Tubes	%	Tubes	%	Tubes	%
1	25	0.45	42	0.76	4	0.07	13	0.23	6	0.11	3	0.05
2	70	1.26	166	3.00	205	3.70	187	3.38	82	1.46	112	2.02
3	111	2.01	47	0.85	56	1.01	50	0.90	101	1.82	28	0.51
4	25	0.45	50	0.90	30	0.54	30	0.54	65	1.17	12	0.22

A relatively early defect of the feed-water distributor inside the SG arose due to accelerated erosion. These elements were replaced at all WWER-440 plants. The heat-exchanging surface of the SG is overdesigned by approximately 15% for the reactors with a design power output of 440 MWe and 10% for the reactors operating at an uprated power level of ≅500 MWe. The acidic crevice pH was a concern regarding the integrity of the heat exchange tubes since the titanium-stabilized stainless steel is susceptible to ODSCC in an acidic environment. However, the plugging statistics were acceptable, and the trends do not show serious operability limitations up to 50 years [40–42].

At several WWER-440 plants, the plugging trends were improved by removing copper alloys from the secondary system and introducing the high-pH secondary water chemistry that essentially slows down the ODSCC. The effect of the power uprate on the heat-exchanging tube integrity is negligible. Nevertheless, the evaluation of operational experiences is needed for establishing an even more effective mitigation technology of titanium-stabilized stainless steel steam generator tubing, mostly in crevices, while at the same time minimizing corrosion product transport and deposition in the steam generators. In accordance with some research results, these efforts are essential for extending over 50 years since this steel might be susceptible to degradation in highly alkaline pH [43,44].

In the future, alternative chemical mitigation materials should be found to replace hydrazine if its use will be restricted due to its health and environmental effects [45].

There has been an isolated case of the cracking of WWER-440 SG primary collector threaded holes exposed to primary water [46,47]. The threaded holes should not be in direct contact with the primary medium, but it was expected that the water had entered during assembly/disassembly. Failure analysis indicated that the primary cause of the cracking was the lubricant, which contained high concentrations of Sulphur and Molybdenum, high stresses, and material impurities. The collector has been repaired by cutting away and replacing the cracked part.

Critical parts of the WWER-440 steam generators are the dissimilar material welds (DMW) connecting the austenitic stainless-steel primary pipe to the K22 ferritic steel body of the steam generator shown in Figure 4.

The WWER-440-related experiences are summarized in several documents and studies, e.g., [48–55]. During the long-term operation of the SG, corrosion damages have been found that were related to intercrystallite corrosion of metal 10Ch16N25AM6, built-up by electrodes EA 395/9 [48]. According to [48], the degradation mechanisms for the flaws in the dissimilar material welds at WWERs are usually a combination of corrosion on the carbon steel side and stress-corrosion-cracking interdendritic character. The factors determining the cracking, i.e., the stresses, the material, and the environment have been thoroughly investigated, e.g., [48–58]. Cases of DMW degradations have been found and repaired at Dukovany NPP and Kola NPP [48,57,58]. The approach of ASME, 2013 Section XI was applied to evaluate allowable flaw sizes in the DMW connecting the austenitic stainless-steel primary pipe to the K22 ferritic steel body of the steam generator [52,53]. A thorough nondestructive examination indicates the performance of these newly repaired DMWs [58]. From the point of view of SG operability for up to 60–70 years, the DMW issue is manageable since verified methods are available for monitoring the degradation and repair. The practical questions are the insurance of the effectiveness of ageing monitoring of DMW that is shown in [49,53]. Moreover, improving the draining, sludge removal is essential, if possible.

The DMW issue is not limited to the steam generator. Since there are several critical locations where DMWs are applied including WWER-440 RPV safe end welds, hydro-accumulator pipe welds, and welds under the pressurizer, the DMW issues are of high priority. In the future, efforts should be made to collect, evaluate, and analyze the operators' operational data and in-service inspection results [44].

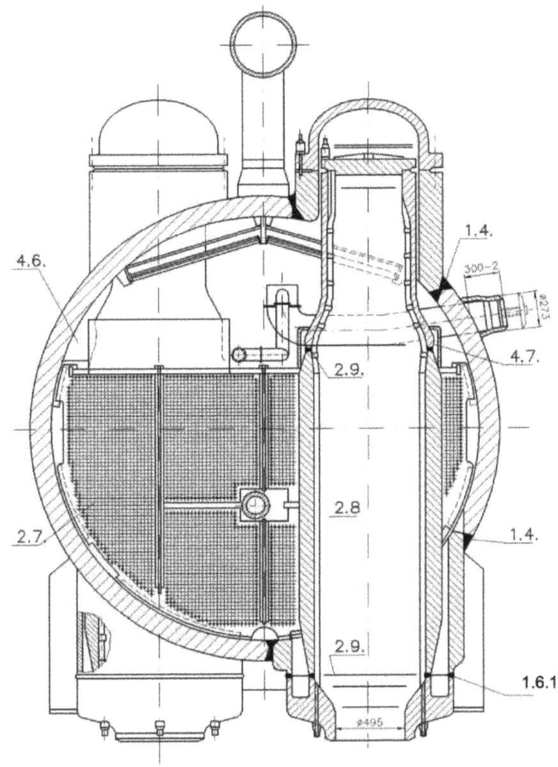

Figure 4. Cross-section of the WWER-440/213 steam generator showing the welds at the collector area. The Dissimilar Material Weld is indicated as 1.6.1.

3. Generic Conditions for Long-Term Operation of WWER-440 Plants

In addition to the above-discussed technical aspects, there are several non-technical aspects related to controlling material degradation and resolving issues arising during operation in the long term. These are as follows:

1. The operating countries established comprehensive regulations on controlling operator ageing management activities and approving long-term operations based on license renewal or the periodic safety review. From the technical point of view, in both cases, the Regulatory Authorities control the information on the ageing of the critical structures and components and the effectiveness of the ageing management programs.
   For example, in the case of Hungary, the regulation includes the basic elements of the US NRC 10 CFR Part 54 [1]. The control of compliance with the current licensing basis is maintained via the annual updating of the Final Safety Analysis Report and its Periodic Safety Review every 10 years. The license renewal itself is a two-step process. First, the LTO Program should be developed and submitted to the regulator at least 4 years before the design life expires, but not before 20 years of operation. Second, the formal license renewal application should be submitted 1 year before the design lifetime expires. Four years of experience implementing the LTO Program should demonstrate that the licensee's Program is effective; it ensures long-term operation and the licensee's assessments regarding the safe lifetime are appropriate. Obtaining the environmental license for an extended term of operation is a precondition to applying for the new operating license.

2. In the WWER-440 operating countries, comprehensive ageing management programs are practiced; see, for example, the case of Paks NPP in [59–61].
3. Supporting the long-term operation research activities are taking place in all WWER-440 operating countries, see the examples for Finland [33,34,36], the Czech Republic [37,50], etc.
4. International cooperation exists for gathering and evaluating operational experiences, e.g., the Pressurized Water Reactor Materials Reliability Program organized by EPRI US [14] or by the IAEA [8]. Guidances and methodologies for the operators are the main products of these activities, e.g., [6,25,26,62].
   Although the national regulatory frames and the applied standards are different, a generalized scientific-technical basis exists for the evaluation and management of the ageing of the critical components thanks to the intensive exchange of the data and operational experiences in the frame of international projects, coordinated by the IAEA, European Union, EPRI. The international benchmarking and comparison of the testing and surveillance methods and qualification and evaluation of the tests are very important. These form the common basis for safe long-term operation.
5. The IAEA SALTO program and services support the operator's practical ageing management activities, e.g., [62–66]. Independent of the national regulation and differences in operator practices, the baseline is defined by the safety requirements, guidelines of the IAEA, and internationally accepted standards and best practices. The IAEA review missions enforce this generic baseline.

4. Discussion and Conclusions

The paper aimed to demonstrate that the expectations for the operability of the WWER-440-type reactors for up to 60–70 years are realistic considering the material aspects of ageing of the main components: The reactor pressure vessel, the reactor pressure vessel internals, and the steam generator. Reviewing the operational experiences and research results related to ageing issues of the WWER-440-type reactors, three main aspects for the justification of long-term operability have been considered:

- Whether the ageing processes and the stressors have been already identified?
- Whether the methods for monitoring and evaluating the material conditions and analysis of time limits operability ensure confidence in the long-term operability of this reactor type?
- Are the already implemented ageing management programs, methods, and mitigative or corrective actions effective?

Regarding the WWER-440 reactor pressure vessels (base metal and welds), irradiation embrittlement is the lifetime-limiting aging effect. The enhanced surveillance and in-service inspection methods are qualified, and the measures for reducing fast neutron fluences and temperature stresses are effective in accordance with experience. Methodology for the analysis of time limits of safe operation could be qualified as proven. As an ultima ratio, the annealing of the RPV welds can be implemented. Decisions on the necessity of annealing can be made based on the evaluation of the material condition.

Regarding the ageing processes of the reactor pressure vessel internals, the operating experience and the related research provided sufficient data and knowledge to prepare for monitoring programs and repair technologies for the ageing-related damages of core baffle-former bolts. The ageing degradation of the reactor pressure vessel internals is receiving more importance during the operation over 40–50 years. The operator's efforts and the supporting research activity should ensure confidence in long-term operability.

The operational performance of the WWER-440 steam generators demonstrates the adequacy of the material selection and the design of steam generators. The critical locations and material degradation processes have been identified, and the operators implemented effective ageing management programs and corrective measures. In-service inspection of heat-exchanging tubes, their plugging technique, also the repair technologies for the dissimilar material welds are proven by experience.

Despite the generic preparedness of the operators for long-term operation and positive experiences regarding the adequacy of material selection and design of the main components of the WWER-440-type reactors, internationally coordinated efforts are needed for gathering and evaluation of operational experiences and data and for ensuring proper scientific basis of long-term operation of the nuclear power plants considered. The ageing phenomena subjects of further investigations have been identified in the paper. The international cooperation will back up the practical efforts of the WWER-440 operators and national regulatory authorities in approving and supervising the long-term operation.

Author Contributions: Conceptualization, T.J.K., Á.B. and S.R.; methodology, T.J.K., Á.B. and S.R.; formal analysis, Á.B.; investigation, T.J.K., Á.B. and S.R.; data curation, Á.B.; writing—original draft preparation, T.J.K.; writing—review and editing, Á.B. and S.R.; visualization, Á.B. and S.R.; supervision, T.J.K. All authors have read and agreed to the published version of the manuscript.

Funding: This research received no external funding.

Data Availability Statement: Except for the information shown in Figure 2, no new data were created or analyzed in this study. Data for Figure 2 are contained within the article. Otherwise, data sharing does not apply to this article.

Conflicts of Interest: The authors declare no conflict of interest.

References

1. *Part 54—Requirements for Renewal of Operating Licenses for Nuclear Power Plants*; Office of Nuclear Reactor Regulation, U.S. Nuclear Regulatory Commission: Washington, DC, USA, 2007. Available online: https://www.nrc.gov/reading-rm/doc-collections/cfr/part054/full-text.html (accessed on 21 March 2023).
2. *Generic Ageing Lessons Learned (GALL) Report*; Final Report (NUREG-1801, Revision 2); Office of Nuclear Reactor Regulation, U.S. Nuclear Regulatory Commission: Washington, DC, USA, 2010. Available online: https://www.nrc.gov/reading-rm/doc-collections/nuregs/staff/sr1801/r2/index.html#pubinfo (accessed on 21 March 2023).
3. *Standard Review Plan for Review of Subsequent License Renewal Applications for Nuclear Power Plant (SRP-SLR)*; Final NUREG-2192; Office of Nuclear Reactor Regulation, U.S. Nuclear Regulatory Commission: Washington, DC, USA, 2017. Available online: https://www.nrc.gov/docs/ML1718/ML17188A158.pdf (accessed on 21 March 2023).
4. *Generic Ageing Lessons Learned for Subsequent License Renewal (GALL-SLR) Report*; Final NUREG-2191; Office of Nuclear Reactor Regulation, U.S. Nuclear Regulatory Commission: Washington, DC, USA, 2017. Available online: https://www.nrc.gov/docs/ML1718/ML17187A031.pdf (accessed on 21 March 2023).
5. IAEA. *Safety Aspects of Long-term Operation of Water Moderated Reactors IAEA-EBP-SALTO*; IAEA: Vienna, Austria, 2007.
6. IAEA. *Ageing Management for Nuclear Power Plants*; IAEA Safety Standards Series No. NS-G-2.12; International Atomic Energy Agency: Vienna, Austria, 2009; ISBN 978-92-0-112408-1.
7. IAEA. *SALTO Guidelines. Guidelines for Peer Review of Long-Term Operation and Ageing Management of Nuclear Power Plants*; IAEA Services Series No. 17; Vienna International Atomic Energy Agency: Vienna, Austria, 2008.
8. IAEA. *Ageing Management for Nuclear Power Plants: International Generic Ageing Lessons Learned (IGALL)*; IAEA Safety Reports Series no. 82 (rev. 1); International Atomic Energy Agency: Vienna, Austria, 2020; ISBN 978-92-0-107419-5.
9. IAEA. *Approaches to Ageing Management for Nuclear Power Plants: International Generic Ageing Lessons Learned (IGALL) Final Report*; International Atomic Energy Agency: Vienna, Austria, 2014; ISBN 978-92-0-104414-3.
10. International Atomic Energy Agency. The Database on Nuclear Power Reactors. Available online: https://pris.iaea.org/pris/home.aspx (accessed on 21 March 2023).
11. IAEA. *Safety Issues and Their Ranking for WWER-440 Model 213 Nuclear Power Plants*; IAEA-EBP-WWER-03; IAEA: Vienna, Austria, 1996.
12. Katona, T.; Bajsz, J. Plex at Paks-making a virtue out of necessity. *Nucl. Eng. Int.* **1992**, *37*, 27–31.
13. Brumovsky, M. *IAEA-NULIFE VERLIFE-Procedure for Integrity and Lifetime Assessment of Components and Piping in WWER NPPs during Operation-Tool for LTO*; IAEA: Vienna, Austria, 2012.
14. Brumovsky, M. Unified Procedure for Lifetime Assessment of Components and Piping in WWER NPPs "VERLIFE", Version 2008. In Proceedings of the ASME/JSME 2004 Pressure Vessels and Piping Conference, San Diego, CA, USA, 25–29 July 2004.
15. Electric Power Research Institute. MRP-471-WWER Issue Management Tables: Identifies Material Research Gaps for WWER Light Water Reactors. Available online: https://www.epri.com/research/summary/000000003002021033 (accessed on 21 March 2023).
16. Katona, T.J. 8—Materials management strategies for WWER reactors. In *Woodhead Publishing Series in Energy, Materials Ageing and Degradation in Light Water Reactors*; Murty, K.L., Ed.; Woodhead Publishing: Cambridge, UK, 2013; pp. 335–384, ISBN 9780857092397. Available online: https://www.sciencedirect.com/science/article/pii/B9780857092397500080 (accessed on 21 March 2023).

17. IAEA. *Integrity of Reactor Pressure Vessels in Nuclear Power Plants: Assessment of Irradiation Embrittlement Effects in Reactor Pressure Vessel Steels*; IAEA Nuclear Energy Series, No. NP-T-3.11; International Atomic Energy Agency: Vienna, Austria, 2009; ISSN 1995-7807, ISBN 978-92-0-101709-3.
18. Brumovsky, M. 11—Irradiation hardening and materials embrittlement in light water reactor (LWR) environments. In *Woodhead Publishing Series in Energy, Understanding and Mitigating Ageing in Nuclear Power Plants*; Philip, G.T., Ed.; Woodhead Publishing: Cambridge, UK, 2010; pp. 357–373, ISBN 9781845695118. Available online: https://www.sciencedirect.com/science/article/pii/B9781845695118500111 (accessed on 21 March 2023).
19. Brumovsky, M. 12—Reactor pressure vessel (RPV) annealing and mitigation in nuclear power plants. In *Woodhead Publishing Series in Energy, Understanding and Mitigating Ageing in Nuclear Power Plants*; Philip, G.T., Ed.; Woodhead Publishing: Cambridge, UK, 2010; pp. 374–386, ISBN 9781845695118. Available online: https://www.sciencedirect.com/science/article/pii/B9781845695118500123 (accessed on 21 March 2023).
20. Katona, T.J.; Rátkai, S.; Jánosiné Bíró, A.; Gősi, P. Time-Limited Ageing Analyses for Justification of Long-Term Operation of Paks NPP. In Proceedings of the ASME 2010 International Mechanical Engineering Congress and Exposition. Volume 11: New Developments in Simulation Methods and Software for Engineering Applications; Safety Engineering, Risk Analysis and Reliability Methods; Transportation Systems, Vancouver, BC, Canada, 12–18 November 2010; pp. 353–360. [CrossRef]
21. Bóna, G.; Elter, J.; Fekete, T.; Keresztúri, A.; Rátkai, S. *Pressurized Thermal Shock Analysis Paks NPP Units 1–4, Summary Report*; Trampus, P., Ed.; EJR No.: 000000A00212 OKA; MVM Paks Nuclear Power Plant Ltd.: Paks, Hungary, 2010.
22. Hordósy, G.; Hegyi, G.; Keresztúri, A.; Maráczy, C.; Temesvári, E.; Vértes, P.; Zsolnay, É. Pressure Vessel Calculations for VVER-440 Reactors. *Radiat. Prot. Dosim.* **2005**, *115*, 100–103. [CrossRef] [PubMed]
23. Hordósy, G. Neutron and Photon Shielding Benchmark Calculations by MCNP on the LR-0 Experimental Facility. *Radiat. Prot. Dosim.* **2005**, *116*, 32–34. [CrossRef] [PubMed]
24. Katona, T.; Rátkai, S.; Pammer, Z. Reconstitution of Time-Limited Ageing Analyses for Justification of Long-Term Operation of Paks NPP. *Nucl. Eng. Des.* **2011**, *241*, 638–643. [CrossRef]
25. IAEA. *Guidelines for Prediction of Irradiation Embrittlement of Operating WWER-440 Reactor Pressure Vessels*; IAEA-TECDOC-1442; IAEA: Vienna, Austria, 2005; ISBN 92-0-105605-2, ISSN 1011-4289.
26. IAEA. *Guidelines on Pressurized Thermal Shock Analysis for WWER Nuclear Power Plants*; IAEA-EBP-WWER-08 (Rev. 1); IAEA: Vienna, Austria, 2006.
27. IAEA. *Pressurized Thermal Shock in Nuclear Power Plants: Good Practices for Assessment*; IAEA TECDOC 1627; IAEA: Vienna, Austria, 2010.
28. PNAE G-7-002-86 (1989): Rules of Strength Calculation for Equipment and Pipelines of Nuclear Power Plants, Energoatomizdat, Moscow. Available online: https://docs.secnrs.ru/documents/pnaes/%D0%9F%D0%9D%D0%90%D0%AD_%D0%93-7-002-86/%D0%9F%D0%9D%D0%90%D0%AD-%D0%93-002-86e.htm (accessed on 21 March 2023).
29. Pištora, V.; Žamboch, M.; Král, P.; Vyskočil, L. PTS Re-Evaluation Project for Czech NPPs. In *Proceedings: Nuclear Power Plant Life Management, Lyon, France, 23–26 October 2017*; Paper: AEA-CN-246-072; IAEAL 21-01402; International Atomic Energy Agency: Vienna, Austria, 2021; ISBN 978-92-0-106921-4.
30. Kryukov, A.; Rubtsov, V.; Lebedinsky, V. Irradiation Embrittlement of WWER RPV Steels Irradiated at High Fluences. *J. Nucl. Technol. Appl. Sci.* **2020**, *8*, 113–118. [CrossRef]
31. Nuclear Regulatory Commission U.S. Baffle-Former Bolts. Available online: https://www.nrc.gov/reactors/operating/ops-experience/baffle-former-bolts.html (accessed on 21 March 2023).
32. PWR Reactor Internals Inspection and Evaluation Guidelines (MRP-227-A). Available online: https://www.nrc.gov/docs/ML1201/ML120170453.html (accessed on 21 March 2023).
33. Ehrnstén, U.; Pakarinen, J.; Karlsen, W.; Keinänen, H. Investigations on Core Basket Bolts from a WWER 440 Power Plant, 15th International Conference on Environmental Degradation of Materials in Nuclear Power Systems-Water Reactors. 2011. Available online: https://cris.vtt.fi/en/publications/investigations-on-core-basket-bolts-from-a-WWER-440-power-plant (accessed on 21 March 2023).
34. Ehrnstén, U.; Pakarinen, J.; Karlsen, W.; Keinänen, H. Investigations on Core Basket Bolts from a WWER 440 Power Plant. *Eng. Fail. Anal.* **2013**, *33*, 55–65. [CrossRef]
35. Ballesteros, A.; Heid, K.; Luostarinen, P. Inspection and Replacement of Baffle Former Bolts in WWER-440 Reactor Type, Transactions. In Proceedings of the SMiRT 19, Toronto, ON, Canada, 12–17 August 2007.
36. Pakarinen, J.; Ehrnstén, U.; Keinänen, H.; Karlsen, W. Microstructural Characterization of Irradiated Baffle Bolts Removed from a Finnish WWER and a French PWR. In Proceedings of the 16th International Conference on Environmental Degradation of Materials in Nuclear Power Systems-Water Reactors, Asheville, NC, USA, 11–15 August 2013; pp. 994–1007.
37. Hojná, A. Environmentally Assisted Cracking Initiation in High-Temperature Water. *Metals* **2021**, *11*, 199. [CrossRef]
38. Kuleshova, E.A.; Fedotova, S.V.; Gurovich, B.A.; Frolov, A.S.; Maltsev, D.A.; Stepanov, N.V.; Margolin, B.Z.; Minkin, A.J.; Sorokin, A.A. Microstructure degradation of austenitic stainless steels after 45 years of operation as WWER-440 reactor internals. *J. Nucl. Mater.* **2020**, *533*, 152124. [CrossRef]
39. IAEA. *Strategy for Assessment of WWER Steam Generator Tube Integrity*; IAEA-TECDOC-1577; IAEA: Vienna, Austria, 2007.
40. Katona, T.; Jánosiné Biró, A.; Rátkai, S.; Tóth, A. Main features of design life extension of WWER-440/213 units NPP Paks Hungary. In Proceedings of the ICONE 11th International Conference on Nuclear Engineering, Tokyo, Japan, 20–23 April 2003.

41. Trunov, N.B.; Denisov, V.V.; Kharchenko, S.A.; Lukasevich, B.I. Consideration of Field Experience in Developing New Projects of Steam Generators for Nuclear Power Stations Equipped with WWER Reactors. *Therm. Eng.* **2006**, *53*, 37–42. [CrossRef]
42. Trunov, N.B.; Stanislav, E.; Davidenko, S.E.; Vladimir, A.; Grigoriev, V.A.; Popadchuk, V.S.; Sergey, I.; Brykov, S.I.; Karzov, G.P. WWER Steam Generators Tubing Performance and Aging Management. In Proceedings of the 14th International Conference on Nuclear Engineering (ICONE 14), Miami, FL, USA, 17–20 July 2006.
43. Kupca, L.; Brezina, M. Analysis of long-term operation influence on the degradation of VVER-440 steam generator piping. In Proceedings of the 7th International Seminar on Horizontal Steam Generators, Podolsk, Russia, 3–5 October 2006.
44. EPRI. *Materials Degradation Matrix*; Revision 4, 3002013781, Final Report; EPRI: Palo Alto, CA, USA, 2018; Available online: https://preview.epri.com/research/products/000000003002013781 (accessed on 21 March 2023).
45. *Update on Hydrazine Alternatives for PWR Secondary Chemistry Control: PWR Chemistry*; Technical Strategy Group Report, 3002010652; EPRI: Palo Alto, CA, USA, 2018.
46. Matocha, K.; Wozniak, J.; Pochman, K. Analysis of WWER-440 SG Primary Collector Bolted Joint Damage. Available online: https://www.osti.gov/etdeweb/servlets/purl/578792 (accessed on 21 March 2023).
47. Postler, M.; Keilova, E.; Burda, J.; Kocik, J.; Charvat, L. Analyses of Cracks in Threaded Holes of SG Collector. In Proceedings of the 6th International Seminar on Horizontal Steam Generators, Podolsk, Russia, 22–24 March 2004; Summaries of Reports. Available online: http://www.gidropress.podolsk.ru/files/proceedings/seminar6/for/nri-postler.pdf (accessed on 21 March 2023).
48. Khodakov, V.D.; Zubchenko, A.S. *Experience of Operation and Repair of Heterogeneous Welds of Equipment and Pipelines Made of Austenitic and Pearlitic Steels, УДК 621.643.4*; In Issues of Atomic Science and Technology, Series: "Ensuring the Safety of Nuclear Power Plants", Scientific and Technical Collection, Issue 23, Reactor Plants with WWER; Publishing House of JSC OKB "GIDROPRESS": Podolsk, Russia, 2008; ISBN 978-5-94883-089-6. Available online: http://www.gidropress.podolsk.ru/files/vant/vant23.pdf (accessed on 21 March 2023).
49. IAEA. *Dissimilar Metal Weld Inspection, Monitoring and Repair Approaches*; IAEA-TECDOC-1852; International Atomic Energy Agency: Vienna, Austria, 2018; ISBN 978-92-0-105618-4.
50. Bystrianský, J.; Ernestová, M.; Haušild, P.; Siegl, J.; Lubomír, J. Conditions for Long-Term Durability of Dissimilar Metal Welds of Power Plants–Environmental Effect. Available online: https://dspace5.zcu.cz/bitstream/11025/50878/2/sbornik2019-131-138.pdf (accessed on 21 March 2023).
51. Szávai, S.; Bézi, Z.; Rózsahegyi, P. Characterization and Numerical Simulation of a Dissimilar Metal Weld. In Proceedings of the 21st European Conference on Fracture, ECF21, Catania, Italy, 20–24 June 2016; pp. 1023–1030.
52. Spisák, B.; Bézi, Z.; Szávai, S. Study of the Stress State of a Dissimilar Metal Weld due to Manufacturing and Operational Conditions. *Period. Polytech. Mech. Eng.* **2022**, *66*, 120–128. [CrossRef]
53. Szávai, S.; Bézi, Z.; Dudra, J.; Takács, C.; Rózsahegyi, P. Material Characterization and Numerical Simulation of a Dissimilar Metal Weld to Support Phased Array Ultrasonic Inspection. In Proceedings of the International Conference on Nuclear Power Plant Life Management, Lyon, France, 23–27 October 2017; IAEA-CN–246-104. Available online: https://inis.iaea.org/search/search.aspx?orig_q=RN:52061982 (accessed on 21 March 2023).
54. Vojna, M. Exposure of Dissimilar Metal Welds to Secondary Water Environment. In Proceedings of the 14th Conference Increasing the Service Life of Energy Equipment Components in Power Plants, Srní, Czech Republic, 24–26 September 2019; pp. 93–98, ISBN 978-80-261-0885-6.
55. Khodakov, D.V. Research and Development of Technology for Repairing Heterogeneous Welded Joints of the Coolant Collector Mounting Unit to the Branch Pipes of the Bodies of Steam Generators PGV-440. Ph.D. Thesis, Moscow, Russia, 2012. Available online: https://www.dissercat.com/content/issledovanie-i-razrabotka-tekhnologii-remonta-raznorodnykh-svarnykh-soedinenii-uzla-krepleni (accessed on 21 March 2023).
56. Horacek, L.; Buldra, I.; Mares, P. UT Qualification and Site Feedback on Dissimilar Metal Welds of WWER Type NPPs in the Czech Republic. Available online: https://www.ndt.net/article/jrc-nde2013/papers/53.pdf (accessed on 21 March 2023).
57. Soukup, T. Nuclear Repair Engineering in Practise: Reconstruction of Feed Water Piping of Steam Generators of WWER TYPE 440/1000 in Dukovany and Temelin NPP. In Proceedings of the 14th Conference Increasing the Service Life of Energy Equipment Components in Power Plants, Srní, Czech Republic, 24–26 September 2019; pp. 195–200, ISBN 978-80-261-0885-6.
58. Khodakov, V.D.; Bazanov, M.A.; Nemchaninova, L.N.; Lukicheva, S.V.; Khodakov, D.V.; Abrosin, A.A.; Fomenko, V.I.; Malyshev, A.G.; Mityashin, V.A.; Zubchenko, A.S. Repair Using Welding of the Weld Unit for the Collectors of the Coolant to the Branch Pipes DN 1100 of the Housings of Steam Generators WWER-440 after Long-Term Operation, In Russian. Available online: http://www.gidropress.podolsk.ru/files/proceedings/seminar8/documents/sgpg2010-027.pdf (accessed on 21 March 2023).
59. Katona, T.; Jánosiné Bíró, A.; Ratkai, S.; Palfi, T.; Toth, A. Equipment Aging Management and Operational Lifetime Extension at the Paks Nuclear Power Plant. In *Aging Management and License Renewal: Presented at the 2004 ASME/JSME Pressure Vessels and Piping Conference, San Diego, NY, USA, 25–29 July 2004*; Bezdikian, G., Shah, V.N., Eds.; American Society of Mechanical Engineers (ASME): New York, NY, USA, 2004; pp. 119–125.
60. Katona, T.; Jánosiné Bíró, A.; Rátkai, S.; Ferenczi, Z. Key Elements of the Ageing Management of the WWER-440/213 type Nuclear Power Plants. In Proceedings of the 18th International Conference on Structural Mechanics in Reactor Technology (SMiRT 18), Beijing, China, 7–12 August 2005. D02-4.

61. Katona, T.J. 19—Plant life management (PLiM) practices for water-cooled water-moderated nuclear reactors (WWERs). In *Woodhead Publishing Series in Energy, Understanding and Mitigating Ageing in Nuclear Power Plants*; Philip, G.T., Ed.; Woodhead Publishing: Cambridge, UK, 2010; pp. 633–705, ISBN 9781845695118. [CrossRef]
62. IAEA. *Plant Life Management for Long Term Operation of Light Water Reactors: Principles and Guidelines*; Technical Reports Series No. 448; IAEA: Vienna, Austria, 2006.
63. Rátkai, S. The Role of the IAEA in the Extension of Operation Life of MVM Paks NPP. In Proceedings of the Fourth International Conference on Nuclear Power Plant Life Management, Lyon, France, 23–27 October 2017; Paper No IAEA-CN-246-009. Available online: https://inis.iaea.org/search/search.aspx?orig_q=RN:52048593 (accessed on 21 March 2023).
64. Krivanek, R.; Fiedler, J. Main corrective measures in an early phase of nuclear power plants' preparation for safe long term operation. *Nucl. Eng. Des.* **2017**, *316*, 125–130. [CrossRef]
65. Krivanek, R. Long term operation of nuclear power plants–IAEA SALTO peer review service and its results. *Nucl. Eng. Des.* **2014**, *280*, 99–104. [CrossRef]
66. Ratkai, S.; Katona, T.J. Experiences Gained During the Development of the License Renewal Application for Unit 1 of Paks NPP. In Proceedings of the ASME 2012 Pressure Vessels and Piping Conference. Volume 7: Operations, Applications and Components, Toronto, ON, Canada, 15–19 July 2012; pp. 53–60. [CrossRef]

energies MDPI

Article

New Optical System for Long Distance Control of Electrical Energy Flows

Vadim Davydov [1,2,*], Bogdan Reznikov [2,*] and Valentin Dudkin [2]

1 Institute of Electronics and Telecommunications, Peter the Great St. Petersburg Polytechnic University, 195251 St. Petersburg, Russia
2 Department of Photonics and Communication Lines, The Bonch-Bruevich Saint Petersburg State University of Telecommunication, 193232 St. Petersburg, Russia
* Correspondence: davydov_vadim66@mail.ru (V.D.); rznkff@gmail.com (B.R.)

Abstract: The problems that arise during the transmission of control commands for key elements and data on their execution via communication systems over long distances at energy facilities are considered. The necessity of a new approach to solving this problem using a fiber-optic communication line (FOCL) is substantiated. A new design of analog FOCL has been developed for transmitting control commands for key elements of substations and switching data. A new method for forming an optical signal in a laser transmitting module using direct current modulation is proposed. The advantages of using a new method for generating commands for control signals and switching data using a sequence of command codes compared with digital signals currently used in control and monitoring systems are substantiated. The main parameters of the developed analog FOCL have been calculated. An experimental prototype of an analog FOCL was assembled, and research was conducted. The results of calculations and experimental data are compared. The directions for further modernization of the developed FOCL are determined to increase the distance for transmitting an optical signal up to 600 km without the use of servers, the maintenance of which in the power transmission line system in some cases is challenging, expensive, and not always possible.

Keywords: electrical substation; switching systems; control; analog optical signal; optical fiber; distance; signal-to-noise ratio; reliability

Citation: Davydov, V.; Reznikov, B.; Dudkin, V. New Optical System for Long Distance Control of Electrical Energy Flows. *Energies* **2023**, *16*, 1040. https://doi.org/10.3390/en16031040

Academic Editor: Tapas Mallick

Received: 12 December 2022
Revised: 13 January 2023
Accepted: 16 January 2023
Published: 17 January 2023

1. Introduction

One of the urgent tasks of modern energy is to increase the reliability of various systems through which electrical power is supplied to consumers [1–3]. The reliability of the operation of electrical energy transmission systems depends on many parameters and various factors [2–6]. One of these factors is the coordinated operation of key elements in the distribution systems of electrical energy between consumers or various units (for example, electric motors at a mobile object with a nuclear power plant and compressors at fuel transfer stations) [7–13]. Key elements also play an important role in the redistribution of electric energy flows in the energy systems of countries with several time zones (for example, USA, Russian Federation, China, Australia, and others) [14–16]. Therefore, more attention is paid to solving problems related to managing and controlling key elements (switches at various transformer substations or high-voltage switches) [17–21].

Fiber optic communication lines (FOCL) are used in substations and switching systems to ensure the reliability of various relays, panel switches, and other equipment [14–17,21–26]. The use of FOCL is the only solution that can significantly reduce the effect of electromagnetic interference on the reliability of the control and monitoring system for electrical energy flows. The developed designs of FOCL make it possible to ensure the transmission of information with control signals over distances of up to 250 km without any problems without optical amplifiers. The use of optical amplifiers in the power transmission line zone, where the FOCL is located, is impossible due to a change in the polarization of laser radiation in the optical amplifier from sparks and lightning. A change in polarization will lead to the transformation of the optical signal with information during optical amplification (the command encoding will change, and another switch will be performed at the substation). It should be noted that the FOCL operates in standby mode. In this mode, the FOCL keeps a low level. Intense electromagnetic radiation with a discharge spark can create a slight surge in the FOCL, amplified by an optical amplifier and identified as a control signal [27–32].

Therefore, to avoid creating such complex problems when transmitting information with control signals to substations or switching systems located at more than 250 km from the control center, a server station is used. In the Russian Federation, ROSSETI PJSC deploys server stations every 200–215 km of the route with FOCL (depending on various conditions). The server station is located on the territory of the automated substation at a small distance from the switching systems in a separate room. In this room, it is necessary to provide a certain temperature regime and conditions for the operation of the server and auxiliary equipment. The server must operate in continuous mode (provide constant transmission of information about the position of key elements on relays and panel switches). Therefore, the server room also hosts a backup set of equipment, which is also maintained in working order. Furthermore, for the same reason, there was a complete rejection of the use of optical amplifiers, which in a certain period cannot ensure uninterrupted transmission of information over FOCL.

It is necessary to conduct preventive maintenance in the area where the server station is located to prevent failures in the operation of the equipment. This requires a substantial amount of money, especially in winter and during the rainy season (huge problems occur in mountainous areas). For example, for several regions of the Russian Federation, especially in the northern regions and some regions of Siberia, these tasks are solved with great difficulty, especially during severe frosts and snowstorms. The departure of PJSC Rosseti employees for more than 1000 km (round trip) becomes a major problem. In addition, it is a highly expensive and sometimes dangerous undertaking. Similar problems can be attributed to several territories in the USA, Canada, Brazil, Kazakhstan, China, Argentina, Chile, and others. Considering that roads were laid to automated substations during their installation, which are in different conditions (roads also require preventive repairs and maintenance), these problems are solved with great difficulty.

This command-and-control system's pain point is when a server substation needs to be installed on the power line route. If the distance between the control center and the automated substation is more than 300 km, it is necessary to install a server substation for stable information transfer. The roads are in a bad state (access to the server substation is extremely difficult). In most cases, you must walk part of the way from the car to the server station on foot with equipment through challenging terrain. This is a difficult job in the presence of snow, heavy precipitation, etc. In addition, the impact of negative weather factors on such a server station is much higher than if it is located on the territory of an automated substation. Vandalism is possible from people (theft of equipment) or wild animals. Now the problem of failure in the operation of such server stations is being solved with great effort (there is no other choice yet). Therefore, reducing the number of such "pain" points in automated substations' control and monitoring system is a highly urgent task.

In addition, it should be noted that the size of the automated remote substation does not affect the preventive maintenance schedule of the server station and the operating conditions of the equipment on it. Service standards are uniform (considering the geographical region's characteristics). Failure to comply with these standards leads to power outages for consumers and accidents, which negatively affect the operation of various consumer equipment [6,7,9,11,13,16,18,19,24,33,34]. Often, the costs of operating small, automated substations exceed revenues, and they are forced to cover them from the entire company's work. Some small problems also make it challenging to operate automated substations using servers. On the other hand, it is extremely difficult to perform this operation without a server station. Therefore, the search for new solutions to this complex problem is highly relevant to the energy systems of many leading countries.

There is a substantial amount of research trying to solve this complex problem, and the previous authors' work is not the only possible solution. Our work proposes a fundamentally new approach to solving the problem noted in comparison with other studies. In the research of other scientists, the main emphasis is on developing a system of strong points using digital FOCL [35,36]. In essence, the tasks are reduced to the creation of shift zones to reduce the departure distance of the maintenance team for both the automatic substation and maintenance work at the server station. In the PJSC Rosseti (Russian Federation) system, these developments were not widely used; in reality, their effectiveness was low. The second main direction in the research of other scientists relates to the possibility of introducing new equipment, for example, electro-optical modulators made based on new materials. These materials are more resistant to electromagnetic radiation and temperature instability [37–39]. Furthermore, optical systems with new electro-optical modulators are planned to be integrated into the power transmission line structure with the rejection of server stations. Modern technologies currently allow the production of optical materials that are less sensitive to a strong electric field than quartz glass and its various modifications [40–45]. In the case of using these materials in communication systems, it is necessary to develop new designs of multiplexers, demultiplexers, and electro-optical modulators, as well as several components for the operation of these elements as part of FOCL. Leading companies in the world (Thorlabs inc., Emcore, Hamamatsu Photonics) are in no hurry to develop and manufacture these devices. Therefore, it is necessary to look for another approach to solve this problem in the development of FOCL. Therefore, it is necessary to look for another approach to solve this problem in the development of FOCL.

2. The Concept of an Analog Fiber-Optic Communication Line and a New Method of Information Transmission

The developed FOCL design proposes not to use direct modulation of laser radiation for current in the transmitting laser module to form analog optical signals with information. This will make it possible to exclude from the design of the FOCL the electro-optical modulators and various elements associated with it, which are expensive and have a high-temperature dependence. It creates additional problems in their operation. For the transmission of modulated laser radiation via FOCL, we propose to use a sinusoidal signal with a subcarrier frequency of 200 MHz. For a control command that switches key elements, a new encoding method based on the formation of a sequence of command codes is used.

The control command is a set of four numbers separated by a hyphen. The first number consists of two, three, or four characters and is, for example, the number of a relay or panel switch. The second number consists of two or three digits and is the number of the key element (its position changes). The third number (two digits) corresponds to the initial position of the key element. The fourth number (two digits) corresponds to the final position of the key element after switching. In this case, it is necessary to steadily transmit four numbers in a certain sequence over the FOCL and indicate the start of the command countdown and its completion. It is proposed that a certain value of the pulse amplitude in relative units (0.950, 0.955, 0.960, . . . , 0.995) corresponds to the numbers: 0, 1, 2, . . . , 9. The amplitude value of 0.94 corresponds to a hyphen (-), i.e. is separating one number from another. In addition, the amplitude value of 1.0 in relative units corresponds to the start of the set of all four numbers and its end in the sequence of command codes. Figure 1 shows, for example, the encoding of the command 1231-47-58-29 using the sequence of command codes proposed by us.

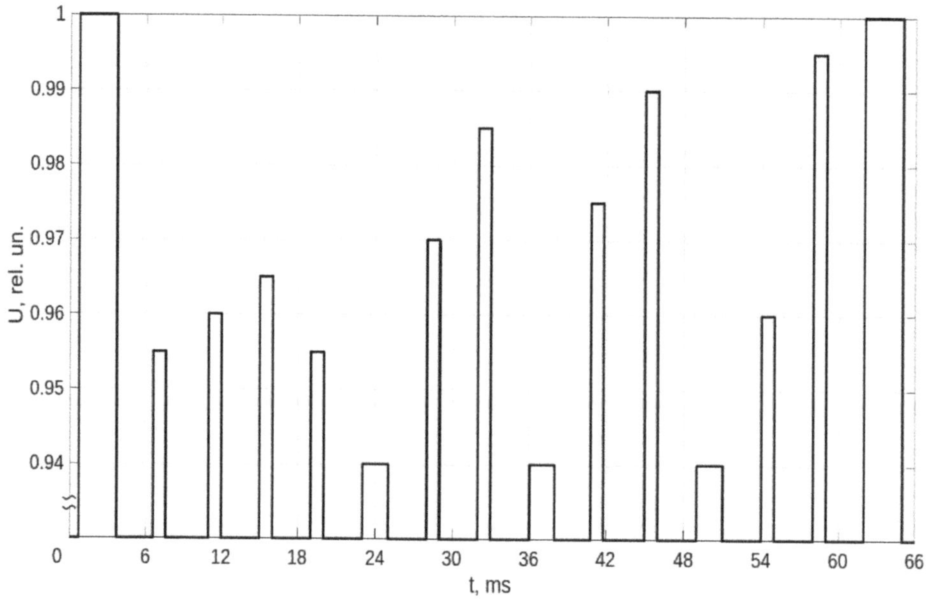

Figure 1. A sequence of command codes in the form of rectangular pulses for transmitting a control command in analog form via FOCL.

In the Russian Federation, one feature is associated with the transfer of control commands. The territory is extensive; it hosts many power plants of various types, so control commands, in some cases, must be transmitted over long distances (up to 500 km).

In addition, in such a situation, the FOCL will be placed on the upper part of the power line to exclude the influence of many technogenic factors on it, especially in the northern regions, steppes, and deserts, as well as in the middle highlands. The same feature is present in the energy systems of the USA, Canada, China, Australia, and several other countries. Under such conditions, the optical signal amplification on the path is excluded due to the great difficulties in its implementation in a strong electromagnetic field and changing climatic conditions (difficulties in ensuring a thermally stable operating mode of the optical amplifier).

Using the developed sequence of command codes makes it possible to transmit information over distances of 500 km via optical fiber. This possibility is due to the fact that laser radiation with a power of $P_{in} \approx$ 100–110 MBT (20.0–20.4 dBm) can be used to transmit these signals. Modern FOCL designs use laser radiation with a power of up to 15 mW (11.8 dBm) to transmit digital signals over various distances. In the near future, it is planned to implement operation of FOCL with a laser power of up to 20 mW. A further increase in power for FOCL, in which digital signals with a high bit-sequence density and channel spectral division multiplexing are transmitted, is limited by the factors number. These factors relate to both technical (the effect of radiation at high power from one channel on the signal in another channel, etc.) and physical phenomena (Kerr nonlinearity, etc.). The stimulated Mandelstam–Brillouin scattering (SMBS) in the case of a pulse duration in a bit sequence much shorter than the photon lifetime has a threshold of more than 20 mW. The Stimulated Raman scattering (SRS) has a threshold of about 600 mW. In this situation, Kerr nonlinearity exerts a significant influence on the information transfer process with an increase in the laser radiation power. The edges of the pulses are distorted (the pulses in the bit sequence are expanded). This expansion of pulses and the incursion of chromatic dispersion leads to the imposition of pulses on each other and the formation of bit errors during signal registration.

In addition, for digital signals, the signal-to-noise ratio at the photodetector must be at least 20 (in dBm). This provides a BER error probability of the order of 0.3×10^{-6}. In the case of overlapping pulse fronts, this signal-to-noise ratio is extremely difficult to provide.

The signal-to-noise ratio on the photodetector for the sequence of command codes should be no worse than 8 (in times). This is ensured by registering the signal on the photodetector in a narrow band, because pulses with a duration of 1 to 3 ms are used. It provides an error probability erfc (BER) which amounts to 10^{-5} [46]. Such a BER value is sufficient to clearly identify a value of pulse amplitude.

It should be noted that with a duration of pulses with information of 1 ms (number) and 2 ms (hyphen), and 3 ms (start and end of the command countdown) with an interval between pulses of 3 ms, the effect of chromatic dispersion will not lead to the superposition of their fronts. The incursion of chromatic dispersion in a single-mode fiber at λ = 1550 nm over a distance L = 550 km is ≈2.3 ns in each pulse edge. For pulses with a duration of 1 ms, these changes are insignificant, and it will not change the width of the signal spectrum.

In contrast to a digital bit sequence, with such pulse durations and the interval between them, the effect of Kerr nonlinearity will be less. In order to introduce an error into signal transmission, this non-linearity must change the upper level of the pulse. This requires more laser power than when transmitting digital signals to distort the fronts. In fact, it is necessary to start the collapse of impulse. In this case, the information will be difficult to identify. The signal-to-noise ratio will decrease (BER will increase). The spectrum of the recorded signal is broadened.

The stimulated Mandelstam–Brillouin scattering will also affect the signal-to-noise ratio and the spectrum of the recorded signal. The threshold for the occurrence of this radiation, in contrast to the case of digital signal transmission, will be higher due to the use of a special fiber and a sequence of command codes. It is difficult to estimate the threshold value of SMBS bias occurrence for the pulse train and special fiber used by us. On the one hand, the spectral width of the pulse is much less than $\Delta\nu_B$. This is considered in the classical theory as continuous pumping. On the other hand, the interval between pulses is 3 ms (that is three times longer than the pulse duration). There is no continuity. In addition, with the signals we use, it is impossible to make assumptions about the random connection of the spectrum of a sequence of zeros and ones, as is accomplished with the transmission of digital signals. In addition, in the signal we use, the ratio between the pulse amplitudes changes only in the upper part of the sequence. Zero is not at the noise level. Therefore, we will evaluate the influence of SMBS bias experimentally by the appearance of nonlinear distortions in the amplitude characteristic of FOCL.

The Stimulated Raman scattering (SRS) for the FOCL design developed by us and the sequence of command codes used has a threshold of about 500–550 mW. Therefore, the SRS and the effects associated with it will not have an impact on the process of changing the amplitude FOCL characteristic.

The proposed method for transmitting control commands is a fundamentally new solution in data transmission using analog signals via FOCL. The sequence of command codes developed by us (Figure 1) makes it possible to use in FOCL for data transmission, which involves various sources of laser radiation with direct modulation of the pump current. This makes it possible to ensure a high stability of laser radiation in terms of power, as well as a change in the depth of modulation over a wide range.

It should be noted that a further increase in P_{in} is inexpedient. This is well illustrated by the following formula:

$$P\ (\text{dBm}) = 10 \cdot \log_{10}(P\ (\text{mW})\ /\ 1\ \text{mW}), \quad (1)$$

An analysis of Formula (1) shows that an increase in P_{in} to 150 mW leads to an increase in power to 21.76 dBm. The signal transmission distance increases slightly. This increases the likelihood of various non-linear distortions and scattering. This will lead to changes in the transmitted pulses shape, leading to errors in decoding the governance command.

The use of pulses with a duration of 1, 2, and 3 ms in the developed sequence of command codes, as well as a time interval between pulses of 3 ms, makes it possible to use a subcarrier frequency F_s in the range from 0.1 to 200 MHz for their transmission over FOCL. This makes it possible to provide a long transmission time ("drift") of the FOCL when transmitting a signal over long distances L. In addition, using such values of subcarrier frequencies F_s makes it possible to record an optical signal on a photodetector in the band from 0.1 to 1 MHz with a small signal-to-noise ratio (S/N), which significantly increases the energy balance of FOCL.

A feature of the system we proposed for transmitting control commands via FOCL is that the pulses in the sequence of command codes (Figure 1) are the envelope for the subcarrier frequency signal, which is fed to the laser transmitting module only at the moments of receipt of these pulses. This makes these signals more resistant to various effects on the FOCL than digital ones. When transmitting digital signals over FOCL, the subcarrier signal is an envelope for a bit sequence consisting of zeros and ones.

The key element switching confirmation command is similarly formed from only three numbers, which are also separated by a hyphen. For example, the operator will receive the confirmation command 1231-47-29 on the central computer in response to the previously sent control command (Figure 1). Figure 2 shows this confirmation command.

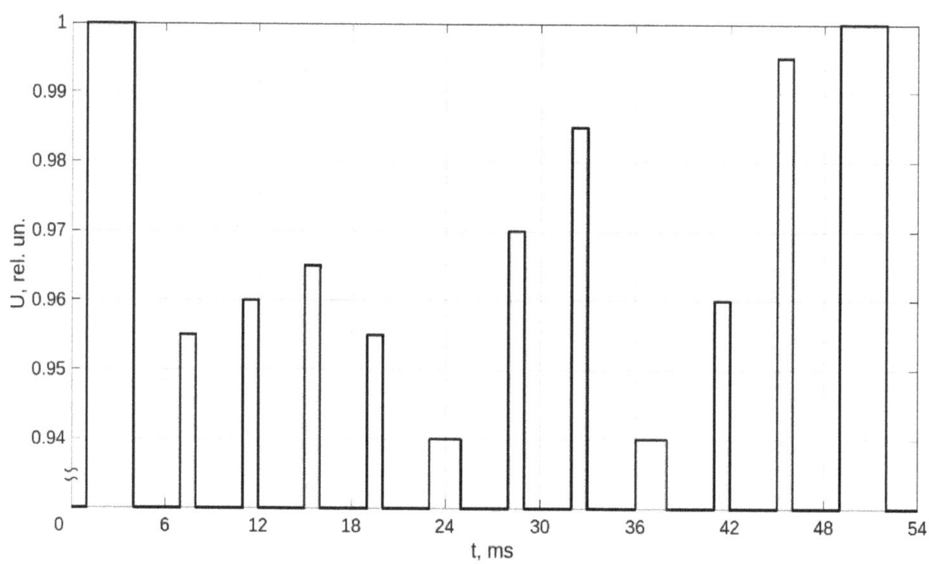

Figure 2. A sequence of command codes in the form of rectangular pulses for transmitting a confirmation command in analog form via FOCL.

To transmit this command, an analog FOCL is also used (its design is identical to the analog FOCL for transmitting control commands) with a transmitting laser module with direct modulation using a subcarrier frequency in the range of 0.1–200 MHz. These two independent FOCLs are placed in one shielded cable on the top of the transmission line. You cannot place two fibers to transmit two commands in one reflective layer. Because the power of laser radiation is large, one channel will negatively impact the other communication channel. It will result in a failure in the transmission of information.

3. The Design of an Analog Fiber-Optic Communication Line and the Calculation of Its Parameters

Figure 3 shows a block diagram of the analog FOCL developed by us for transmitting control commands and confirmation signals about the completed switching.

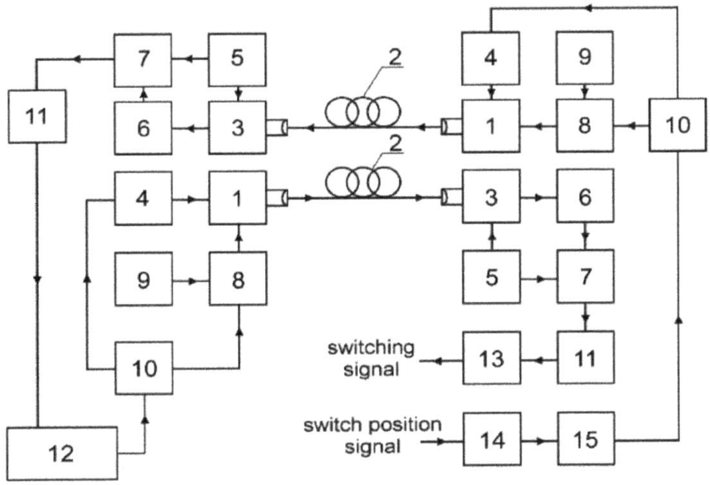

Figure 3. Structural diagram of an analog FOCL for controlling and monitoring the operation of key elements: 1—semiconductor laser; 2—optical fiber; 3—photodetector module; 4—laser power supply unit with direct current modulation; 5—multifunctional power supply; 6—tunable LC filter; 7—analog-to-digital converter (ADC); 8—electronic key; 9—subcarrier frequency signal generator; 10—the device for forming a sequence of rectangular pulses; 11—the device for processing information obtained using a sequence of command codes; 12—dispatcher's central computer; 13—device for generating switching signals of key elements; 14—device for generating data from key elements; and 15—information processing device.

In the developed design of the FOCL, in contrast to the classical schemes for transmitting analog signals, the subcarrier frequency F_s is used. This information transmission feature is because low-frequency signals (sequence of command codes) are transmitted over long distances via FOCL. Laser radiation always contains flicker noise, which is also low frequency. The conducted studies have shown this phenomenon in laser radiation with a power of 150 mW or more (the noise level is about (−62 dBm) or more).

These noises can form on the upper part of a rectangular pulse during direct modulation of laser radiation and affect the signal-to-noise ratio during the registration of an optical signal. The conducted studies have shown that the use of a subcarrier frequency signal F_s with a higher frequency than the transmitted signals, for which rectangular pulses are the envelope, makes it possible to make the effect of noise flicker on the registration of the modulated optical signal insignificant. To achieve this, also in the developed design of the FOCL, an electronic key 8 is used. The subcarrier signal enters the laser 1 only when supplying rectangular pulses with information to its power element.

Such formation of an optical information signal makes it possible to place a tunable LC filter 6 after the photodetector 3 (Figure 3), which cuts out the subcarrier signal. The information processing device 11 after the ADC 7 receives only rectangular pulses with information. Control device 10 is designed to convert information from sensors into command codes. This device ensures that they arrive at the control input of the electronic key 6 and the laser power supply 4 in the specified mode. After processing device 11, the information enters either the central computer 12 or the switching signal generation device 13, which generates commands for switching key elements. After switching the key element, information about its position through 14 is sent to the processing device 15, which converts the received data into the required format for transferring it to 10. In device 10, a command code is generated about the switching performed, which is sent to 8 and 4 to transmit it via FOCL to the central computer 12.

Difficulties arise with registering an optical signal transmitted over long distances and significantly decreases in power. Currently, various models of photodetectors for recording analog signals are produced (for example, PDA400 (company "Thorlabs Inc." (New York, NY, USA)) or the sensitive InGaAs photodiode PDINCH300 (Company "Emcore"(Lion, France)) in the range of changes in the detected power ΔP_r of laser radiation, which is ≈90–95 dB. Different limits determine different models of photodetector modules ΔP_r (for example, from 10 dBm to −80 dBm or from 0 dBm to −95 dBm).

Because it is necessary to transmit information over distances of more than 500 km, selecting a receiver with a lower value of the registration range of minus 95 dBm is necessary. At such information transmission distances, additional noise is formed in the optical fiber in addition to photodetector noise and others. The experience of operating FOCL in normal mode shows that all these noises do not exceed 2–3 dBm. In addition, 2 dBm is lost when laser radiation is injected into the optical fiber. In this case, for the FOCL developed by us, the value of ΔP_r is more than 110 dBm. The value of L, when all the factors are considered, can be estimated using the following formula:

$$L = \Delta P_r / \alpha_{1550} > 550 \text{ km}, \tag{2}$$

where α_{1550} is the power loss in the optical fiber at λ = 1550 nm (standard losses $\alpha_{1550} \approx 0.195$ dB/km).

The result obtained shows the possibility of transmitting an optical signal and its stable registration at distances $L \approx 550$ km. When transmitting over such long distances, there must be a margin for losses (in case of repair and other situations).

Because information is transmitted over FOCL over more than 200 km, it is necessary to calculate the most important parameters of the developed FOCL. These are the rise time of the optical system τ_s, the time of the signal transmitted through the optical fiber τ_0, and the energy balance a_e. These parameters show the capabilities of the developed FOCL design for transmitting an analog signal over a distance of L.

The following data are used to calculate the developed FOCL. Information is transmitted at a wavelength λ = 1550 nm. The optical power P_t of the transmitting laser module 1 (Emcore) is 150 mW, and the modulation depth is 70%. For a given P_t, the width of the laser emission spectrum is ΔF_1 = 600 MHz. The subcarrier frequency is F_s = 100 MHz. A specialized single-mode optical fiber of the G.652 standard with a shifted zero dispersion (triangular profile) M = 0.3 ps/(nm·km) is used to transmit information. A photodetector module is used to receive an optical signal (with the following parameters: bandwidth ΔF_2 = 1 GHz, NEP = 10^{-14} W·Hz$^{1/2}$; optical fiber length L = 550 km).

When using high-power laser radiation to transmit command codes, there is a problem with determining the line width $\Delta\lambda$. As the value of P_t increases, the value of $\Delta\lambda$ also changes (the manufacturer of the laser transmitters only specifies the maximum permissible value). Therefore, we conducted additional studies of the change in the value of $\Delta\lambda$ from P_t (Figure 4).

The obtained results showed that the value of $\Delta\lambda$ slightly changes when the value of P_t changes to 160 mW. Furthermore, the value of $\Delta\lambda$ changes nonlinearly to 0.227 nm at a power P_t = 250 mW. For P_t = 150 mW, the linewidth is $\Delta\lambda$ = 0.112 nm. The following formulas are used to calculate the values of τ_s and τ_0:

$$\tau_0 = \frac{B}{F_s}, \tag{3}$$

where B = 0.35 is a factor that considers the nature of the linear analog signal.

$$\tau_s = \sqrt{{\tau_1}^2 + {\tau_2}^2 + {\tau_3}^2}, \tag{4}$$

where τ_1 is the rise time of the transmitter, τ_2 is the rise time of the receiver, and τ_3 is the rise time of the optical fiber are defined as follows:

$$\tau_1 = \frac{B}{\Delta F_1}, \tag{5}$$

$$\tau_2 = \frac{B}{\Delta F_2}, \quad (6)$$

$$\tau_3 = \frac{B}{\Delta F_3}, \quad (7)$$

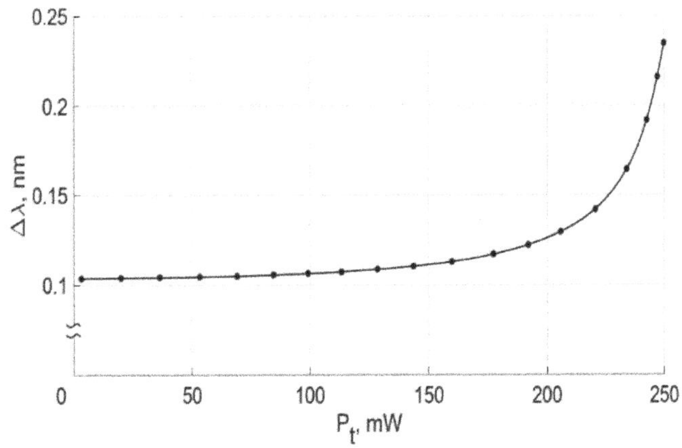

Figure 4. Dependence of the change in the linewidth $\Delta\lambda$ of laser radiation on the power P_t at a temperature T = 293.2 K.

All values in Relations (5) and (6) are determined by the manufacturers of the receiving and transmitting modules. For calculations, it is necessary to determine ΔF_3 from the following relationship:

$$\Delta F_3 = 0.35/(M \cdot \Delta\lambda \cdot L), \quad (8)$$

As a result of calculations, the following values were obtained: τ_0 = 3.5 ns, τ_1 = 0.58 ns, τ_2 = 0.35 ns, τ_3 = 0.018 ns, τ_s = 0.678 ns. An analysis of the obtained time values showed that the relation $\tau_0 > \tau_s$ is satisfied in the developed FOCL design. Information about the switching command will be transmitted over 550 km.

The following formulas are used to calculate the energy balance a_e:

$$a_e = a_1 - (a_2 + N \cdot a_3 + a_4 + a_5), \quad (9)$$

where a_1 is the loss margin in the optical fiber, a_2 is the attenuation of the optical signal over the entire length of the fiber, a_3 is the attenuation at various connections (welding is used), N is the number of connections, a_4 is the loss at the modulation depth in the transmitting optical signal, and a_5 is the attenuation on classic connections (connectors).

Let us determine for the developed FOCL all the values included in Equation (9):

$$a_1 = P_{in} - P_R, \quad (10)$$

where P_R is the minimum optical power recorded on the photodetector module.

$$P_R - 10\log\left(\frac{\text{NEP}}{10^{-3}}\right) + 5\log(\Delta F_t) + 0.5 \cdot \text{SNR(dB)}, \quad (11)$$

The following values are used to calculate the P_R value. The optical signal is recorded with a signal-to-noise ratio (SNR) equal to 8 (in times) with an information transmission bandwidth ΔF_t = 100 kHz.

$$a_2 = \alpha_{1550} \cdot L,\ a_3 = 0.05\ \text{dB}, N = 99, a_4 = 1.5\ \text{dB},\ a_5 = 0.24\ \text{dB}, \quad (12)$$

As a result of calculations, the following values were obtained: P_R = −84.55 dBm, a_1 = 104.55 dB, a_2 = 97.50 dB, and a_e = 0.36 dB.

The obtained result shows that the developed FOCL can stably transmit information over distances up to 500 km. The further transmission of information is not recommended because the value of a_e is slightly greater than 0 dB with a recommended margin of at least 1 dB. The distance L = 500 km is critical.

4. Results of Experimental Investigations and Discussion

The dynamic range is the main characteristic of FOCL for the transmission of analog signals. In the conditions of the research laboratory of Bonch-Bruevich Saint Petersburg State University of Telecommunication, with the support of company LLC "T8" and company PJSC ROSSETI, the laboratory layout was assembled for research of FOCL developed by us (the block diagram of laboratory layout is shown in Figure 3). A during developing analog FOCLs with using laser radiation a power of more than 15 mW, modeling of line characteristics is not very promising work due to the difficulty of considering all non-linear distortions and scattering that affect the amplitude and shape of the transmitted signal. It is more reliable to measure everything. Without the results of experimental research, companies do not accept the development for implementation.

In Figure 5 shows the results of measuring the output power Pout from a change in the laser power P_{in}, which enters the optical fiber for various values of L.

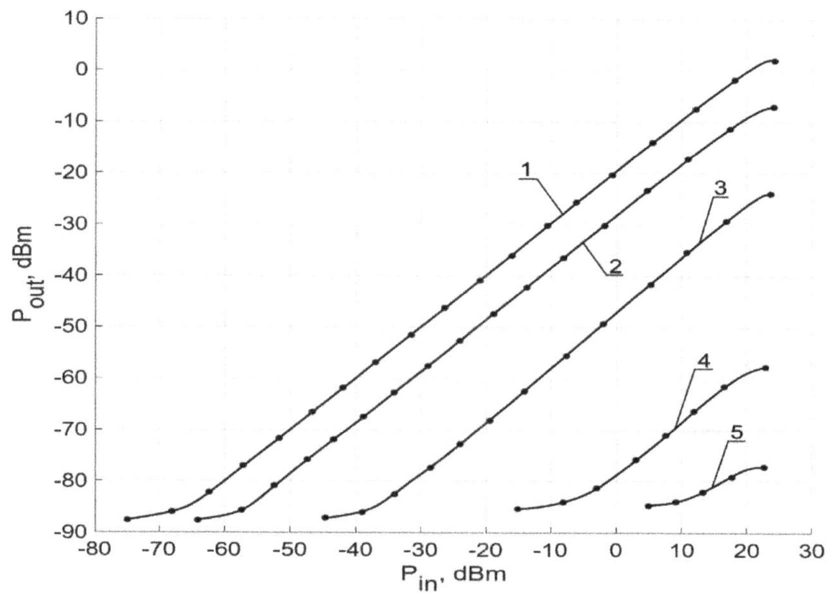

Figure 5. Amplitude characteristic of FOCL. Graphs 1, 2, 3, 4, and 5 correspond to L in km: 100; 150; 250; 400; and 500.

The conducted studies of the change in the dynamic range of the developed design of the analog FOCL showed that stable transmission of control commands to key elements and control of their position over distances of up to 500 km is ensured. The presence of a nonlinear section on the amplitude characteristic in the lower part is explained by the fact that the amplitude of the recorded signal has become comparable to the intrinsic noise of the photodetector module. Moreover, with an increase in L, the number of welded joints increases (the construction length of a solid fiber is 5000 m), which increases the overall loss of the optical signal in the fiber. Therefore, there are changes in the nonlinearity of the amplitude response with increasing L.

Nonlinear distortions in the amplitude characteristic in the upper part can be explained by the influence of SMBS bias and Kerr nonlinearity on the transmitted signal parameters.

As a result of experiments, we found that the parameters of laser transmitting modules do not affect the formation of this nonlinearity. During the research, various options for the rectangular pulse formation were used (using a semiconductor laser tunable in power and equipment of company LLC "T8" using an amplifier). Changes in the amplitude characteristics measured for two cases turned out to be insignificant.

As an example, Figure 6 shows the results of a study of the transmission of rectangular pulses with a duration of 1 ms (digit) over the developed analog FOCL with a pulse interval of 3 ms for various values of L.

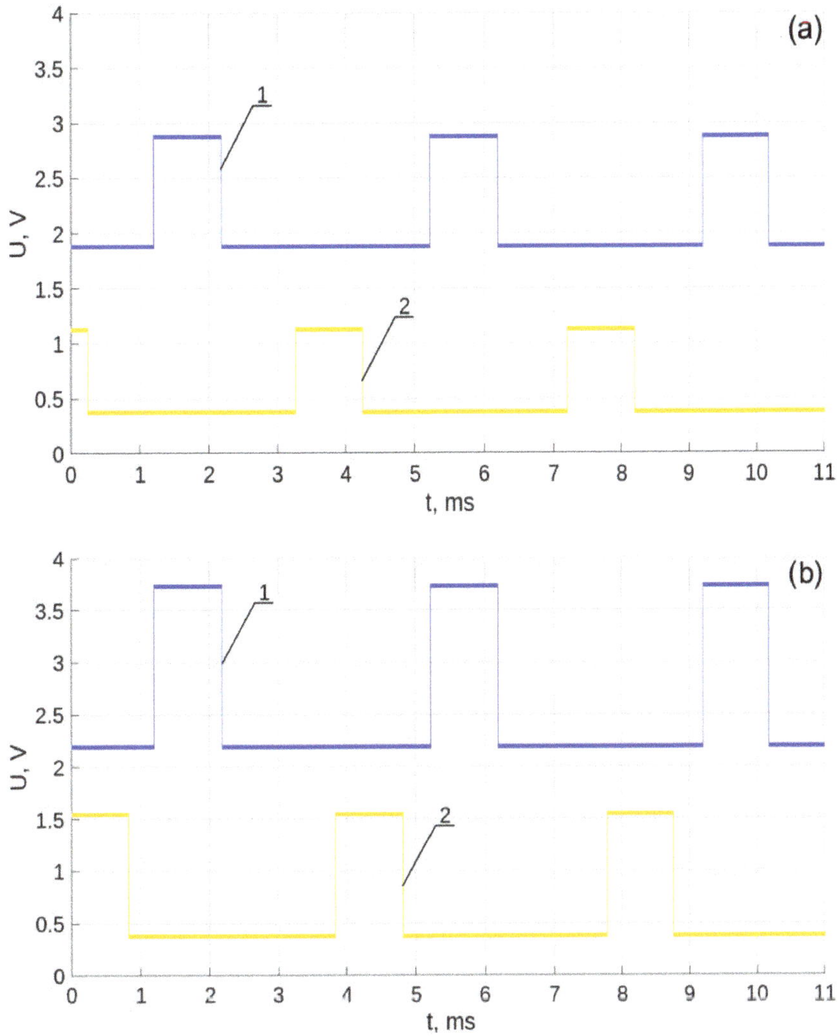

Figure 6. Rectangular pulses for the transmission of digits in numbers as a sequence of command codes: graph 1 at the input to 2, graph 2 at the output 3. Studies are presented for various values of L in km: (**a**) −200; (**b**) −250.

As an example, Figure 7 shows the results of studying the transmission of rectangular pulses with a duration of 1 ms (digit) over the developed analog FOCL with a duty cycle of 100 MHz with an interval between pulses of 3 ms for various values of L.

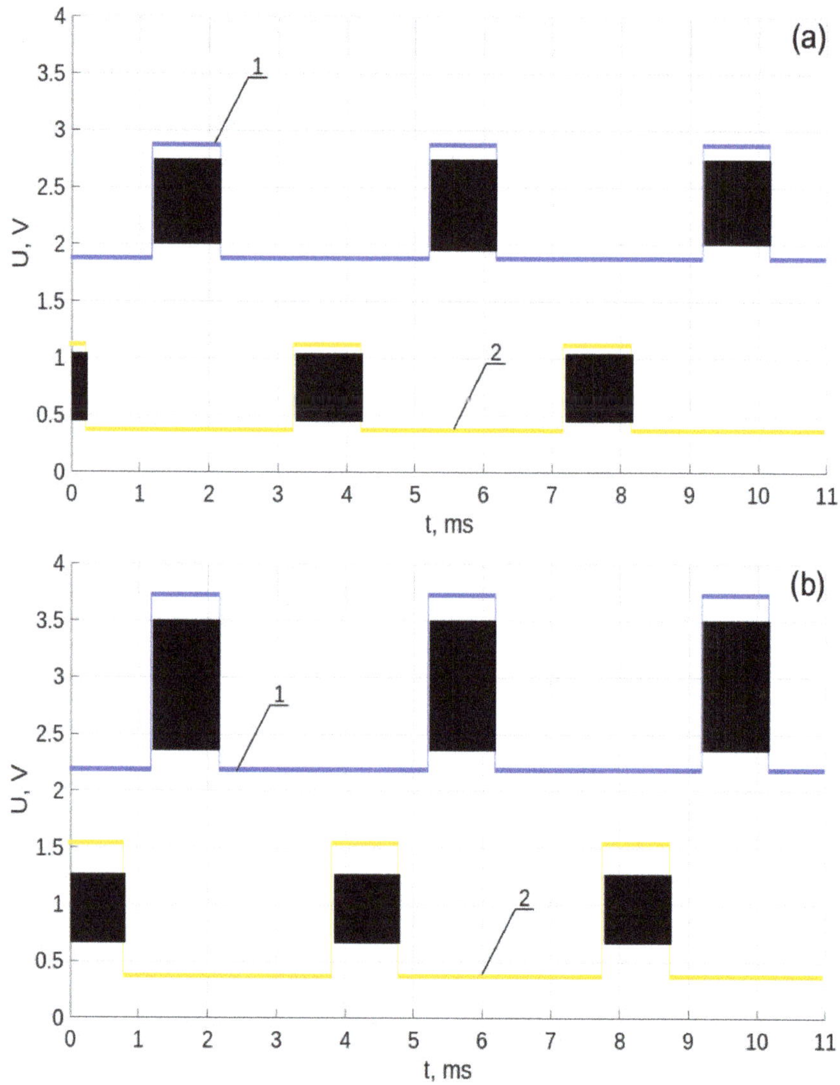

Figure 7. Rectangular pulses for the transmission of digits in numbers as a sequence of command codes: graph 1 at the input to 2, graph 2 at the output 3. Studies are presented for various values of *L* in km: (**a**) −200; (**b**) −250.

As an example, Figure 8 shows the results of the registration of rectangular pulses transmitted over the developed analog FOCL with a duration of 1 ms (digit) with a duty cycle of 100 MHz with an interval between pulses of 3 ms for various values of *L*.

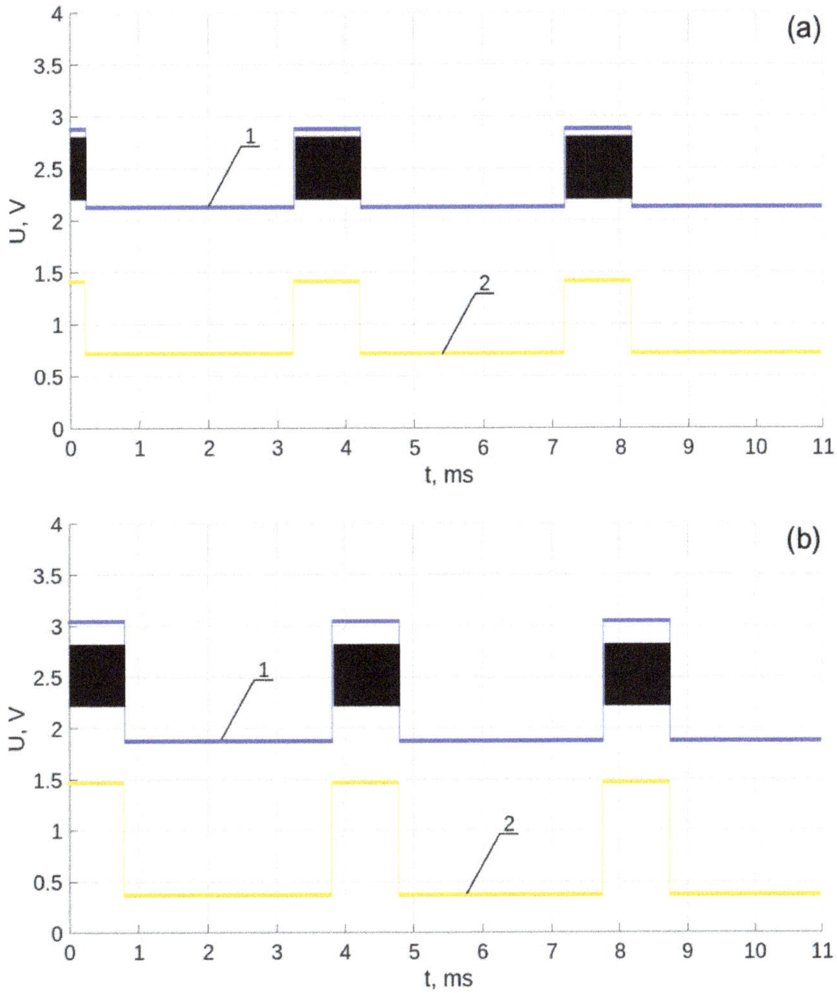

Figure 8. Rectangular pulses for transmitting digits in numbers as a sequence of command codes. Graph 1 corresponds to the signal at the input to 2. Graph 2 corresponds to the signal at output 6 for different values of L in km: (**a**) 200; (**b**) −250.

An analysis of the results obtained shows stable operation of the developed analog FOCL when transmitting information in the form of a sequence of command codes at a power of P_{in} = 100 mW over various distances.

The absence of significant distortions in the fronts and upper part of the recorded pulses transmitted over the FOCL developed by us shows that for this laser radiation power, the influence of the SMBS bias and Kerr nonlinearity is insignificant.

The investigations conducted made it possible to establish the insignificant effect of changing the subcarrier frequency F_s in the range from 1 to 200 MHz on the amplitude and shape of recorded analog signal transmitted over the FOCL.

5. Conclusions

An analysis of the obtained experimental and calculated data confirms the adequacy of our proposed developments in the implementation of the design of an analog FOCL for transmitting control commands for key elements in various power systems and for obtaining reliable information about the switching performed (the position of panel switches, etc.). The experiments and calculations confirmed the possibility of reliable transmission of control commands and other information over distances of 500 km. It is at least 5–6 times more than in operating control systems operating on other physical principles.

It should be noted that the developed analog FOCL is easily integrated into the power transmission lines in operation. For its placement and functioning, it is not necessary to make any changes and improvements to the structures of power lines installed on the routes. In addition, using the developed FOCL makes it possible to exclude two server stations from the control and monitoring system. The service life of the FOCL developed by us is more than 20 years, and the service life of the server station in the PJSC ROSSETI (Russian Federation) system is ten years at maximum. For example, preventive maintenance on the developed FOCL is performed as needed (possibly never before the expiration of its service life). At the server station in the northern regions of the Russian Federation in winter, preventive maintenance is performed at least once a week and, if necessary, more often (due to snow drifts, icicles, etc.). Such exploitation requires time and significant human and financial resources. Using our development significantly reduces these costs in all areas. These conclusions are made by the authors of this article on the basis of many years of cooperation with company PJSC ROSSETI (formerly with company RJS "Unified Energy Systems of Russia (UES Russia)" until 2008). Specialists of the company PJSC ROSSETI are involved in the development of this project. This project is planned to be implemented in practice in the divisions of this company. At the request of the article authors, an expert specialist group of PJSC ROSSETI performed (at the moment) an economic assessment of operational efficiency for one developed FOCL within 20 years in a power transmission line system compared to existing equipment. Economic efficiency is from 5 to 20 million US dollars. Such a spread is determined by climatic conditions and inflationary processes over the entire period of FOCL operation. The risk of work in the power transmission line system in difficult weather conditions is not included in this amount.

Increasing the range of information transmission (more than 500 km) without the use of optical amplifiers, which are extremely difficult to integrate into existing power transmission lines, is a challenging issue. Its solution requires further research. This is because the classical methods of increasing L (the choice of a photodetector with a lower NEP is associated with the production of such elements, and the choice of an optical fiber with a lower attenuation coefficient α_{1550}) are either technically difficult to implement or sharply increase the cost of the FOCL design (fiber with a core made of pure quartz α_{1550} = 0.171 dB/km). When using this type of optical fiber, the transmission distance of control commands and necessary information is increased up to 600 km.

It should be noted that the analog FOCLs developed by us are lines for unidirectional data transmission (one way). This eliminates the possibility of unauthorized connection to the FOCL by unknown persons without its identification. When using server stations, these connections are possible. Such circumstances further increase the reliability of the operation of the FOCL developed by us for the management and control of the operation of key elements in power systems.

Considering all of these factors, as well as the circumstances in which the experimental FOCL layout developed by us is assembled using equipment and components from leading world companies, the following statement can be made. The FOCL developed by us can be implemented in the energy systems of the USA, Canada, China, and other countries with minor modifications.

Author Contributions: Conceptualization, V.D. (Vadim Davydov) and V.D. (Valentin Dudkin); methodology, V.D. (Vadim Davydov); software, B.R.; validation, V.D. (Vadim Davydov), V.D. (Valentin Dudkin) and B.R.; formal analysis, B.R. and V.D. (Valentin Dudkin); investigation, V.D. (Valentin Dudkin); resources, V.D. (Vadim Davydov) and V.D. (Valentin Dudkin); data curation, B.R.; writing—original draft preparation, V.D. (Vadim Davydov); writing—review and editing, V.D. (Valentin Dudkin); visualization, B.R.; supervision, V.D. (Valentin Dudkin); project administration, V.D. (Vadim Davydov); funding acquisition, V.D. (Vadim Davydov). All authors have read and agreed to the published version of the manuscript.

Funding: This research received no external funding.

Data Availability Statement: Not applicable.

Conflicts of Interest: The authors declare no conflict of interest.

References

1. Agafonova, N.D.; Egorov, M.Y.; Sergeev, V.V.; Gotovskii, M.A.; Kruglikov, P.A.; Lebedev, M.E.; Sudakov, A.V.; Fedorovich, E.D.; Fokin, B.S. Heat-and-Mass Transfer Intensification in Saturated-Steam Generators in NPP with VVER as a Means for Increasing Efficiency and Reliability. *At. Energy* **2018**, *123*, 154–158. [CrossRef]
2. Gulevich, A.V.; Dekusar, V.M.; Chebeskov, A.N.; Kuchinov, V.P.; Voloshin, N.P. Possibility of Fast-Reactor Exportation Under an International Nuclear Non-Proliferation Regime. *At. Energy* **2020**, *127*, 192–195. [CrossRef]
3. Dong, Z.; Liu, M.; Jiang, D.; Huang, X.; Zhang, Y.; Zhang, Z. Automatic Generation Control of Nuclear Heating Reactor Power Plants. *Energies* **2018**, *11*, 2782. [CrossRef]
4. Patil, P.R.; Tanavade, S.; Dinesh, M.N. Analysis of Power Loss in Forward Converter Transformer Using a Novel Machine Learning Based Optimization Framework. *Technol. Econ. Smart Grids Sustain. Energy* **2022**, *7*, 21. [CrossRef]
5. Abbas, G.; Asad, M.U.; Gu, J.; Alelyani, S.; Balas, V.E.; Rashid Hussain, M.; Farooq, U.; Chang, C. Multivariable unconstrained pattern search method for optimizing digital pid controllers applied to isolated forward converter. *Energies* **2021**, *14*, 77. [CrossRef]
6. Başoğlu, M.E. Forward converter-based distributed global maximum power point tracking in partial shading conditions. *SN Appl. Sci.* **2020**, *2*, 248. [CrossRef]
7. Jiang, W.; Liu, X.; Zhang, X.; Chen, H.; Li, P. Reliability Analysis of Fast Transfer Switch under Multiple Degradation Factors. *J. Mech. Eng.* **2022**, *58*, 105–115. [CrossRef]
8. Javaid, U.; Freijedo, F.D.; Dujic, D. MVDC supply technologies for marine electrical distribution systems. *CPSS Trans. Power Electron. Appl.* **2018**, *3*, 65–76. [CrossRef]
9. Rajora, G.L.; Sanz-Bobi, M.A.; Domingo, C.M. Application of Machine Learning Methods for Asset Management on Power Distribution Networks. *Emerg. Sci. J.* **2022**, *6*, 905–920. [CrossRef]
10. Johal, H.; Ren, W.; Pan, Y.; Krok, M. An integrated approach for controlling and optimizing the operation of a power distribution system. In Proceedings of the IEEE PES Innovative Smart Grid Technologies Conference Europe (ISGT Europe), Gothenburg, Sweden, 11–13 October 2010; p. 5638859. [CrossRef]
11. Northcote-Green, J.; Wilson, R. Control and automation of electrical power distribution systems. *Control Autom. Electr. Power Distrib. Syst.* **2017**, *1*, 464. [CrossRef]
12. Davydov, R.; Logunov, S.; Bobyl, A. Optical Method for Controlling the Flow Rate of the Coolant in Nuclear Reactors. In Proceedings of the 2021 International Conference on Electrical Engineering and Photonics, EExPolytech, St. Petersburg, Russia, 14–15 October 2021; pp. 179–183. [CrossRef]
13. Davydov, V.V.; Velichko, E.N.; Karseev, A.Y. Fiber-optic system for simulating accidents in the cooling circuits of a nuclear power plant. *J. Opt. Technol.* **2015**, *82*, 132–135. [CrossRef]
14. Tor, O.; Shahidehpour, M. Power distribution asset management. In Proceedings of the 2006 IEEE Power Engineering Society General Meeting (PES), Montreal, QC, Canada, 18–22 June 2006; p. 1709234. [CrossRef]
15. Opathella, C.; Elkasrawy, A.; Adel Mohamed, A.; Venkatesh, B. MILP formulation for generation and storage asset sizing and sitting for reliability constrained system planning. *Int. J. Electr. Power Energy Syst.* **2021**, *116*, 105529. [CrossRef]
16. Koziel, S.; Hilber, P.; Westerlund, P.; Shayesteh, E. Investments in data quality: Evaluating impacts of faulty data on asset management in power systems. *Appl. Energy* **2021**, *281*, 116057. [CrossRef]
17. Martínez-Galán, P.; Crespo, A.; de la Fuente, A.; Guillén, A. A new model to compare intelligent asset management platforms (IAMP). *IFAC-Pap.* **2020**, *53*, 13–18. [CrossRef]
18. Napoleone, A.; Roda, I.; Macchi, M. The implications of condition monitoring on asset-related decision-making in the Italian power distribution sector. *IFAC-Pap.* **2016**, *49*, 108–113. [CrossRef]
19. Mirhosseini, M.; Keynia, F. Asset management and maintenance programming for power distribution systems: A review. *IET Gener. Transm. Distrib.* **2021**, *15*, 2287–2297. [CrossRef]
20. Gavrikova, E.; Volkova, I.; Burda, Y. Strategic aspects of asset management: An overview of current research. *Sustainability* **2020**, *12*, 5955. [CrossRef]

21. Koksal, A.; Ozdemir, A. Improved transformer maintenance plan for reliability centred asset management of power transmission system. *IET Gener. Transm. Distrib.* **2016**, *10*, 1976–1983. [CrossRef]
22. Hao, Q.; Yan, H.; Tian, B.; Zhang, A. Research on intelligent photovoltaic control and protection switch based on fuzzy PID controller. *J. Phys. Conf. Ser.* **2022**, *2310*, 012047. [CrossRef]
23. Kumar Gupta, M.; Kumar Gupta, P.; Kumar Giri, R.; Gupta, A. Smart electric control system using PLC & HMI. *Int. J. Mech. Eng. Technol.* **2018**, *9*, 548–555.
24. Tan, H.G.R.; Tan, A.C.; Iriana, M.; Mok, V.H. Power socket programmable circuit breaker system. In Proceedings of the IEEE 2nd International Power and Energy Conference, Johor Bahru, Malaysia, 1–3 December 2008; pp. 574–577. [CrossRef]
25. Nistane, K.S.; Raut, P.V. Power generators using automatic transfer switch. *Int. Res. J. Eng. Technol. (IRJET)* **2020**, *7*, 112–114. [CrossRef]
26. Gupta, A.K.; Gupta, A. A Novel Design of Smart Train. In Proceedings of the 2018 International Conference on Smart Electric Drives and Power System (ICSEDPS 2018), Nagpur, India, 12–13 June 2018; pp. 69–74. [CrossRef]
27. Kumar, S.; Sharma, N. Emerging Military Applications of Free Space Optical Communication Technology: A Detailed Review. *J. Phys. Conf. Ser.* **2022**, *2161*, 012011. [CrossRef]
28. Xu, R.; Song, G.; Hou, J.; Zhang, C.; Zhang, Y. Pilot Protection for High Voltage Direct Current Transmission Line Based on Polarity Characteristics of Converter Control Signal Deviation. *J. Xi'an Jiaotong Univ.* **2021**, *55*, 143–153. [CrossRef]
29. Song, X.; Shen, H.; Huang, S.; Han, L.; Xiao, Z. Effect of distributed capacitance on differential current protection and relevant countermeasures. *Dianli Zidonghua Shebei/Electr. Power Autom. Equip.* **2018**, *38*, 169–172. [CrossRef]
30. Song, G.; Gao, S.; Cai, X.; Zhang, J.; Rao, J.; Suonan, J. Survey of relay protection technology for HVDC transmission lines. *Autom. Electr. Power Syst.* **2012**, *36*, 123–129. [CrossRef]
31. Zhang, C.; Song, G.; Dong, X. A Non-unit Adaptive Traveling Wave Protection Method for High Impedance Faults. *Proc. Chin. Soc. Electr. Eng.* **2021**, *40*, 3548–3556. [CrossRef]
32. Jia, K.; Li, M.; Bi, T.; Wang, C.; Qi, H.; Yang, Q. Energy Distribution-Based Differential Protection for VSC-DC Distribution Lines. *Power Syst. Technol.* **2017**, *41*, 3058–3065. [CrossRef]
33. Davydov, R.V.; Gureeva, I.M. Coolant Flow Rate Measurement Technique for Nuclear Power Reactor Control Systems. *Meas. Tech.* **2022**, *64*, 895–902. [CrossRef]
34. Davydov, R.; Logunov, S.; Nikolaev, D. Fiber Optic System for Monitoring Coolant Parameters in Nuclear Power Plants. In *Lecture Notes in Computer Science (Including Subseries Lecture Notes in Artificial Intelligence and Lecture Notes in Bioinformatics)*; Springer: Cham, Switzerland, 2022; Volume 13158, pp. 221–229. [CrossRef]
35. Davydov, V.V.; Ermak, S.V.; Karseev, A.U.; Petrov, A.A.; Velichko, E.N. Fiber-optic super-high-frequency signal transmission system for sea-based radar station. In *Lecture Notes in Computer Science (Including Subseries Lecture Notes in Artificial Intelligence and Lecture Notes in Bioinformatics)*; Springer: Cham, Switzerland, 2014; Volume 8638, pp. 694–702. [CrossRef]
36. Chapalo, I.; Petrov, A.; Bozhko, D.; Bisyarin, M.; Kotov, O. Averaging methods for a multimode fiber interferometer: Experimental and interpretation. *J. Light. Technol.* **2020**, *38*, 5809–5816. [CrossRef]
37. Filatov, D.L.; Galichina, A.A.; Vysoczky, M.G. Features of transmission at analog intermediate frequency signals on fiber—Optical communication lines in radar station. *J. Phys. Conf. Ser.* **2017**, *917*, 082005. [CrossRef]
38. Shurupov, D.; Pchelkin, G. Development of high-speed fiber-optic communication lines for transferring large amounts of information. In Proceedings of the ACM International Conference Proceeding Series, New York, NY, USA, 15–16 December 2021; pp. 577–580. [CrossRef]
39. Kotov, O.I.; Chapalo, I.E.; Petrov, A.V. Signals of an intermodal fiber interferometer induced by laser frequency modulation. *Tech. Phys. Lett.* **2016**, *42*, 11–14. [CrossRef]
40. Yang, X.; Tang, B.; Cao, X. Association study between the optical-photocatalysis behaviors and the carrier transfer characteristics of g-$C_3N_4/Sr_2MgSi_2O_7$:(Eu,Dy). *J. Environ. Chem. Eng.* **2022**, *10*, 108820. [CrossRef]
41. Maqbool, H.; Rafique, A.; Bhatti, I.A.; Ayub, K.; Iqbal, J. Novel endohedrally and exohedrally metals (Li, Na, and K, Ag) doped (15-crown-5) with remarkable electronic, static and dynamic NLO response. *Optik* **2022**, *271*, 17016. [CrossRef]
42. Alioueche, A.; Benmerkhi, A.; Bouchemat, M. Optical photonic crystal sensor based on U-shaped ring resonator. *Opt. Quantum Electron.* **2022**, *54*, 831. [CrossRef]
43. Zhuang, F.; Hao, G.; Zhou, R.; Li, H.; Zhang, R. Electric fields and strains effect on the electronic and optical properties of Zr_2CO_2/MoSSe van der Waals heterostructure. *Mater. Today Commun.* **2022**, *33*, 104295. [CrossRef]
44. Jiang, R.; Wang, R.; Guo, J.; Wu, Y.; Li, Y. Design and analysis of a two-dimensional large-scale silicon-photonic optical phased array. *Opt. Laser Technol.* **2022**, *156*, 108551. [CrossRef]
45. Davydov, V.; Nikolaev, D.; Moroz, A.; Dmitrieva, D.; Pilipova, V. Research of fatigue damage in organic glass by speckle images. *AIP Conf. Proc.* **2020**, *2308*, 060005. [CrossRef]
46. Gowar, J. Optical communication systems. In *Lecturer in Electronic Engineering*; Universlty of Brlstol (Prentice—Hall International): London, UK, 1984.

energies

Article

Developing a Reactivity-Equivalent Physical Transformation to Simulate an Axially Heterogeneous Boiling Water Reactor

Robert Mossop [1,*], Bruno Merk [1], Maulik Patel [1], Tetsushi Hino [2] and Aiden Peakman [1,3]

1 School of Engineering, University of Liverpool, Liverpool L69 3BX, UK
2 Hitachi, Ltd., 7-1-1, Omika-cho, Hitachi-shi 319-1292, Ibaraki-ken, Japan
3 National Nuclear Laboratory, Chadwick House, Warrington WA3 6AE, UK
* Correspondence: r.mossop@liverpool.ac.uk

Abstract: Hitachi is advancing their designs for a conceptual reactor called the resource-renewable boiling water reactor (RBWR), a concept reactor similar to the advanced boiling water reactor with a harder neutron spectrum. This design aims to minimise construction costs and waste production as well as to utilise separated plutonium and minor actinide fuel. However, the axial heterogeneity of the design poses calculation difficulties. The aim of this work is to use a known method, reactivity-equivalent physical transformation (RPT), for calculating fuel with double heterogeneity and apply it to a BWR-type fuel pin. This could reduce the calculation time needed for optimisation of the design of the RBWR. The objective of the study is to use SCALE 6.2 to produce an equivalent axial pin model by comparison with the burnup and neutron spectra of a radial model of the fuel. This model can then be used for 2D burnup calculations, and in future work will be used for the generation of two-group and multigroup cross-sections for further deterministic calculations as part of a two-step approach for analysis of the RBWR. The RPT method has been extensively tested on spherical fuel, and SCALE is a standard industry code. The initial radial model is a hexagonal assembly with 20% enriched UO_2 fuel in a zircaloy cladding, surrounded by light water moderator. The derived axial model has a water density distribution taken from Hitachi's RBWR designs. Criticality over 70 GWd/tU burnup is estimated using the model. The application of the RPT to the BWR pin was shown to be possible, but to have limitations with the introduction of additional radial complexity. For a single pin, excellent agreement between the radial and axial models could be found across a range of water densities, but in the case of an assembly level calculation distinct equivalence models were required for each water density. In addition, the produced RPT model is validated using SCALE's 3D Monte Carlo module, KENO.

Keywords: waste burner; axial heterogeneity; deterministic; RBWR

check for updates

Citation: Mossop, R.; Merk, B.; Patel, M.; Hino, T.; Peakman, A. Developing a Reactivity-Equivalent Physical Transformation to Simulate an Axially Heterogeneous Boiling Water Reactor. *Energies* **2023**, *16*, 3359. https://doi.org/10.3390/en16083359

Academic Editor: Roman Vadimovich Davydov

Received: 23 February 2023
Revised: 1 April 2023
Accepted: 4 April 2023
Published: 11 April 2023

1. Introduction

The Generation IV reactor program seeks to improve several aspects of nuclear reactors, including safety, cost, and efficiency [1]. Improving sustainability also stands as one of the key goals for the successful large-scale deployment of future reactors. Generating energy sustainably and promoting long-term availability of nuclear fuel are extremely important in any scenario where the coverage of nuclear power is significantly expanded, and in response to questions about geological disposal and the cost of permanent storage of spent fuel, minimising long-term nuclear waste and reducing the stewardship burden are at the forefront of the conversation around civil nuclear power [2].

Hitachi's resource-renewable boiling water reactor (RBWR) is 'designed to improve the use of resources and reduce the load imposed on the environment by spent fuel' [3], while the concept is still based on a classical BWR plant design, which should reduce challenges with the implementation of a new technology by relying on several mature technologies from existing commercial BWR systems. By hardening the neutron spectrum compared to a conventional BWR, the RBWR can reuse transuranic elements as fuel [4,5]. A report by Hitachi to the ICAPP 2019 [6] suggests that due to a discrepancy in results of both Monte Carlo and deterministic calculations performed on the RBWR core by the JAEA, several US universities, and Hitachi, a 'more sophisticated' understanding and an improved level of modelling of the heterogeneity in the axial direction of the fuel pin are required [7].

In a 2012 report by the University of Michigan, the following is noted: 'An inherent deficiency of the neutronics modelling described above is that it does not take accurate account of the axial variation of the neutron spectrum that is very significant in the RBWR cores and their interface with the reflectors. This is because the generation of the group cross-sections was performed using 2-D lattice calculations. The computational method used by Hitachi for the RBWR core designs has a similar deficiency' [8].

Thus, there is a strong demand to develop a new approach for a more robust cross-section preparation procedure that is able to handle the axial heterogeneity. One approach could be to move to a full 3D cross-section preparation using Monte Carlo methods, which would lead to significantly increased computational demand. While this approach provides the most accurate cross-sections, it is not practical or necessary for design activities requiring parametric studies. Based on this request, the approach should be applicable in a standard lattice calculation as is typical for the first step of the two-step procedure of light water reactor analysis.

The problem in the RBWR seems, therefore, comparable to the double heterogeneity problem of high temperature reactor (HTR) fuel based on tristructural isotropic (TRISO) fuel particles. This problem was discussed in a paper by Kim et al. (2004) as part of a preliminary assessment of lattice physics for HTR analysis by Argonne National Lab [9]. Kim and Baek (2005) of KAIST proposed a novel method called the reactivity-equivalent physical transformation for solving the double heterogeneity problem [10]. A general conceptual diagram of the method can be seen in Figure 1. For an HTR, the first step is achieved by smearing TRISO particle materials (which contain graphite and fuel) in a compact graphite region at the centre of the model to eliminate the complex single structures. This region is surrounded by pure graphite moderator (such as a pebble or block in a real case). This step will lead to an incorrect criticality result due to loss of the self-shielding effect caused by homogenization. In a second step, the radius of the smeared fuel region is adjusted to account for this and to match the criticality of the now-homogenized and densified fuel region and the surrounding graphite to the criticality of the heterogeneous composition. Key to the process is the fact that the material configuration is not changed—the number of particles in the region stays constant and only the geometric dimensions are adopted in the new reactivity-equivalent physical transformation (RPT) model to mimic the self-shielding. KAERI (Korean Atomic Energy Research Institute) developed a computer code for analysis of the HTR for treatment of the 'unique' HTR double heterogeneity problem, based on HELIOS/MASTER in 2008 [11]. Rohde et al. (2012) expanded the RPT method by modifying the way in which the TRISO particles were smeared in the transformed model, completely separating the different material regions in the particle [12]. This allowed for separate temperature assignment to different materials in the model, further improving the usefulness of the transformation for the cross-section preparation in reactor lattice codes. As of 2022, progress on the RPT method continues for TRISO spherical fuel. In their proposed Ring RPT method, Lou et al. (2020) correct the method for inclusion of burnable poisons [13]. Handling the problem of the fuel's double heterogeneity through the RPT and solving it in a lattice code provides significant benefit in terms of computational time, due to the nature of the difference between the deterministic method used in this study and a Monte Carlo (MC) route, noting that MC calculation routes are inherently computationally intensive [12,14,15]. The RPT has not yet been applied to fuel types outside of the area of TRISO based HTR fuel.

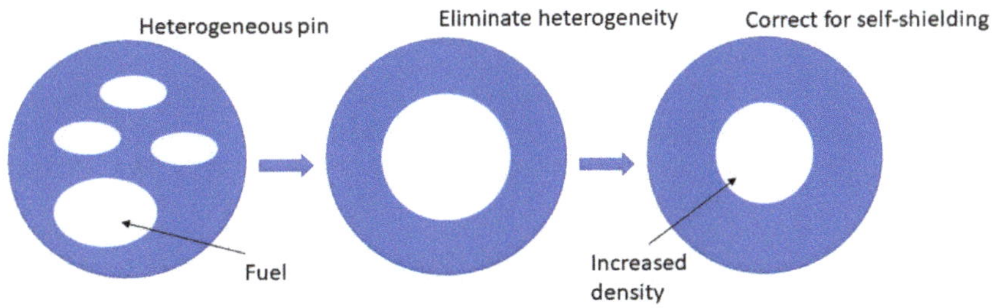

Figure 1. A diagrammatic representation of the core principle of previous uses for the RPT. A highly heterogeneous pin/fuel pellet is simplified to allow lattice calculations in 2D, and then the self-shielding differences which result from the geometric manipulation are corrected for by a manipulation of the density and radius of the fuel.

The current study investigates the applicability of the RPT method to the RBWR fuel pin, specifically focusing on the level of discrepancy between models of the pin produced for the axial and radial direction in a 2D calculation. An axial model of both a single pin and assembly, based on the RBWR design, are produced. The density and size of the fuel and moderator regions are changed, keeping the pitch and total mass of fuel and moderator constant. By using this adapted method of correction for self-shielding, heterogeneity in the axial direction, which is characteristic for RBWR fuel assemblies, could be calculated deterministically in a classical lattice code. As in the case of TRISO, the RPT method has the potential to reduce computation time for the design phase of the RBWR (or other axially heterogeneous reactor types) by expanding the applicability of deterministic codes through easing the production of homogenised and condensed cross-section libraries, an essential step for coupled neutronic/thermal hydraulic design and transient studies. The well-known lattice and burnup code system SCALE 6.2 is used for both models (classical radial as well as RPT corrected axial) to evaluate the question, 'Can the RPT be applied to an axially heterogeneous pin model of a BWR?'

2. Code Descriptions

Steady-state transport calculations and burnup calculations were performed using SCALE 6.2. SCALE 6.2 is a comprehensive modelling and simulation code package for criticality safety, reactor physics, radiation shielding, radioactive source term characterisation, and sensitivity and uncertainty analysis developed by the Oak Ridge Nuclear Laboratory (ORNL, Oak Ridge, TN, USA). Multigroup neutron energy libraries based on ENDF/B-VII.0 and ENDF/B-VII.1 data were used throughout this work. The specific structure used for these calculations is the 238-energy group structure. All calculations were performed with the TRITON module of SCALE 6.2. TRITON draws from several modules to perform calculations. T-XSEC is called to prepare multigroup cross-section libraries accounting for spatial and energy-dependent self-shielding. Libraries for each problem consist of microscopic cross-sections for each nuclide defined in the TRITON script. To build on the cross-section data, TRITON calls NEWT (New ESC-based Weighting Transport code), for the determination of the 2D neutron flux distribution. NEWT is 'a two-dimensional discrete ordinates transport code developed based on the Extended Step Characteristic (ESC) approach' [16].

Depletion calculations are performed using the ORIGEN depletion module, which is called by TRITON for time-dependent transmutation of each material. ORIGEN requires the neutron flux distribution and cross-sections as well as the user-defined material specifications. It provides new isotopic concentrations by performing depletion operations, cross-section collapsing, and flux normalisation.

Validation calculations were performed using the SCALE module KENO, specifically KENO-VI, which is a multi-group, flexible geometry 3D Monte Carlo transport calculator. In each validation case the energy groups used are kept the same between TRITON and KENO.

3. Model Descriptions for BWR Burnup Calculations

Unit or pin cell studies based on SCALE are performed for several scenarios relating to the production and implementation of a 2D 'RPT' transformed model. In all cases an RBWR-analogous uranium breeder reactor is used, consisting of a hexagonal fuel assembly with 20% enriched UO_2 fuel (see Figure 2). For burnup calculations applying the transformed model, effective multiplication factors were compared for different seed/blanket geometries to a burnup of 70 GWd/tU. Isotopic and spectral analyses were also carried out using the developed RPT model. For the RPT development, two models were created. The first is a hexagonal pin cell model of a uranium-breeder based on the geometry of the RBWR-TB2. The second is an axial model of a pin, made up of 40 sections of 10 cm height (see Figure 3). In the case of the calculations used to produce the RPT model, the fuel pin consists of 20% enriched UO_2 similar to the radial model, but for later calculations applying the equivalent model these sections can be filled either with the 20% enriched 'seed' or natural uranium 'blanket' as foreseen in the RBWR design, to test for the efficiency of different breeder reactor geometries.

Figure 2. SCALE 6.2 model of radial RBWR-analogue pin-cell.

Figure 3. SCALE 6.2 model of a section of the axial pin model used for applying the RPT to an RBWR-analogue reactor.

3.1. RBWR-Analogue Radial Pin-Cell Model

The radial pin-cell of a uranium-breeder based on the geometry of the RBWR-TB2 was produced to calculate reference values of neutron multiplication and energy distribution of the neutron flux, to which RPT-transformed models will be compared. Table 1 shows the main geometrical and operational parameters used in the simulations. Figure 2 displays the SCALE 6.2 model of the RBWR-analogue hexagonal pin cell which consists of fuel and cladding, surrounded by moderator. Reflective boundary conditions were applied to the system.

Table 1. RBWR-analogue pin-cell parameters.

Parameter	Data
Fuel radius, mm	3.00
Zircaloy clad radius, mm	3.60
Pitch, mm	9.10
Fuel temperature, K	850
Natural Zr clad and moderator temperature, K	620
Fuel enrichment, wt%	20
Fuel density, g/cm^3	10.2540
Moderator densities, g/cm^3	0.3–0.7

3.2. Axial Uranium Breeder Pin/Burnup

Calculations on the axial direction were performed using a 400 cm long reactor pin discretized into sections of 10 cm height as a basis for later studies. In the first step, these sections consisted entirely of 20% enriched UO_2 fuel for the approach to produce the equivalent model, and thereafter consisted of either 20% enriched UO_2 or natural uranium in order to see the effect of different seed/blanket arrangements as proposed in the RBWR. For the later usage, the water density curve is important when investigating an axial cut of a boiling water reactor as it varies considerably along the height of the core. The opportunity to take this into account is foreseen through splitting the outer moderator region of the model into four 100 cm height sections, each with the opportunity to provide different water densities. The choice of four sections was determined, based on the current approaches for BWRs, to deliver a sufficient level of detail to gain useful insights into the behaviour of different sections of the pin without significantly increasing the computational time. Table 2 provides the data for the equivalent axial pin model, as a basis for the reactivity equivalent physical transformation to be applied.

Table 2. RBWR-analogue equivalence axial model parameters.

Parameter	Data
Modified Fuel radius	To be determined
Modified Zircaloy clad radius	To be determined
Pitch, mm	9.10
Fuel temperature, K	850
Natural Zr clad and moderator temperature, K	620
Fuel enrichment, wt%	20
Axial region size for self-shielding calculation, cm	10
Calculated region sizes of seed/blanket structure, cm	20/100
Region sizes for water density distribution, cm	100

3.3. Spatial Self-Shielding Effect

Spatial self-shielding occurs when the fuel geometrically shields itself from neutron penetration due to the heterogeneous nature of the pin cell, resulting in a lower neutron flux inside the fuel rod compared to near the surface. As this effect is dependent on geometry, the process required by this paper's application of the RPT must account for a change in the self-shielding due to a change in the geometry.

Figure 4 shows that by reducing the radius and increasing the density of the fuel in the axial pin RPT model, there is an increase in the criticality. This is due, in part, to the reduction in self-shielded neutrons, but there may also be a spectral effect due to the respective increase in moderator region volume, as shown in Equations (1)–(5). The contribution of these two effects could be worthy of additional analysis, by separating the pin into several smaller regions to compare fluxes. However, this analysis was deemed unnecessary for an exploration of the applicability of the RPT in this case.

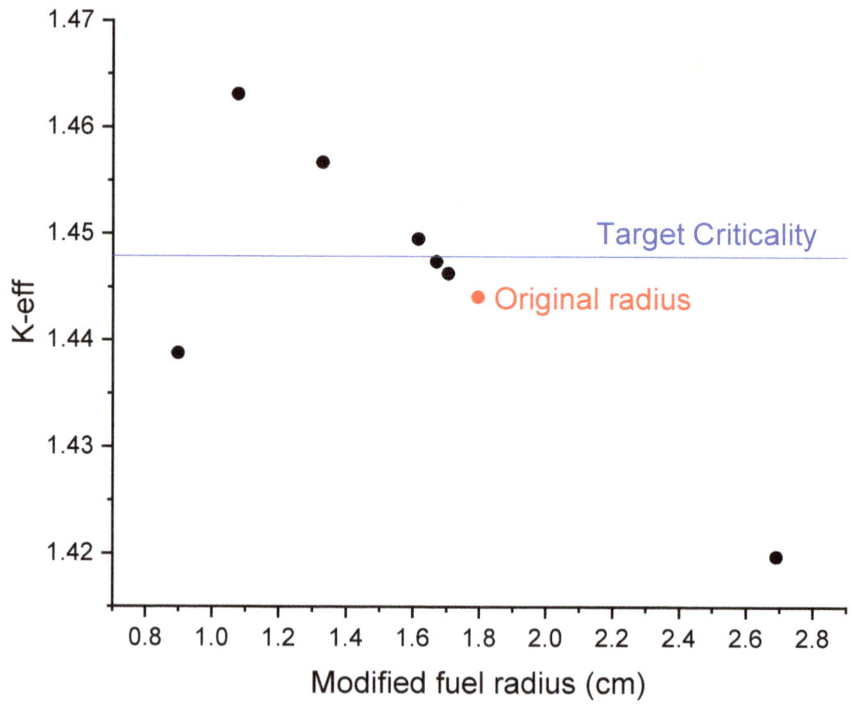

Figure 4. Effect of modification of the fuel radius with compensating density modification across several different modified fuel radii. It can be seen that the behaviour is predictable for most degrees of modification with higher criticality being calculated for denser fuel, due to the reduction in the effect of self-shielding.

3.4. Monte Carlo Validation

Using the same unit cell model as was created for TRITON calculations (shown in Figure 2), multiple validation calculations were run using SCALE KENO-VI. The first point of validation was a direct comparison of the radial unit cell model to a 3D Monte Carlo extrapolation of the same unit cell. The Monte Carlo model was created with a single fuel material for this comparison, in order to evaluate the size of the difference between the two calculation methods before introducing axial heterogeneity through a seed and blanket structure.

When axial heterogeneity is added it is pivotal that the RPT-produced axial model is validated against a 3D calculation, as without this the only equivalence that is proven is for two one-dimensional models. Therefore, 3D Monte Carlo calculations for several seed/blanket axial geometries and varying void fractions were completed and compared against RPT-axial models, for various configurations relevant to future calculations for cross section production.

4. Results and Discussion

4.1. Producing the Equivalence Model

4.1.1. Effective Multiplication Factor

The first point of evaluation to determine the RPT radius is matching of the reference and the RPT model regarding the effective multiplication factor at a fresh steady-state, without burnup. This factor, k_{eff}, is the key value for the comparison between the two models to be matched in the RPT process. Δpcm is the measure of the difference between the k_{eff} of the pin model (RPT model) and of the radial unit cell model (reference), given by the equation:

$$\Delta\rho[pcm] = \frac{k_{eff}(axial) - k_{eff}(radial)}{k_{eff}(radial) * k_{eff}(axial)} * 10^5 \quad (1)$$

In order to keep both the mass of the fuel and moderator constant, as well as the pitch, a fuel diameter/width modification factor (α) is defined from which the density and width of each of the moderator, clad, and fuel can be defined. Therefore, trivially from α the following can be defined:

$$x_f = \alpha x_{f0} \quad (2)$$

$$x_m = P - x_{clad} - \alpha x_{f0} \quad (3)$$

$$\rho_f = \frac{\rho_{f0}}{\alpha} \quad (4)$$

$$\rho_m = \frac{x_{m0}\rho_{m0}}{P - x_{clad} - \alpha x_{f0}} \quad (5)$$

where x_{f0}, ρ_{f0}, x_{m0}, ρ_{m0} are the initial widths and densities of the fuel and moderator, respectively, and x_f, ρ_f, x_m, ρ_{m0} are the modified values [17–19]. These values are defined only by geometrically transforming the radial unit cell model to an axial pin model conserving the mass and number of particles of the fuel and moderator and conserving the pitch of the radial model. This is a straightforward geometric transformation, shown in the first step of Figure 5, using the ratios of the areas of the material regions in the hexagonal model and converting them directly into widths in the pin model. P is the pitch, and the size, x_{clad}, and density of the cladding region of the model were kept constant, as varying it was found to have a negligible effect on results.

Figure 6 shows the discrepancy in k_{eff} between the axial and radial models for four different water density calculations, each at steady state at zero burnup. The water density values were chosen to maximise similarity in the two models over a range which is relevant to RBWR/ABWR water density curves. The results obtained at a value of α between 0.9 and 0.95 are within a pcm difference of −200 to +100 for all water densities. For the value of α = 0.93, all pcm differences are <±50 pcm, showing a very good standard of agreement between the two models, at least for steady state criticality. However, it is necessary to expand on the investigation for this value to further implement the new equivalent model.

The next step of the analysis required to calibrate the RPT is to compare values of k_{eff} over a burnup cycle. This is performed for a burnup up to 70 GWd/tU. As can be seen from Figure 7 in the case of 50% void, and as was seen for comparisons performed over the same burnup for 60% and 70% void, the pcm difference between the models remains consistent with what is found in the steady state analysis until a burnup of 20 GWd/tU or higher. At this point the discrepancy between the two models begins to increase, which is a consequence of the change in the fissile material due to the breeding of plutonium and the related change in the neutron spectrum. This effect produces a discrepancy of more than 100 pcm at very high burnup. In the 60% void case there was a maximum difference of 77 pcm, and for 70% void this was 62 pcm across the same burnup length.

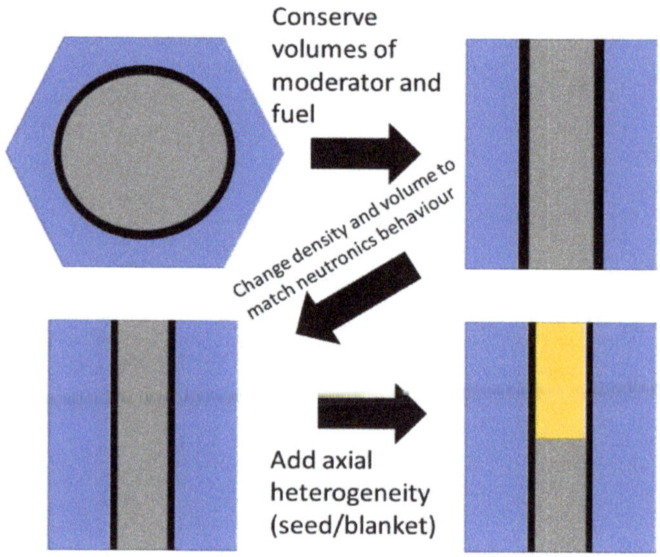

Figure 5. Simple diagrammatic representation of the processes undertaken as part of this work.

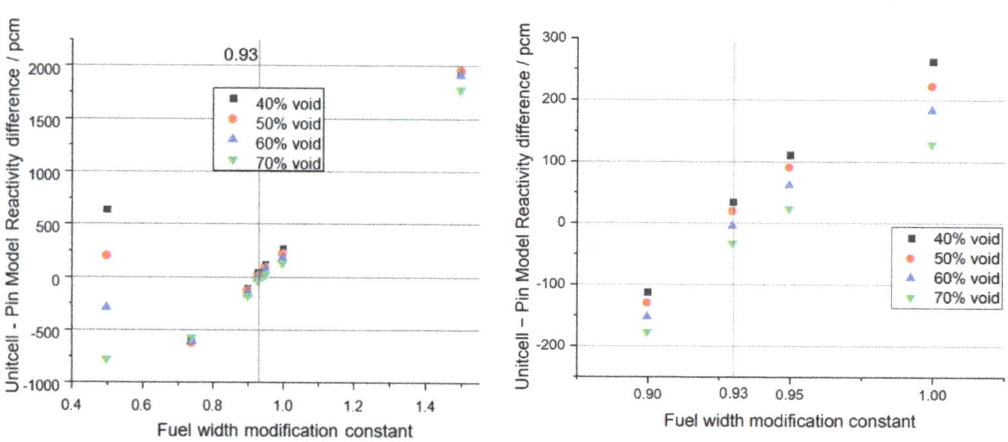

Figure 6. Reactivity difference between the axial and radial models of the pin in pcm units. When α is significantly below 1, with highly dense fuel represented in the model, the behavior of the pin model is inconsistent across different void fractions. The best agreement is found at a width reduction of 0.93, where the difference between the transformed pin model and the unit cell radial model is below 30 pcm across all void fractions calculated. Figures (**left**) and (**right**) are the same set of data, with different axis scaling.

4.1.2. Spectral Comparison

Following the investigation of the k_{eff} development over burnup, the next, more detailed step will be the comparison of the neutron spectra. The neutron spectra of the two models were investigated as a follow-up metric by which to evaluate the RPT. Due to the high enrichment of the fuel and relatively small pitch in the fuel assembly of the RBWR, the model produces a much harder neutron spectrum than a classical LWR, as shown in Figure 8. Figure 9 shows that for the fast spectrum, there is a strong correlation between the two models, as there are no energy brackets of the SCALE-defined 238-group structure within this range where the number of interactions in either model varies from the other by more than 1%. At lower, thermal energies within the two models there are larger differences, even up to around 3% at ~1 eV. The normalisation of lower absolute numbers of interactions leads to these comparably large values.

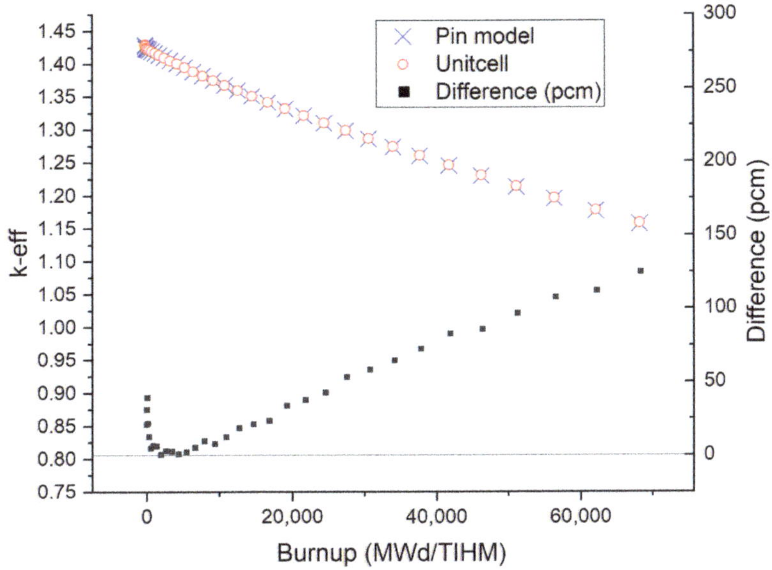

Figure 7. Difference in k-eff, the effective neutron multiplication in the model, between the axial RPT model of the pin and the radial unit cell model across a standard burnup cycle. As the pin reached high burnup (>50 GWd/tU), the difference in criticality between the two models increased to >100 pcm. This calculation was performed with a void fraction of 50%. In the cases of 60% and 70% void, the axial model showed more similar behaviour at high burnup.

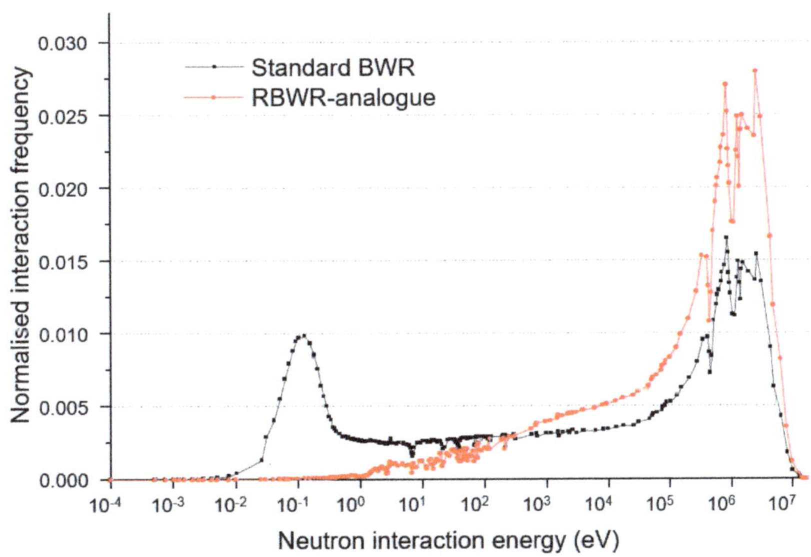

Figure 8. Comparison of neutron energy distribution between the RBWR-analogue calculated reactor used in this work and a standard BWR as calculated from a simple example case from the SCALE user guide. Both calculations were run at steady state before any depletion.

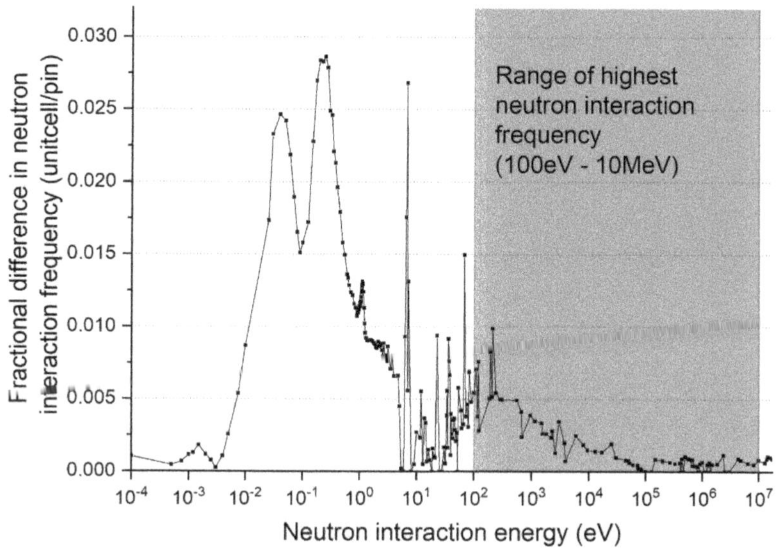

Figure 9. Fractional difference in neutron interaction frequency between the unit cell and RPT pin models calculated across a 238-energy group spectrum. In the energy range of highest interaction frequency for this pin, agreement of within 1% is seen between the axial and transformed radial models. Outside of this most-frequent range, higher discrepancies between the models can be found but very few interactions between neutrons of energies lower than 100 eV take place in the breeder reactor, as can be seen in Figure 8. Fractional difference is calculated using $\frac{2(unitcell-pin)}{unitcell+pin}$ to normalise values for energy groups of lower interaction frequency.

4.2. Applying the Equivalence Model

The RPT model was used to investigate the effect of different levels of heterogeneity on the breeding capabilities of the RBWR-analogue reactor with varying seed and blanket arrangements along the length of the pin. This was achieved by changing the placement of some parts of the fuel within the pin and running a burnup calculation as before. In each case, the amount of highly enriched uranium fuel was kept constant, but the geometry of seed and blanket regions across the height of the pin were changed. Calculations were run for repeated regions of seed and blanket of 100 cm, 40 cm, 20 cm, and 10 cm height. For the evaluation TRITON is used for observing the number densities of several main actinides, particularly fissile plutonium isotopes which were tracked at the same set of burnup points.

At a first glance of Figure 10 it is clear that by the end of the burnup cycle the smaller seed and blanket regions lead to a pin which can produce more fissile plutonium. This is also seen in the comparison between Figures 11 and 12, where the same comparison of 100 cm/20 cm seed and blanket regions is seen, but with the data split into different regions of the reactor. There are different observations to make. First the increase in the plutonium amount in the seed is slightly higher in the 100 cm case than in the 20 cm case. Second, in the case of the 100 cm height blanket regions, the natural uranium at the centre of the region, 50 cm from its closest highly enriched uranium region, produces a negligible amount of fissile plutonium. Thus, the observation in the blanket is clear—there are simply not enough neutrons reaching all areas of the breeding blanket to create a significant amount of neutron capture in U-238 to breed Pu-239 when the blanket area is 100 cm long. The other observation needs a bit more of an explanation but thinking about the different structure of power production in the seed and in the blanket gives an answer. In the 20 cm case, the power production in the blanket region will already be significant due to the strong breeding of Pu which typically dominates fissile content after a burnup of 30 GWd/tU and higher in an LWR. Thus, in the 20 cm case, the local burnup in the seed will be lower since some burnup is already created in the blanket. In contrast, in the 100 cm case, due to the weak breeding, almost all burnup is accumulated in the seed and there is a clear correlation between the neutron flux, or the power produced in a fuel, and the plutonium bred due to the exposure to the neutron flux. It can be observed in all calculations of the seed that the Pu production increases almost linearly with the burnup. Based on this observation, the slightly increased Pu content in the seed of the 100 cm case can be explained with the higher local burnup there, while the global value is still lower, as in Figure 10.

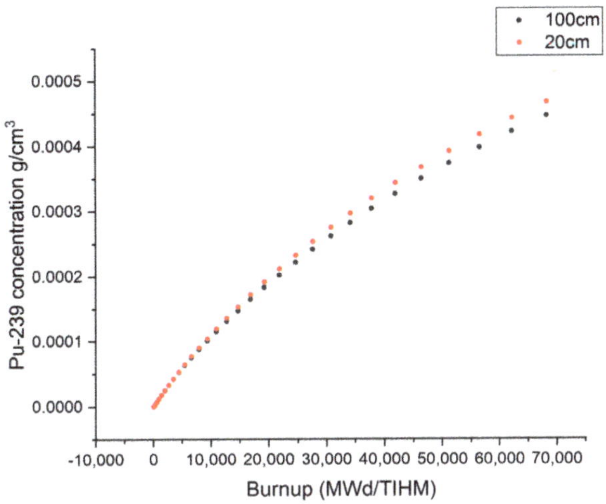

Figure 10. Pu-239 concentration across a burnup cycle ending at 70 GWd/tU, averaged across the whole pin. The two different models are referred to by their seed and blanket region sizes of 20 cm and 100 cm. The difference in plutonium production is similar at low burnup, but at higher burnup the 20 cm seed/blanket model produces more Pu-239.

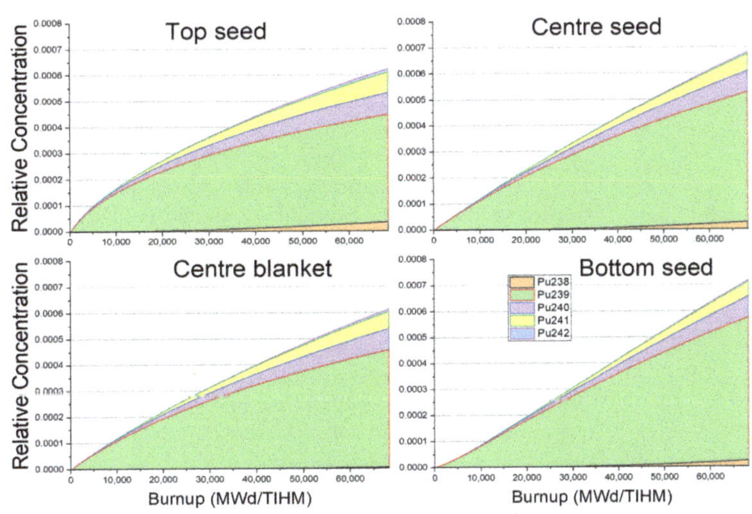

Figure 11. Concentrations of different plutonium isotopes throughout a burnup cycle, for a seed and blanket structured pin with regions of 20 cm size. The production of fissile plutonium (Pu-239 and Pu-241) in the top seed is faster at the start of the burnup cycle, in the region where the water density is lowest. The opposite is true for the bottom of the reactor pin. The central blanket region produces less plutonium than any seed region, but still yields over 70% of the plutonium of the central seed region.

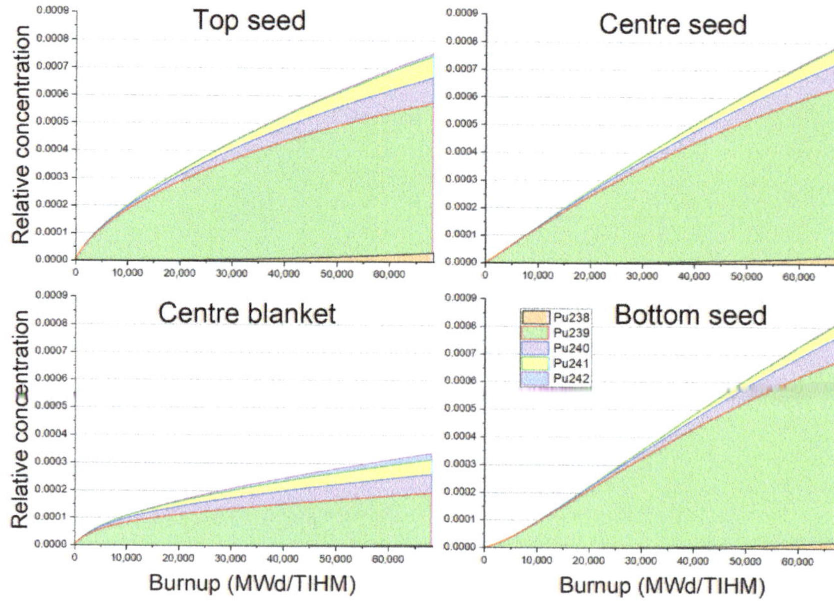

Figure 12. Concentrations of different plutonium isotopes throughout a burnup cycle, for a seed and blanket structured pin with regions of 100 cm size. The production of fissile plutonium (Pu-239 and Pu-241) in the top seed is faster at the start of the burnup cycle, in the region where the water density is lowest. The opposite is true for the bottom of the reactor pin. The central blanket region produces less plutonium than any seed region.

In addition, Figures 10 and 11 show that the different water densities across the height of the reactor have a noticeable effect on the plutonium production, and particularly by looking at the differences between the top and bottom seed regions for the 100 cm seed/blanket model, it can be observed that the effect is for both the quantity and quality of the plutonium produced. In the case of the top seed region, very high-quality plutonium is produced quickly at low burnup, but this is not extended to high burnup points.

4.3. Reapplying the Results to a Full Assembly Model

After using the radial unit-cell model as a proof of concept for the equivalence procedure as a first step, it is possible to apply the process to a more realistic full assembly model. The new model includes areas of improved moderation (see Figure 13) which change the fuel width modification needed to produce an axial model with similar behaviour, but this time for the full fuel assembly. The effect on criticality is given in Figure 14 for different water densities. In order to produce cross-sections for the equivalent full assembly models, different width modifications in the RPT model are required for each water density due to the larger effect of the radial heterogeneity, paired with the effect of different spatial self-shielding. The results of the procedure are shown in Figure 15. α values for water densities of 0.3, 0.4, 0.5, and 0.6 were found to be 0.90, 0.74, 0.74, and 0.60, respectively.

4.4. Validation Results

Figure 16 shows the pcm difference over the same burnup cycle as used elsewhere in this work, in this case between a 3D Monte Carlo calculation (as described briefly in Section 3.4) and a single material axially homogeneous RPT-transformed pin model. The maximum pcm difference is around 100, showing a very strong level of agreement between the two models, as would be expected.

Figures 17 and 18 show the results of the validation of the 2D RPT axial model with a seed and blanket structure. In all of these cases, the RPT model was 'calibrated' using a case with 50% void fraction and 50 cm repeating seed and blanket axial regions. The results clearly show that for the calibration case there is very good agreement between 3D Monte Carlo and RPT. The results also include comparisons for void fractions and seed and blanket geometries which are not the calibration case, to show the robustness of the RPT method. It is shown that the RPT calibration is relatively robust to changes to void fraction in the model, but responds far less well to changes in the axial heterogeneous geometry.

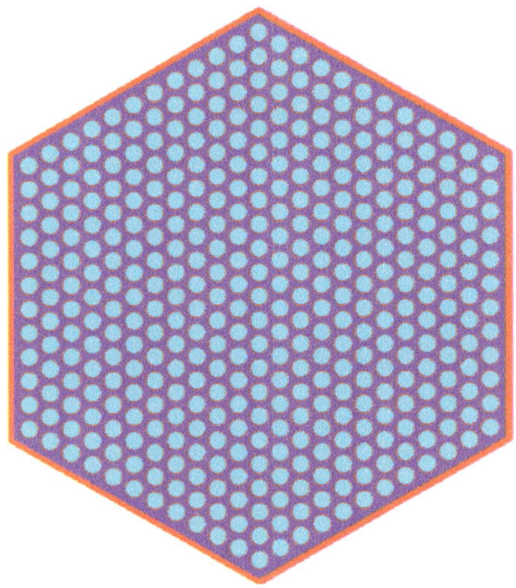

Figure 13. SCALE 6.2 model of radial RBWR-analogue assembly.

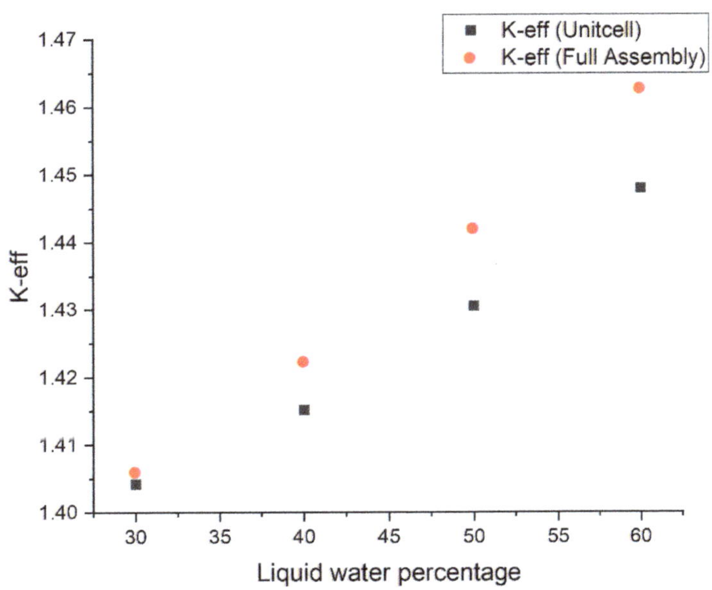

Figure 14. Effect of areas of increased moderation on the criticality of the radial model. At lower void fractions the effect is shown to be greater, up to a difference around 1000 pcm at the lowest void fraction of 60% water density.

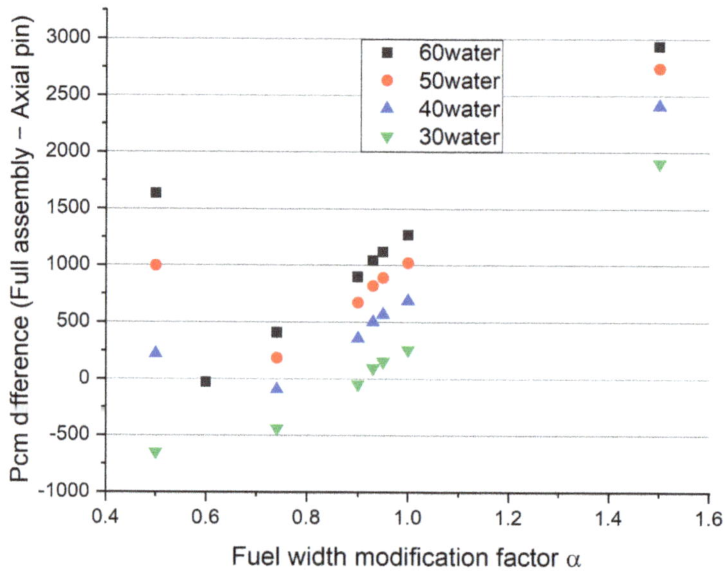

Figure 15. Difference in pcm units between the k-eff values of the full assembly radial model and an equivalent axial pin model modified by the given fuel width modification factor. Different factors for modification are required to minimise the difference in behaviour between the two models, with the largest modification required for the lowest void fraction, as expected.

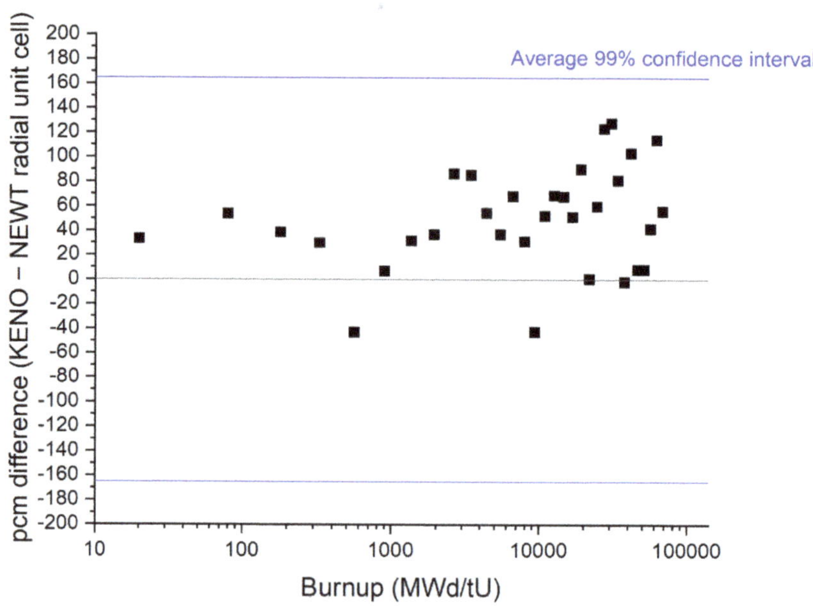

Figure 16. Difference in pcm units between the k-eff values over ~75 GWd/tU burnup between the radial unit cell model from which the RPT is calibrated, and a 3D SCALE KENO model of the same geometry and material composition. In both cases there is only a single fuel material in the pin. For the entirety of the burnup cycle the difference remains within the 165 pcm 99% confidence interval determined by the Monte Carlo statistics used in the KENO 3D model.

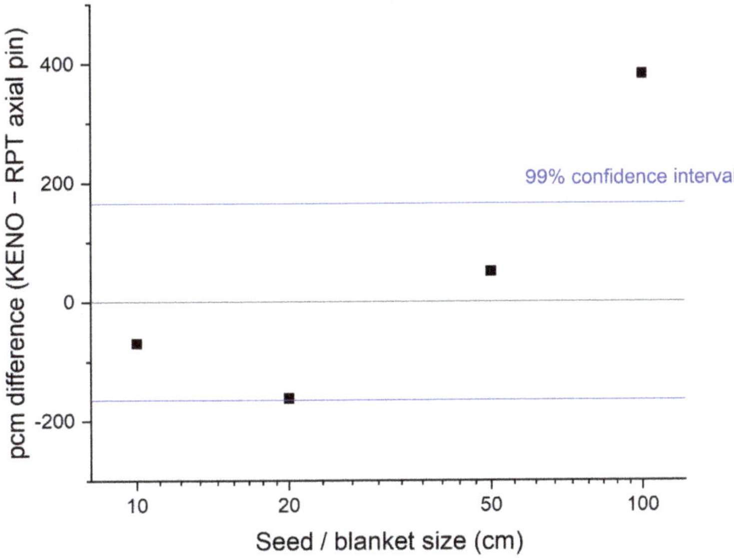

Figure 17. Difference in pcm units between the k-eff values of the 3D SCALE KENO and an equivalent axial pin model modified by the given fuel width modification factor. In this case, a seed and blanket geometry of 50 cm regions and a void fraction of 50% is used to calibrate the RPT. The results show that, firstly and most importantly, axial heterogeneity can be successfully modelled by the RPT method in 2D, and that the method requires recalibration for different axial geometries to be well-validated.

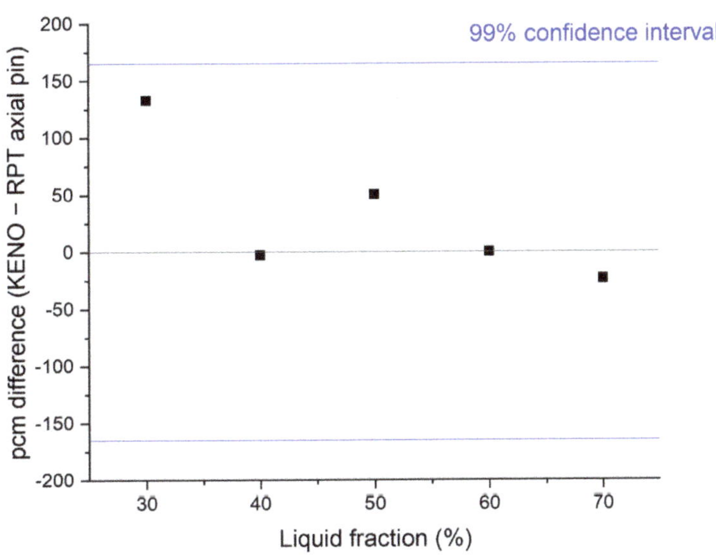

Figure 18. Difference in pcm units between the k-eff values of the 3D SCALE KENO and an equivalent axial pin model modified by the given fuel width modification factor. In this case, a seed and blanket geometry of 50 cm regions and a void fraction of 50% is used to calibrate the RPT. The results show that, firstly and most importantly, axial heterogeneity can be successfully modelled by the RPT method in 2D, and that the method is relatively robust against requiring recalibration for changes in water density.

5. Discussion and Conclusions

Development of reactor designs such as the RBWR is important for progression of the nuclear industry towards providing solutions to one of its most difficult problems—the demand of improving the fuel usage—as well as to develop promising solutions for the waste management. Both are required to improve the already good sustainability of nuclear energy to a new level that will allow widespread use of nuclear energy as a core contributor to a future net zero society [20]. Lattice physics computation is a computationally less-intensive tool when compared to Monte Carlo simulations, which for a reactor design early in the iteration stages can be extremely useful. In addition, lattice calculations build the backbone of all later core simulations and transient studies since the lattice simulations are the key step for the cross-section preparation for the widely used nodal core simulators, which are used for neutronic/thermal hydraulic coupled analysis. To support the later application of these coupled core simulators, the application of the reactivity-equivalent physical transformation process to an axially heterogeneous boiling water reactor pin was adapted from another application and investigated. Two models, one a radial representation of an RBWR-analogue pin and the other an axial representation, were produced. Through several stages of analysis and comparison using the code system SCALE 6.2, it was possible to apply the transformation process to these models and produce an equivalence model which could then be used to examine some properties of axially heterogeneous pins without the need for 3D Monte Carlo simulations.

The equivalence model was produced by varying the density and size of the moderator and fuel regions of the axial model and then comparing with the radial model, using several different properties as comparators [21]. The first comparison was the steady-state criticality. In this case, it was found that for void fractions of 30–70%, a fuel width reduction constant of 0.93 provided an equivalence model with <50 pcm difference to the radial model. The models were then compared across burnup, which provided further evidence of the efficacy of the reactivity equivalent physical transformation, with a maximum of ~100 pcm difference between the models even at very high burnup points. A spectral analysis was then performed, showing that the models were very consistent across all energy groups for neutron interaction frequency, and particularly consistent in the energy region of most interest between 1 keV and 100 MeV.

There were limitations to the model in both the burnup and spectral comparisons. At very high burnup an increasing level of disparity between the models was observed. This may have been caused by the Pu production and the appearance of fission products in the fuel, which would have had an amplified impact on the criticality and the evolution of the self-shielding of the modified model which would require further adjustment or an optimization not singularly based on fresh fuel. In the case of the spectral analysis, normalisation accounts for some of the larger discrepancies at low energy, wherein there are far fewer interactions, and the effect of small discrete differences between the models' interaction frequency appears as large fractional differences.

The model was then used to investigate the effect of heterogeneity, and more specifically the seed/blanket breeder structure of concept reactors such as the RBWR, on plutonium production. Applying the equivalence model, it was found that for larger areas of seed/blanket, blanket regions would not produce as much fissile plutonium as for smaller seed/blanket geometries. This is due to the lack of neutron penetration into large blanket regions past a few mean free paths.

Based on all of the above, it is clear that the application of the reactivity-equivalent physical transformation to fuel geometries, outside of its current use cases, has the potential to be a very powerful tool. However, the RBWR-analogue reactor which is used for all of the calculations in this work is not a realistic reactor. In the case of more complex and continuous water density distributions, the model may require such a level of detail that the computational advantage is lessened [22], but there is still a great opportunity to use this approach for the essential cross-section preparation required for coupled 3D core analysis and transient studies [23].

Author Contributions: Conceptualization, R.M. and B.M.; methodology, R.M.; validation, R.M.; investigation, R.M.; resources, B.M. and T.H.; data curation, R.M.; writing—original draft preparation, R.M.; writing—review and editing, M.P., A.P. and B.M.; supervision, M.P., A.P. and B.M.; funding acquisition, A.P. and T.H. All authors have read and agreed to the published version of the manuscript.

Funding: This research was funded by Hitachi Ltd., specifically through a funded studentship from the Centre for Technology Innovation—Energy, Research and Development group.

Data Availability Statement: The data presented in this study are available on request from the corresponding author. The data are not publicly available due to some data sets being proprietary.

Conflicts of Interest: The authors declare no conflict of interest.

References

1. Merk, B.; Litskevich, D.; Bankhead, M. iMAGINE-A disruptive change to nuclear or how can we make more out of the existing spent nuclear fuel and what has to be done to make it possible in the UK? *ATW Int. J. Nucl. Power* **2019**, *64*, 353–359.
2. Merk, B.; Litskevich, D.; Whittle, K.; Bankhead, M.; Taylor, R.; Mathers, D. On a Long Term Strategy for the Success of Nuclear Power. *Energies* **2017**, *10*, 867. [CrossRef]
3. Kito, K.; Hino, T.; Matsumura, K.; Matsuura, M. Hitachi's Vision for Nuclear Power and Development of New Reactors. *Hitachi's Nucl. Power Bus.* **2020**, *69*, 156–162.
4. Hino, T.; Nagayoshi, T.; Mitsuyasu, T.; Ohtsuka, M.; Moriya, K. Development of RBWR (Resource-renewable BWR) for Recycling and Transmutation of Transuranium Elements (1)—Overview and core concept. In Proceedings of the ICAPP2015, Nice, France, 3–6 May 2015.
5. Hino, T.; Miwa, J.; Mitsuyasu, T.; Ishii, Y.; Ohtsuka, M.; Moriya, K.; Shirvan, K.; Seker, V.; Hall, A.; Downar, T.; et al. Core Design and Analysis of Axially Heterogeneous Boiling Water Reactor for burning Transuranium Elements. *Nucl. Sci. Eng.* **2017**, *187*, 213–239. [CrossRef]
6. Hino, T.; Miwa, J.; Murakami, Y.; Mitsuyasu, T.; Soneda, H.; Lindley, B.; Tollit, B.; Smith, P.; Shirvan, K.; Downar, T.; et al. Core Design of RBWR (Resource-renewable Boiling Water Reactor) and Benchmark Calculation of Core Analysis Tools. In Proceedings of the ICAPP2019, Juan-les-Pins, France, 12–15 May 2019.
7. Hino, T. Research needs for BWR core and core analysis. In Proceedings of the Conference and Workshop on the Research Needs of Boiling Water Reactors, Bangor, UK, 25–27 October 2016.
8. Lui, Y. A Full Core Resonance Self-Shielding Method Accounting for Temperature-Independent Fuel Subregions and Resonance Interference. Ph.D. Thesis, University of Michigan, Ann Arbor, MI, USA, 2015.
9. Kim, T.; Yang, W.; Taiwo, T.; Smith, M. Preliminary assessment of lattice physics capabilities for VHTR analysis. In Proceedings of the ANS Winter Meeting, Washington, DC, USA, 14–18 November 2004.
10. Kim, Y.; Baek, M. Elimination of double-heterogeneity through a reactivity-equivalent physical transformation. In Proceedings of the International Conference on Nuclear Energy Systems for Future Generation and Global Sustainability, Tsukuba, Japan, 9–13 October 2005.
11. Noh, J.; Kim, K.; Kim, Y.; Lee, H. Development of a computer code system for the analysis of prism and pebble type VHTR cores. *Ann. Nucl. Energy* **2008**, *35*, 1919–1928. [CrossRef]
12. Rohde, U.; Merk, B.; Baier, S.; Fridman, E.; Dürigen, S.; Kliem, S.; Weiss, F. Development of a reactor dynamics code for block-type HTR at Forschungszentrum Dresden-Rossendorf. In Proceedings of the Fachtag der KTG: "Aktuelle Themen der Reaktorsicherheitsforschung in Deutschland", Dresden, Germany, 7–8 October 2010.
13. Kim, Y.; Kim, K.; Noh, J. Preservation of Fuel Characteristics in the RPT Method. In Proceedings of the Spring Meeting of the Korean Nuclear Society, Gapyoung, Republic of Korea, 22 December 2006.
14. Miwa, J.; Hino, T.; Mitsuyasu, T.; Nagaya, Y. Whole-core Monte Carlo burnup calculation for RBWR by parallel computing. In Proceedings of the Joint International Conference on Supercomputing in Nuclear Applications + Monte Carlo 2020, Chiba, Japan, 18–22 May 2020.
15. Fridman, E.; Merk, B. Modification of the Reactivity Equivalent Physical Transformation Method for HTGR Fuel Element Analysis. In Proceedings of the 5th International Topical Meeting on High Temperature Reactor Technology, Prague, Czech Republic, 18–20 October 2010.
16. Rearden, B.; Jessee, M. *SCALE Code System, Manual*; Oak Ridge National Laboratory: Oak Ridge, TN, USA, 2017.
17. Lynch, S. Reactivity-Equivalent Physical Transformation Model or Pin Cell Arrays. Bachelor's Thesis, Massachusetts Institute of Technology, Cambridge, MA, USA, 2010.
18. Goluoglu, S.; Williams, M.L. Modeling Doubly Heterogeneous Systems in SCALE. *Trans. Am. Nucl. Soc.* **2005**, *93*, 963–965.
19. Rohde, U.; Baier, S.; Duerigan, S.; Fridman, E.; Kliem, S.; Merk, B. Development and verification of the coupled 3D neutron kinetics/thermal-hydraulics code DYN3D-HTR for the simulation of transients in block-type HTGR. *Nucl. Eng. Des.* **2012**, *251*, 412–422. [CrossRef]
20. Baldova, D. Feasibility Study on High-Conversion Th-U233 Fuel Cycle for Current Generation of PWRs. Ph.D. Thesis, Czech Technical University in Prague, Prague, Czech Republic, 2014.

21. Lou, L.; Chai, X.; Yao, D.; Peng, X.; Chen, L.; Li, M.; Yu, Y.; Wang, L. Research of ring RPT method on spherical and cylindrical Double-Heterogeneous systems. *Ann. Nucl. Energy* **2020**, *147*, 107741. [CrossRef]
22. Kulikowska, T. *Reactor Lattice Codes*; Lecture; Institute of Atomic Energy: Otwock, Poland, 2000.
23. Downar, T.; Hall, A.; Jabaay, D.; Ward, A.; Greenspan, E.; Ganda, F.; Bartoloni, F.; Bergmann, R.; Varela, C.; DiSanzo, C. *Technical Evaluation of the HITACHI Resource—Renewable BWR (RBWR) Design Concept*; Technical Report; Electric Power Research Institute: Palo Alto, CA, USA, 2012.

energies
MDPI

Article

Photosensitivity of Nanostructured Schottky Barriers Based on GaP for Solar Energy Applications

Vasily Rud [1,2,*], Doulbay Melebaev [3], Viktor Krasnoshchekov [4], Ilya Ilyin [2], Eugeny Terukov [1,5], Maksim Diuldin [4], Alexey Andreev [2], Maral Shamuhammedowa [3] and Vadim Davydov [4,6]

[1] Ioffe Institute, 195251 St. Petersburg, Russia
[2] Faculty of Global Studies, M.V. Lomonosov Moscow State University, 119992 Moscow, Russia
[3] Department of Physics, Turkmen State University Named for Magtymguly, Ashgabat 744000, Turkmenistan
[4] Higher School of International Educational Program, Peter the Great Saint-Petersburg Polytechnic University, 195251 St. Petersburg, Russia
[5] Department of Electronics, Saint Petersburg Electrotechnical University "LETI", 197022 St. Petersburg, Russia
[6] Department of Photonics and Communication Lines, The Bonch-Bruevich Saint Petersburg State University of Telecommunication, 193232 St. Petersburg, Russia
* Correspondence: rudvas.spb@gmail.com

Abstract: This work investigates the surface-barrier photoelectric properties of Au-palladium-n-GaP structures. Research into the visible spectrum region, under the action of both linearly polarized and natural radiation, provides us with new information about the height of the barrier, the interface m-s section, and the GaP band structure. SBs based on GaP (p- and n-type) are helpful for researchers in developing advantageous structures for creating various photovoltaic devices—photodetectors for fiber-optic control of energy systems or possible structures for solar energy. Despite many years of research, issues concerning the band structure of semiconductors based on the phenomenon of photoelectroactive absorption in such surface-barrier structures' m-s remain urgent in the creation of new high-performance devices. Such structures may also be interesting for creating solar energy systems. They create a thin insulating dielectric layer (usually an oxide layer) in solar cells on SBs between the m and the semiconductor substrate. The advantage of solar cells based on m dielectric semiconductor structures is the strong electric field near the surface of the semiconductor that usually has a direction favoring the collection of carriers created by short-wavelength light. Diffusion of impurities usually results in crystal defects in the active region. There are no such defects in the studied elements. This is also the difference between solar cells on m dielectric structures and elements with diffusion in p-n junctions. We studied the PS of Au-Pd-n-GaP nanostructures to determine the height of the potential barrier $q\varphi_{Bo}$ and obtained accurate data on the zone structure of the n-GaP. The PS of nanostructured Au-Pd-n-GaP structures was studied in the visible region of the spectrum. Essential information about the semiconductor's potential barrier parameters and band structure was obtained. The intermediate Pd nanolayer between Au and GaP has specific effects on the Au-Pd-n-GaP nanostructure, which are of considerable practical and scientific significance for future needs.

Keywords: Schottky barrier; nanostructure; surface-barrier structures; photocurrent spectrum; spectral dependences; PS

Citation: Rud, V.; Melebaev, D.; Krasnoshchekov, V.; Ilyin, I.; Terukov, E.; Diuldin, M.; Andreev, A.; Shamuhammedowa, M.; Davydov, V. Photosensitivity of Nanostructured Schottky Barriers Based on GaP for Solar Energy Applications. *Energies* **2023**, *16*, 2319. https://doi.org/10.3390/en16052319

Academic Editor: James M. Gardner

Received: 29 December 2022
Revised: 21 February 2023
Accepted: 24 February 2023
Published: 28 February 2023

1. Introduction

Researchers have been concentrating for a long time on the study of Schottky barriers (SBs) based on GaP. They are interesting as structures for developing different radiation-resistant photovoltaic and electronic instruments for high-frequency power electronics, ultraviolet (UV) photoelectronics, and solar energy [1–8]. The m-s-structures with SBs based on gallium phosphide are promising in the development of UV short-wave radiation photodetectors. Such devices are easy and inexpensive to produce and robust in operation, as the industry has mastered GaP as a wide-band-gap (high-energy-gap) semiconductor. The devices' high PS in the UV region is provided by the high threshold energy (E_o = 2.8 eV, 300 K) of direct GaP optical transitions. The PS of an m-GaP structure with a Schottky barrier in the UV region (hν = 3.1 ÷ 6.0 eV) was studied more thoroughly than the PS in the visible light region (hν = 1.5 ÷ 3.1 eV). The study of the m-GaP structure visible light region provides important information on Schottky barriers and the fundamental parameters of GaP band structures.

As generally known, the Schottky barrier height ($q\varphi_{Bo}$) is a fundamental parameter of m-semiconductor (MS) and m-dielectric-semiconductor (MIS) devices used in state-of-the-art optoelectronics and microelectronics. Therein, the most accurate method for determining the $q\varphi_{Bo}$ is considered the photoelectric method (following Fowler's theory) [9,10]. With respect to a rectifying non-point MS contact, the Fowler's semi-phenomenological theory was first applied in [9–12]. Those works proposed a photovoltaic method based on measuring the PS spectra of surface-barrier structures to determine the $q\varphi_{Bo}$. In those works, a method of determining the $q\varphi_{Bo}$ for m-s structures under irradiation in the irradiation geometry from the MS junction, both from the side of the metal and from the side of the semiconductor, was offered. If the photon energy hν was more than the height of the barrier but less than the energy-gap width of semiconductor E_g, then electron photoemission from metal to semiconductor occurred. Following Fowler's theory, the relationship between photon energy and short-circuit photocurrent I_{f0} per equal number of incident photons can be described, as in [13], at $h\nu - q\varphi_{Bo} > 3$ kT:

$$I_{f0} \sim \left(h\nu - q\varphi_{B_o}\right)^2, \tag{1}$$

where T is the absolute temperature and k is the Boltzmann constant.

Therefore, photocurrent $I_{f0}^{1/2}(h\nu)$ dependence should be linear, and the extrapolation of it to the energy axis will provide values of the $q\varphi_{Bo}$. The values of the $q\varphi_{Bo}$ for Au-n-GaP structures were experimentally determined by the photoelectric method in the Fowler region of the photocurrent spectrum ($q\varphi_{Bo} < h\nu < E_g$). When the SB was irradiated from the side of a semitransparent m layer [2,13–19], the $q\varphi_{Bo}$ value at room temperature was determined within the range of 1.20 ÷ 1.45 eV. The noticeable difference between these values and the $q\varphi_{Bo}$ values was explained by the influence of the intermediate layer on the diode capacitance value, since such a layer can modify the very dependence of the charge distribution in the diode on the bias. The broadband effect of PS in SBs with respect to natural light is well known [16–19]. The obtained structures show a high PS for this type of photoconverter. The PS spectra are broadband in nature, and their total width at half maximum $\delta^{1/2}$, when irradiated from the barrier contact side, is high.

Studies of obtained structures in linearly polarized radiation (LPR) showed that the polarization PS in SBs on bulk crystals and epitaxial layers, regardless of the nature of the metals used and the technological method of their deposition on the surface of isotropic semiconductors, begins to manifest itself only when the LPR beam direction deviates from the normal position to the surface of the barrier. The azimuthal dependences of the photocurrent of all barriers under conditions when the angle of incidence differs from 0° correspond to the periodic law [14], and the inequality $I_P > I_S$, which follows from the analysis of the processes of passing through the air/SB interface based on the Fresnel formulas [15], is valid in the entire PS region.

2. Materials and Methods

The surface-barrier structure is a semiconductor plate. To develop an SB, an ohmic contact is firstl created on a semiconductor wafer. Then, the semiconductor surface for creating the contact is prepared. A fairly clean surface area of $A^{III}B^{V}$ semiconductors is obtained after etching them in a mixture of bromine (0.5 ÷ 10%) and methanol (99.5 ÷ 90%) and subsequent washing them in pure methanol [20].

An m-s barrier contact can be created by the following methods: vacuum deposition, gas-transport chemical reactions, electrochemical deposition, and chemical deposition of m [18,21–25].

The primary benefit of the chemical deposition method is the capability to easily obtain structures with properties close to the ideal model. This method creates high-quality diodes that meet today's micro- and optoelectronics requirements.

In 1971, researchers from the laboratory of Professor Dmitry Nasledov of the Ioffe Physical-Technical Institute developed a technique for metal film chemical deposition on the surface of a semiconductor to create Schottky surface-barrier diodes [26]. This technique is widely used in making various m-s diode structures. Our work was mainly aimed at improving the chemical method of Yury Goldberg [19], using the idea of nanotechnology.

The object of our study was nanostructured SB Au-Pd-n-GaP. We used [$n = (0.1 \div 5) \times 10^{17}$ cm^{-3}, T = 300 K] wafers oriented in the (100) crystallographic plane with a 350–400 μm thickness, which were constructed by the Czochralski method, as an initial material for making the structures. On one GaP surface (97% In + 3% Te), ohmic contact was created. A palladium (Pd) tunnel-transparent intermediate layer 20–30 Å thick and, then, a gold (Au) barrier contact nanolayer $100 \div 120$ Å thick, were formed on the other surface. The Pd intermediate barrier contact (BC) and layer were created by the chemical method [26,27]. A thin Pd nanolayer located between GaP and Au worked as a barrier. It improved the quality of the m-s interface and prevented diffusion of Au towards GaP. The Pd intermediate nanolayers had different thicknesses; the maximum thickness was 45 Å. The palladium intermediate layer thickness (d) was measured using an LEM-ZM ellipsometer on a number of GaP-Pd layer structures before the formation of a barrier contact by the ellipsometric method. The chemical deposition of the Au layer at room temperature from an aqueous solution of $HAuCl_4$ (4 g/L) + HF (100 mL/L), according to the procedure set out in [21,23], was used to create semitransparent barrier contact. Before Au deposition or before Pd nanolayer formation, the GaP surface was treated in 4% Br_2 + 96% CH_3OH [23]. According to the ellipsometric data, the palladium intermediate layer's thickness (d) reached 20–45 Å.

For the obtained barrier structures, the barrier contact area (S) was 0.1–1.0 cm^2. Silver current leads of 0.05 mm in diameter were soldered to the barrier and ohmic contacts, after which capacitance-voltage characteristics (C-U) and current-voltage characteristics (I-U) were measured. We also collected information about short-circuit photocurrent spectra $I_{f0} = f(h\nu)$. All measurements were performed at T = 300 K. Au-Pd-n-GaP SBs photoelectric properties were studied experimentally in two stages. In the first stage, non-polarized (natural) light irradiated the studied structures from the Au layer side (Figure 1, tap a). In the second stage, the same structures were irradiated from the same side with polarized light. The I_{f0} spectrum of the short-circuit photocurrent of the structures was measured using a DMR-4 monochromator with quartz prisms; SI 8-200 and DRT-375 lamps were used as light sources. A high PS of the SB was observed in the interval of $1 \div 2.8$ eV. The absolute value of the obtained structure's quantum efficiency in the interval of $h\nu = 1 \div 5.4$ eV was determined using reference photodetectors calibrated in the interval of $1 \div 5.2$ eV.

Based on the analysis of SBs' electric and photoelectric properties, the following GaP crystal parameters were determined: the free electron concentration $N = N_d - N_a$, the indirect (E_g, E_L), and the direct (E_0) interband transition energy. In addition, the BS energy diagram parameters were estimated: the space-charge layer width W_0, the maximum electric field E_{m0} at the interface under zero bias, and the barrier height $q\varphi_{Bo}$. The direct dark current I dependence on U was also measured, which in a current densities interval of 10^{-7}–10^{-2} A/cm^2 is exponential:

$$I = I_s \cdot e^{qU/\beta kT},\qquad (2)$$

where I_s is the saturation current.

This formula helps determine the dependence of structure ideality coefficient β on the thickness of the intermediate dielectric layer [28,29]. We used two distinct methods (photoelectric and the method of capacitance-voltage characteristics at a frequency of 10 kHz) to find the Schottky barrier's height and compared the results when $\beta < 1.3$ practically coincided (see Table 1). At this stage of the study, to determine the $q\varphi_{Bo}$ and the E_0, the PS spectra were measured in the photon energy interval $h\nu = 1.5 \div 3.2$ eV under conditions of Au-Pd-n-GaP SB irradiation from the Au side. The measured characteristic parameters of typical SBs are given in the Table 1. The volt-farad characteristics of the Au-Pd-n-GaP structures at different frequencies were measured using an E7-20 instrument. In the spectral measurements, the irradiated area of the resulting m-s junctions was $\approx$0.01 cm^2, and to receive equal number of incident photons, the photocurrent was reduced (Figure 2).

Table 1. Parameters of BS energy diagrams of Au-Pd-n-GaP at T = 300 K.

№	S, cm^2	$N_d - N_a$, cm^{-3}	E_{m0}, V/cm	qU_D, eV	β	$q\varphi_{Bo}$, eV ($I_{f0} - h\upsilon$)	$q\varphi_B$, eV ($C - U$)
1	0.230	5.6×10^{17}	4.4×10^5	1.425	1.35	1.28	1.49
2	0.197	1.1×10^{17}	2.1×10^5	1.285	1.28	1.38	1.41
3	0.165	5.9×10^{17}	4.9×10^5	1.345	1.39	1.21	1.43
4	0.110	3.8×10^{17}	3.8×10^5	1.285	1.19	1.33	1.38
5	0.123	2.6×10^{17}	3.2×10^5	1.385	1.25	1.35	1.48

Figure 1 shows typical PS spectra of Au-Pd-n-GaP BS in the spectral interval 1.5–3.2 eV when irradiated from the side of the Au nanolayers. The photocurrent of Au-Pd-n-GaP surface-barrier structures at photon energies more than the barrier height of the m-s structure ($q\varphi_{Bo} = 1.3$ eV), but less than the gallium phosphide energy gap width ($E_g = 2.26$ eV), was determined by the metal–semiconductor photoemission of electrons, and at photon energies close to and greater than the energy-gap width it was determined by the nonequilibrium current carriers' excitation in the semiconductor and their separation (by the field of the surface potential barrier) [14].

The photocurrent spectrum at photon energies $q\varphi_{Bo} < h\nu < E_g$ was studied in detail in this paper. It was in this spectrum window that the $q\varphi_{Bo}$, a fundamental parameter of an MS structure, was determined. This parameter corresponds to the properties of Schottky diodes occurring at the MS interface. Typical spectral characteristics of Au-Pd-n-GaP structures under irradiation from the side of the Au nanolayer in the photon energy interval 1.5–2.0 eV are shown in Figure 2.

As we know, Au nanoparticles exhibit plasmonic properties, and plasmon resonance is a reason for their absorption of a light quantum [30]. Their properties strongly depend on their size. We studied the plasma resonance absorption spectrum of the Au colloidal solution by the dynamic light scattering method. It was established that plasmonic properties of nanoparticles occurred in the λ = 480–600 nm spectral interval (where $h\nu$ = 2.06–2.58 eV). For example, the 15.27 nm Au nanoparticle diameter corresponded to the plasma resonance peak at λm = 2525 nm wavelength (where $h\nu$ = 2.36 eV) [31]. The results indicate that as the diameter of the Au nanoparticle in the colloid decreases, the shorter wavelengths correspond to an absorption peak. Our experiments showed that the studied Au-Pd-n-GaP structures have no plasmon resonance in the Fowler spectrum region ($h\nu$ = 1.5–2.0 eV, λ = 620–826 nm).

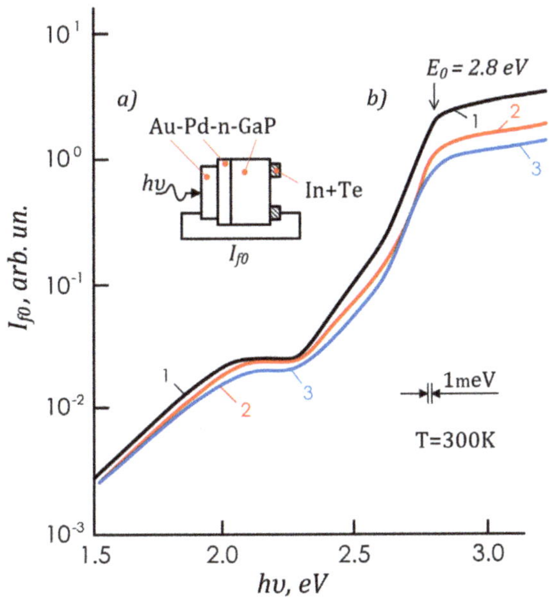

Figure 1. Diagram of a photodetector with lighting geometry (tab a); photocurrent spectra of surface-barrier structures Au-Pd-n-GaP with different thicknesses of the interlayer of palladium d. d: 1—40 Å (β = 1.35); 2—32 Å (β = 1.28); 3—45 Å (β = 1.39) (tab b). The photocurrent related to an equality of incident photons.

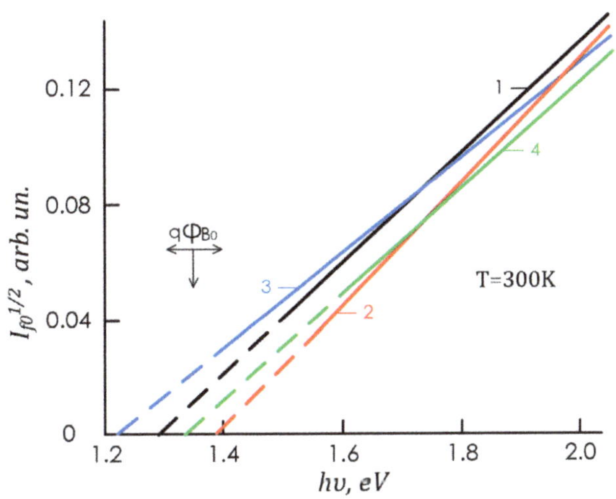

Figure 2. Dependencies of photocurrent $I_{f0}^{1/2}$ related to equality of incident photons on photon energy hν for four Au-Pd-n-GaP surface-barrier structures with different thicknesses of the intermediate palladium layer d. d: 1—40 Å, 2—32 Å, 3—45 Å, and 4—22 Å. Here and below, we indicate sample numbers near the curves and correspond to the table.

3. Investigations of Photoelectrical Properties in Non-Polarized Light

Typical spectral characteristics of Au-Pd-n-GaP nanostructures with various Pd intermediate nanolayer thicknesses are shown in Figure 2. The $I_{f0}(h\nu)$ dependence in the interval $h\nu$ = 1.5–2.0 eV obeys the Fowler law (1).

The experimental $I_{f0}^{1/2}$ dependence on $h\nu$ (Figure 2, curves 1–4) turned out to be linear. The $q\varphi_{Bo}$ value was determined by extrapolating the dependence to the $I_{f0}^{1/2} = 0$. In the studied structures, this value was between 1.21–1.38 eV. For various structures, the $q\varphi_{Bo}$ value was different (see the Table 1).

Figure 2 shows that the β coefficient depends on the palladium layer thickness d. The presence of a thin palladium layer (d = 22–45 Å; β = 1.19–1.39) between the GaP and Au at the m-s interface in the SB changed the interface properties. It caused changes in the height of the Schottky barrier (Figure 2; see the Table 1). These studies showed that by measuring the barrier height $q\varphi_{Bo}$, one can obtain important information about the properties of the m-s and m-dielectric (oxide)-s interface. In the series of studies performed, it was for the first time experimentally established that the use of a thin m Pd layer with a thickness from 20 to 35 Å between a GaP and Au does not reduce the Au-n-GaP structures potential barrier height.

A good agreement with the C-U data (Figures 3–5, Table 1) was obtained for the $q\varphi_{Bo}$ value of the Au-Pd-n-GaP structures in the Fowler photocurrent spectrum region ($h\nu$ = 1.5–2.0 eV).

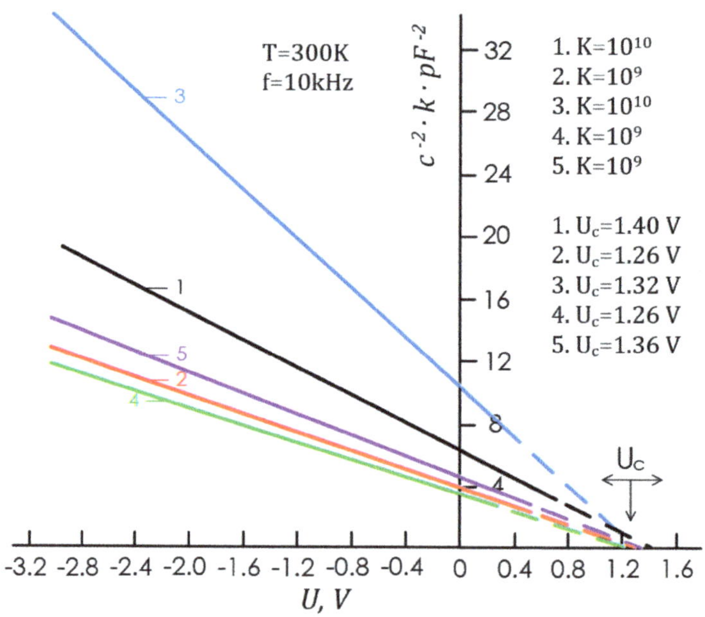

Figure 3. The differential capacitance dependence on the bias voltage of Au-Pd-n-GaP SB with various thicknesses of intermediate palladium layer d: 1—40 Å (β = 1.35); 2—32 Å (β = 1.28); 3—45 Å (β = 1.39); 4—22 Å (β = 1.19); 5—29 Å (β = 1.25).

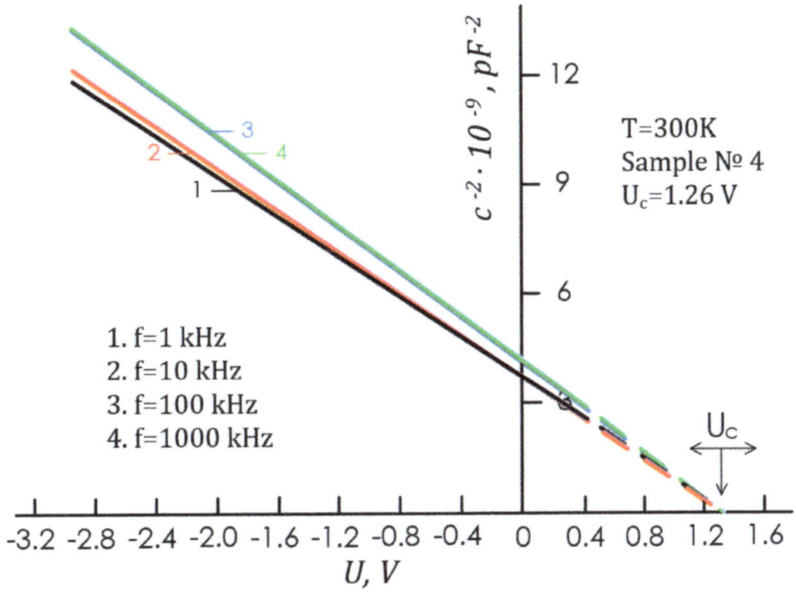

Figure 4. Differential capacitance dependence on the bias voltage for Au-Pd-n-GaP structures (Sample No. 4, β = 1.19) at different frequencies (f).

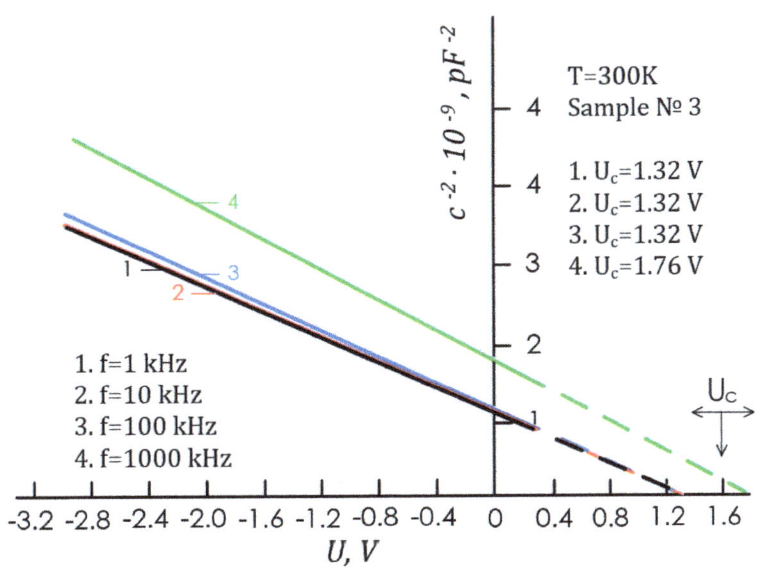

Figure 5. Differential capacitance dependence on the bias voltage for studied structures (No. 3, β = 1.39) at different frequencies (f).

The capacitance-voltage characteristics of the Au-Pd-n-GaP structures were measured. For all the studied structures, the differential barrier capacitance dependence on the voltage at the potential barrier in the coordinates $1/C^2 = f(U)$ at $U = -3.0 \div 0.4$ V obeyed a linear law (Figures 3–5), which was in agreement with Schottky theory and allowed the direct determination of the $q\varphi_{Bo}$ from the expression $q\varphi_{Bo} = qU_D - \mu = qU_C + kT - \mu$, where U is the diffusion potential, U_C is the capacitive cutoff voltage, and μ is the level energy counted from the conduction band Fermi in the GaP bulk [29–34].

At the temperature T = 300 K, we obtained the cutoff voltage from the intersection of the experimental dependences C = f(U) with an abscissa axis for different structures of $U_0{}^C = U_C \approx 1.26$–1.40 V (Figure 3). For the optimal Pd intermediate layer thickness d ≈ 20–35 Å, the $q\varphi_{Bo}$ values found by the two methods (I_{f0}~hν, C^{-2}-U) turned out to be practically equal (see Table 1). In Figure 4, almost ideal diodes (sample No. 4) at different frequencies (f = 1÷ 1000 kHz) show the observed dependence of C^{-2} on U. It was found that in an ideal diode in the C^{-2}-U characteristic, the value of U_C remains constant. This means that in an almost ideal diode in a space charge layer, the internal electric field (samples No. 2, No. 4, and No. 5) does not change with frequency and the $q\varphi_{Bo}$ value remains unchanged. Figure 5 shows the dependence in C^{-2} coordinates on U for non-ideal diodes (No. 3, d ≈ 45 Å; β = 1.39) at different frequencies. It was found that in non-ideal diodes of the Pd intermediate layer with d > 35 Å the value of U_C does not retain a constant, changing with the frequency (see Figure 5).

It was found that the average value of the Schottky barrier height was $q\varphi_{Bo} = 1.35$ eV (300 K) for almost ideal Au-Pd-n-GaP structures in the Fowler region (samples No. 2, No. 4, and No. 5, hν = 1.5 ÷ 2.0 eV). This was new received data for this diode structures, which confirmed the $I_{f0}{}^{1/2}$~hν and C^{-2}-U measurements. The reference literature contains $q\varphi_{Bo}$ data for Au-n-GaP (structures measured in different samples by different methods). We measured I-U, C-U, I_{f0}~hν in the same sample using three methods. As far as we know, such a result for Au-Pd-n-GaP surface-barrier structures was obtained for the first time.

To study the band structures of GaP, Mead and Spitzer [23,25] proposed depositing a metal film on the semiconductor surface and observing the photovoltaic effect. Photo-electromotive force was proportional to optical absorption if the region was depleted of carriers and the thickness of the intermediate palladium layer d.

In the resulting transition, the diffusion length of minority carriers was much smaller than the penetration depth of light. This eliminated the difficulties of very thin samples' preparation required in light-absorption experiments. In the monograph by Sze [3] and in other works [2,14,28], it was shown that the photoelectric measurement method is the most accurate method for determining the barrier height of an m-s structure surface-barrier. Later, to determine the band parameters of both graded-gap and homo-gap semiconductors, a modernized contact photoelectric method was proposed [34–38]. The photocurrent of Au-Pd-n-GaP surface-barrier structures at photon energy hν = 2.2 ÷ 2.8 eV reflected the gallium phosphide band structure properties (Figures 6–8).

In the visible-light region, the photocurrent in the hν = 2.3–2.65 eV interval was due to electron transitions from the valence band to the X valley of the conduction band, while the photocurrent in the hν > 2.65 eV interval was due to electron transitions from the valence band to the "G" valley of the conduction band ($E_0 = 2.8$ eV) and was located in the violet and ultraviolet regions (Figures 6–8).

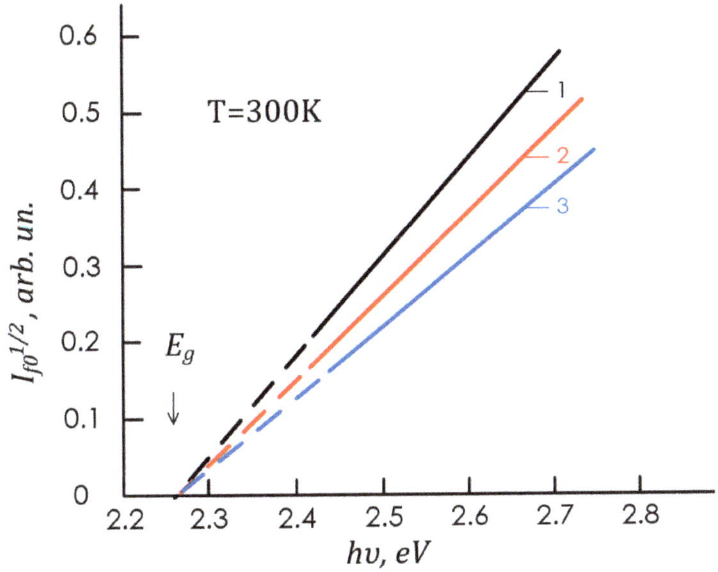

Figure 6. Dependence of the photocurrent $I_{f0}^{1/2}$ related to equality of incident photons on the hν for three Au-Pd-n-GaP surface-barrier structures with different d: 1—40 Å, 2—32 Å, 3—45 Å.

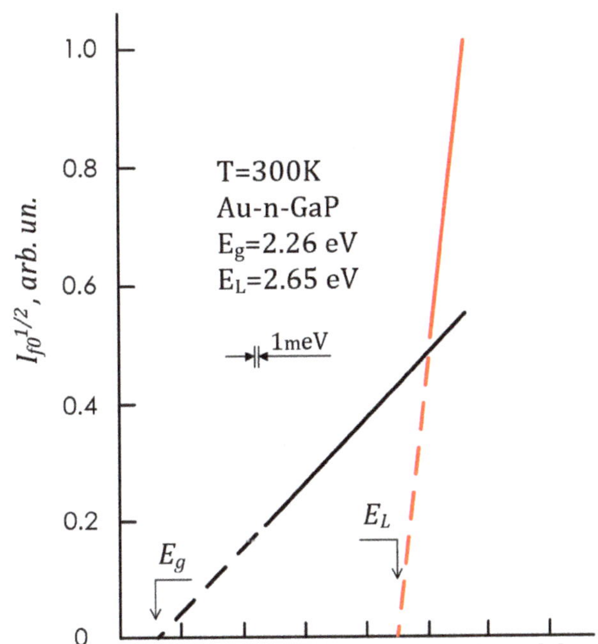

Figure 7. Dependence of the photocurrent $I_{f0}^{1/2}$ related to an equality of incident photons on the hν for four Au-Pd-n-GaP surface-barrier structures (Sample No. 4, β = 1.19).

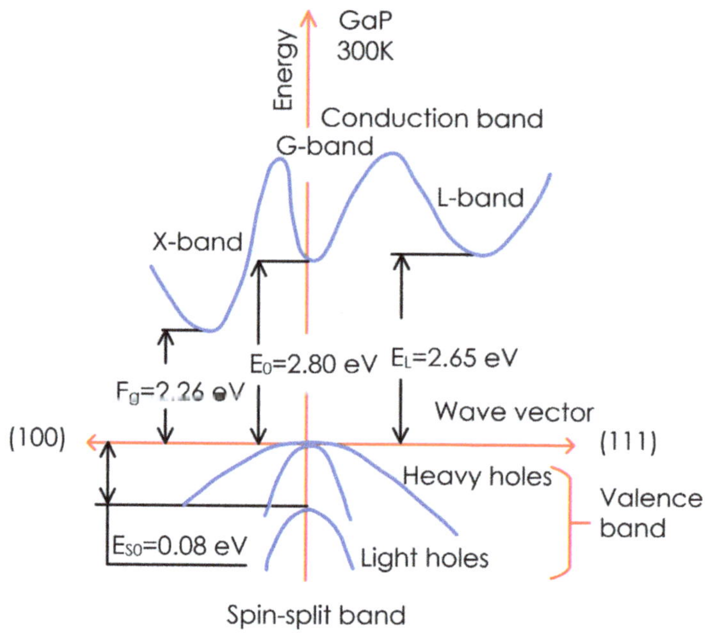

Figure 8. Scheme of optical transitions in the gallium phosphide band structure. E_g, E_0, and E_L are energies of optical transitions determined from the photocurrent spectrum of SBs.

When analyzing the experimental dependence of I_{f0} on hν (Figure 1, b) using the Spitzer and Mead method [11]: I_{f0}~$(h\nu\text{-}E_g)^2$, where hν < Eo, EG was determined; for GaP, it was 2.26 eV at 300 K (Figure 6). To determine the threshold energy of direct optical transitions for GaP, a modernized photoelectric method was used [36]. By comparing the absorption spectrum of GaP [38–40] with the photocurrent spectra of Au-Pd-n-GaP structures in the ultraviolet regions, it was established (Figure 1, b and Figures 9 and 10) that the photocurrent I_{f0} in m-s structures at hν < E_0 usually increases exponentially with hν, and at hν > E_0 it increases first linearly and then sublinearly. The linear section begins at hν = E_0, which allows one to determine E_0 directly from the photocurrent spectrum I_{f0} of the m-s structure when the m-s junction is irradiated through the Au nanolayer, both under natural (Figure 1, b) and polarized (Figure 10) radiation. The E_o for GaP was determined by this method; it was 2.80 eV at 300 K.

4. Investigation of PS in Linearly Polarized Light

The basic results of the performed polarization studies are presented in Figures 9–11. The regularities of photocurrent polarization measurements are as follows. When the LP light ($\ominus \neq 0°$) falls from the side of Au, a broadband polarimetric effect arises in the GAP structures. In the region of the spectrum under study, the photocurrent of the structures depends on the azimuthal angle between the light wave electric vector $\vec{E}$ and the incidence plane.

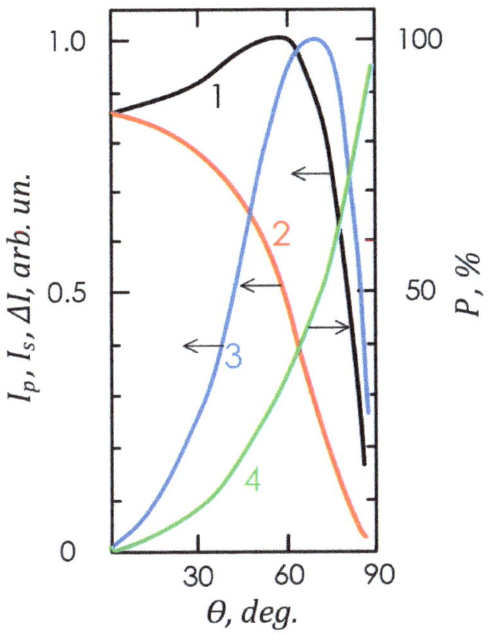

Figure 9. Dependences of photocurrents (1 corresponds I_P, 2 corresponds I_S, 3 corresponds $\Delta I = I_P - I_S$) and PPL coefficient (4 corresponds $P = \frac{\Delta I}{I_P - I_S} \cdot 100$ %) for surface-barrier structures (Sample No. 5) on the angle of incidence of linearly polarized radiation (λ = 0.437 μm, illumination is from the side of the barrier contact, T = 300 K).

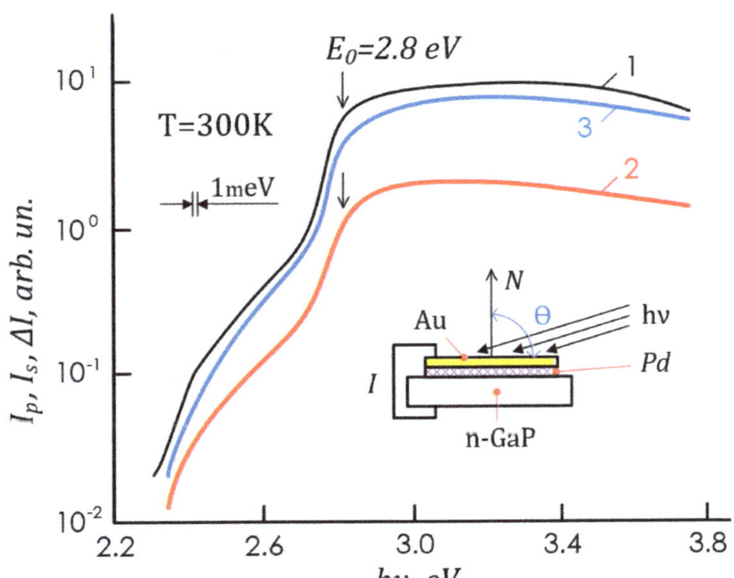

Figure 10. Spectral dependences of short-circuit photocurrent $I^{\parallel}$ (1), photocurrent polarization difference (2), short-circuit photocurrent $I^{\perp}$ (3), for surface-barrier structures (Sample No. 5). The inset shows the scheme of the experiment, ⊖ = 80°.

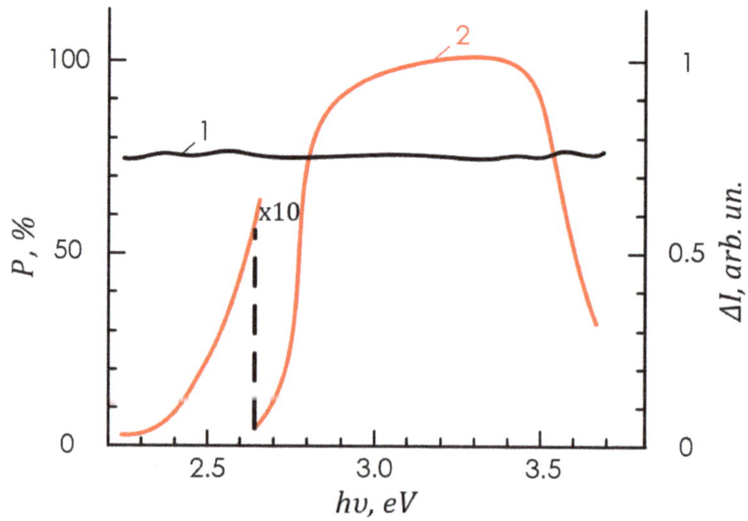

Figure 11. Spectral dependences of the PPL coefficient (1) and the polarization difference of photocurrents (2) of Au-Pd-n-GaP surface-barrier structures (Sample No 5, T = 300 K, ⊖ = 80°).

Figure 9 presents the dependences of the photocurrents (1—I_P, 2—I_S, 3—$\Delta I = I_P - I_S$) and the coefficient of induced PPL (4—$P = \frac{\Delta I}{I_P - I_S} \cdot 100\ \%$) on the angle of incidence of radiation ⊖ to the RP of the structures (the photocurrent I_P and I_S correspond to the irradiation of the structures when RP parallel or perpendicular to $\vec{E}$). For example, if the structures were illuminated with light with $\vec{E}$||RP polarization, the photocurrent I_P with increasing ⊖ increases at first, as can be seen from Figure 9, reaching a maximum in the vicinity of ⊖ ≈ 60°, then decreases quickly.

It should be emphasized that the increase in I_P with increases in the ⊖ angle is due to a decrease in reflection losses from the Au surface and takes place only at E//PR. Our studies showed that the increasing IP effect is related to both the interface and Au layer surface quality. For best-quality structures, $I_{⊖=60}/I_{⊖=0}$ = 1.3. For lower-quality structures, the dependence $I_P(⊖)$ resembles $I_S(⊖)$ (Figure 9) and I_S-curve 2 (Figure 10), i.e., upon increasing the photocurrent effect, I_P disappears. The analysis of the $I_P(⊖)$ and $I_S(⊖)$ experimental dependences should take into account the above regularities, and they are useful for controlling metal layer deposition quality on semiconductor surfaces and m-s interface quality.

Figure 10 shows the induced PPL spectral dependence in typical SB structures at angles of incidence ⊖ = 80°. The features of the studied structures connected with the long-wave photocurrent edge of I_P and I_S. According to the dependences of the polarization photocurrent I_P on hν and I_S on hν, the E_g and E_0 for GaP were determined using the well-known method [35–39]. The maximum value of the polarization difference of photocurrents ΔI (Figure 9, curve 3) in the obtained Au-GaP structures was achieved at angles of ⊖ = 70°. At ⊖ = 0°, following the selection rules for interband optical transitions in GaP [39], the polarization difference of photocurrents ΔI, and the PPL coefficient P = [ΔI/(I_P + I_S)], 100% of the studied structures turned out to be equal to zero, i.e., there was no PS to polarization. With an increase in ⊖, the PPL coefficient increased according to the law P~$⊖^2$, in accordance with [26,35]. The maximum value P = 74% was reached at ⊖ = 80° and, according to [26], corresponded to the refractive index n = 3.2 characteristic of GaP [39].

Spectral dependences of polarization parameters for a typical structure at ⊖ = 80° are shown in Figure 11. The main regularity of these studies—the nonselective nature of the PPL coefficient—was observed in the entire PS region, which corresponded to the results of the analysis [35,38]. In a wide range of incident photon energies of 2.3 ÷ 3.7 eV, the obtained structures had a constant value of P, which could be quickly controlled by choosing the value of ⊖ (Figure 9, curve 4). The polarization difference ΔI (Figure 11, curve 2) had a spectral contour corresponding to the spectral dependence of the photocurrent of these structures in unpolarized radiation. The maximum value of ΔI fell on the fundamental absorption region of GaP and was realized for the obtained structures in the interval 2.9 ÷ 3.4 eV. Since the maximum azimuthal PS of a polarimetric photodetector is $\Phi_i \sim \Delta I$ [35], it was obvious that the obtained structures could be used as photoanalyzers in the wavelength interval 0.36 ÷ 0.42 μm for use in energy systems. The maximum value of the azimuthal PS of Au-GaP structures was Φ_i = 0.2 ÷ 0.21 A/W·deg at T = 300 K.

5. Conclusions

A comprehensive detailed study of the GaP surface-barrier structures based on Au with palladium intermediate layer showed that, upon the formation of high-quality interface, the GaP surface stabilized and, in this case, the potential barrier height tended to a value of 1.35 eV at 300 K.

A thin palladium layer (d < 40 Å) performed two functions in the studied GaP surface-barrier structures. First, this tunnel-transparent intermediate nanolayer made the GaP surfaces chemically stable. Second, it prevented gold atoms' movement to the semiconductor, providing a high-quality, stable interface with the GaP semiconductor.

Investigation of PS in the visible region provided valuable information about semiconductor band structure parameters and the potential barrier. For the first time, it was experimentally established that the presence between GaP and Au of a thin metallic Pd layer with a thickness of 20 to 35 Å does not reduce the potential barrier height of the Au-n-GaP structures.

Thus, an intermediate palladium nanolayer between GaP and Au with a thickness of 20–30 Å forms specific effects in the Au-Pd-n-GaP nanostructure, which are of considerable practical and scientific significance for future needs.

Author Contributions: Conceptualization, V.R., E.T. and D.M.; methodology, V.R. and D.M.; software, V.K., M.S. and A.A.; validation, M.D.; formal analysis, V.K.; investigation, V.R. and I.I.; resources, V.R. and I.I.; data curation, V.K. and M.D.; writing—original draft preparation, V.R. and M.D.; writing—review and editing, M.D. and V.K.; visualization, M.D. and V.D.; supervision, V.R.; project administration, V.K. and E.T. All authors have read and agreed to the published version of the manuscript.

Funding: The research was partially funded by the Ministry of Science and Higher Education of the Russian Federation under the strategic academic leadership program "Priority 2030" (agreement 075-15-2021-1333, dated 30 September 2021).

Data Availability Statement: Not applicable.

Conflicts of Interest: The authors declare no conflict of interest.

Abbreviations

SB	Schottky barriers
PS	photosensitivity
PPL	photopleochroism
UV	ultraviolet
m	metal
s	semiconductor
$q\varphi_{Bo}$	Schottky barrier height
E_g	width of forbidden band
I_{f0}	short circuit photocurrent

T	absolute temperature
BC	barrier contact
I_s	saturation current
β	the structure ideality coefficient
U_D	diffusion potential
U_C	capacitive cutoff voltage
$\ominus$	angle of incidence of radiation
P	coefficient of photopleochroism
Φ_i	azimuthal photosensitivity of a polarimetric detector
RP	receiving plane

References

1. Nannichi, V.; Pearson, G.L. Properties of GaP Schottky barrier diodes at elevated temperatures. *Solid-State Electron.* **1969**, *12*, 341–348. [CrossRef]
2. Khvostikov, V.P.; Sorokina, S.V.; Khvostikova, O.A.; Nakhimovich, M.V.; Shvarts, M.Z. GaSb-Based Thermophotovoltaic Converters of IR Selective Emitter Radiation. *Semiconductors* **2021**, *55*, 840–843. [CrossRef]
3. Sze, S.M.; Kwok, K.; Ng, K.K. *Physics of Semiconductor Devices*, 3rd ed.; John Wiley & Sons, Inc.: New York, NY, USA, 2007; p. 815.
4. Sorokina, S.V.; Soldatenkov, F.Y.; Potapovich, N.S.; Shvarts, M.Z.; Khvostikov, V.P. Au- and Ag-Containing Contacts to GaSb-Photovoltaic Converters. *IEEE Electron. Device Lett.* **2022**, *43*, 418–421. [CrossRef]
5. Salamov, I.; Bobyl, A.; Mekhilef, S. The efficiency of the on-grid solar power plant in the Chechen Republic. *IOP Conf. Ser. Earth Environ. Sci.* **2020**, *578*, 012044. [CrossRef]
6. Ken, O.S.; Zhukov, E.A.; Akimov, I.A.; Korenev, V.L.; Kopteva, N.E.; Kalitukha, I.V.; Sapega, V.F.; Wieck, A.D.; Ludwig, A.; Schott, R.; et al. Effect of electric current on the optical orientation of interface electrons in AlGaAs/GaAs heterostructures. *Phys. Rev. B* **2020**, *102*, 045302. [CrossRef]
7. Nekrasov, S.V.; Akimov, I.A.; Kusrayev, Y.G.; Yakovlev, D.R.; Bayer, M. Effect of nuclear quadrupole interaction on spin beats in photoluminescence polarization dynamics of charged excitons in InP/(In,Ga)P quantum dots. *Phys. Rev. B* **2019**, *100*, 235415. [CrossRef]
8. Bodnar', I.V.; Osipova, M.A.; Rud', V.Y.; Rud', Y.V.; Bairamov, B.K. Photoelectric properties of In/CdP_2 surface barrier structures. *J. Appl. Spectrosc.* **2010**, *77*, 148–151. [CrossRef]
9. Lubyankina, E.A.; Toporov, V.V.; Mizerov, A.M.; Timoshnev, S.N.; Shubina, K.Y.; Bairamov, B.H.; Bouravleuv, A.D. Raman Spectroscopy of GaN Epitaxial Layers Synthesized on Si(111) by Molecular Beam Epitaxy with Nitridation. *Semiconductors* **2020**, *54*, 1847–1849. [CrossRef]
10. Cowley, M.; Heffner, H. Gallium Phosphide-Gold Surface Barrier. *J. Appl. Phys.* **1964**, *35*, 255–256. [CrossRef]
11. Mead, C.A. Metal-semiconductor surface barriers. *Solid-State Electron.* **1966**, *9*, 1023–1033. [CrossRef]
12. Smith, B.L. Near ideal Au-GaP Schottky diodes. *J. Appl. Phys.* **1969**, *40*, 4675–4676. [CrossRef]
13. White, H.G.; Logan, R.A. GaP surface-barrier diodes. *J. Appl. Phys.* **1963**, *34*, 1990–1997. [CrossRef]
14. Goldberg, Y.A.; Posse, E.A.; Tsarencov, B.V. Ideal GaP surface—Barrier diodes. *Electron. Lett.* **1971**, *7*, 601–602. [CrossRef]
15. Kesamanly, F.P.; Rud', V.Y.; Rud', Y.V. Induced photopleochroism in semiconductors Review. *Semiconductors* **1999**, *33*, 483–503. [CrossRef]
16. Brillson, L.J.; Viturro, R.E.; Slade, M.L.; Chiarada, P.; Kilday, D.; Kelly, M.K.; Margaritondo, G. Near-ideal Schottky barrier formation at metal—GaP interfaces. *Appl. Phys. Lett.* **1987**, *50*, 1379–1381. [CrossRef]
17. Sullivan, M.V.; Kolb, G.A. Chemical polishing of GaAs in bromine and methanol mixture. *J. Electrochem. Soc.* **1963**, *110*, 585–587. [CrossRef]
18. Etmers, G.W.; Stevens, E.H. Properties of Co-GaAs, Ni-GaAs diodes fabricated by electroless plating. *Solid-State Electr.* **1972**, *15*, 721–727. [CrossRef]
19. Goldberg, Y.A. Semiconductor near-ultraviolet photoelectronics. *Semicond. Sci. Technol.* **1999**, *14*, R41. [CrossRef]
20. Konnikov, S.G.; Rud, V.Y.; Rud, Y.V.; Melebaev, D.; Berkeliev, A.; Serginov, M.; Tilevov, S. Photopleochroism of GaPxAs1-x Surface-Barrier Structures. *Jpn. J. Appl. Phys.* **1993**, *32*, 515–517. [CrossRef]
21. Moulin, E.; Luo, P.; Pieters, B.; Sukmanowski, J.; Kirchhoff, J.; Reetz, W.; Müller, T.; Carius, R.; Royer, F.X.; Stiebig, H. Photoresponse enhancement in the near infrared wavelength range of ultrathin amorphous silicon photosensitive devices by integration of silver nanoparticles. *Appl. Phys. Lett.* **2009**, *95*, 033505. [CrossRef]
22. Amanmadov, A.; Durdyyev, R.; Kotyrov, M.; Melebaev, D. Synthesis of gold nanoparticles via citrate reduction and their characterization. *Eurasian Union Sci.* **2019**, *12*, 34–42. [CrossRef]
23. Spitzer, W.G.; Mead, C.A. Barrier Height Studies on Metal-Semiconductor Systems. *J. Appl. Phys.* **1963**, *34*, 3061–3069. [CrossRef]
24. Mead, C.A.; Spitzer, W.G. Conduction band minima at AlAs and AlSb. *Phys. Rev. Lett.* **1963**, *11*, 358–361. [CrossRef]
25. Spitzer, W.G.; Mead, C.A. Conduction Band Minima of Ga(As_(1−x)P_x). *Phys. Rev.* **1964**, *133*, A872–A875. [CrossRef]
26. Spitzer, W.G.; Gershenzon, M.; Frosch, C.J.; Gibbs, D.E. Optical absorption in n-type gallium phosphide. *J. Phys. Chem. Sol.* **1959**, *11*, 339–341. [CrossRef]

27. Davydov, R.; Davydov, V.; Dudkin, V. The Nuclear Magnetic Flowmeter for Monitoring the Consumption and Composition of Oil and Its Complex Mixtures in Real-Time. *Energies* **2022**, *15*, 3259. [CrossRef]
28. Mikhailova, M.P.; Moiseev, K.D.; Yakovlev, Y.P. Discovery of III–V Semiconductors: Physical Properties and Application. *Semiconductors* **2019**, *53*, 273–290. [CrossRef]
29. Aspnes, D.E.; Studna, A.A. Dielectric functions and optical parameters of Si, Ge, GaP, GaAs, GaSb, InP, InAs and InSb from 1.5 to 6.00 eV. *Phys. Rev. B.* **1983**, *27*, 985–1009. [CrossRef]
30. Hudges, R.C.; Zipperian, T.E.; Dawson, L.R.; Biefeld, R.M.; Walko, R.J.; Dvorak, A.M. Gallum phosphide junctions with low leakage for energy conversion and near ultraviolet detectors. *J. Appl. Phys.* **1991**, *69*, 6500–6505.
31. Seredin, P.V.; Goloshchapov, D.L.; Arsentyev, I.N.; Nikolaev, D.N.; Pikhtin, N.A.; Slipchenko, S.O. Spectroscopic Studies of Integrated GaAs/Si Heterostructures. *Semiconductors* **2021**, *55*, 44–50. [CrossRef]
32. Diuldin, M.V.; Melebayew, D.; Terukov, E.; Hogland, W.; Kosolapov, V.M.; Bobyl, A.V.; Pashikova, T.D.; Garadzha, N.; Shamuhammedowa, M. Highly sensitive photodetectors on the basis of Au-oxide-n-GaP0.4As0.6. In *IOP Conference Series: Earth and Environmental Science*; IOP Publishing: Bristol, UK, 2022; Volume 1096, p. 012005.
33. Davydov, R.; Antonov, V.; Makeev, S.; Batov, Y.; Dudkin, V.; Myazin, N. New high-speed system for controlling the parameters of a nuclear reactor in a nuclear power plant. In Proceedings of the E3S Web of Conferences, St. Petersburg, Russia, 25–27 June 2019; Volume 140, p. 02001.
34. Davydov, R.; Antonov, V.; Angelina, M. Parameter Control System for a Nuclear Power Plant Based on Fiber-Optic Sensors and Communication Lines. In Proceedings of the 2019 IEEE International Conference on Electrical Engineering and Photonics, EExPolytech, St. Petersburg, Russia, 17–18 October 2019; pp. 42–45.
35. Boudjemila, L.; Krasnoshchekov, V.; Olimov, S.; Diuldin, M.; Kasimakhunova, A. Some features of photoelectrical properties of highly efficient solar cells based on Si. In Proceedings of the 2020 IEEE International Conference on Electrical Engineering and Photonics, EExPolytech, St. Petersburg, Russia, 15–16 October 2020; pp. 223–226.
36. Rud, Y.V. Photopleochroism and the physical principles of the preparation of semiconductor polarimetric photodetectors. *Sov. Phys. J.* **1986**, *29*, 638–650. [CrossRef]
37. Murat, S.; Fahrettin, Y. Photovoltaic and interface state density properties of the Au/n-GaAs Schottky barrier solar cell. *Thin Solid Film.* **2011**, *519*, 1950–1954.
38. Stirn, R.J.; Yeh, Y.C.M. A 15% efficient antireflectioncoated metal-oxide-semiconductor solar cell. *Appl. Phys. Lett.* **1975**, *27*, 95–98. [CrossRef]
39. Rud, V.Y.; Rud, Y.V.; Terukov, E.I. Polarization studies of the photoelectric properties of II–IV–V_2-semiconductor-compound–electrolyte systems. *J. Opt. Technol.* **2016**, *83*, 275–278. [CrossRef]
40. Rud, V.; Melebaev, D.; Trapeznikova, I.; Diuldin, M.; Yakusheva, M.; Bykova, N. Features of Metal Dielectric Structures Based on GaP with Avalanche Multiplication Effect. In Proceedings of the 2022 International Conference on Electrical Engineering and Photonics (EExPolytech), St. Petersburg, Russia, 20–21 October 2022; pp. 379–382.

energies

MDPI

Article

Phase-Homogeneous $LiFePO_4$ Powders with Crystallites Protected by Ferric-Graphite-Graphene Composite

Dmitry Agafonov [1], Aleksandr Bobyl [2,*], Aleksandr Kamzin [2], Alexey Nashchekin [2], Evgeniy Ershenko [2], Arseniy Ushakov [3], Igor Kasatkin [4], Vladimir Levitskii [5], Mikhail Trenikhin [6] and Evgeniy Terukov [7]

1 Department of Electrochemical Production Technology, St. Petersburg State Institute of Technology, Moskovski Ave. 26, 190013 St. Petersburg, Russia
2 Division of Solid State Physics, Ioffe Institute, Politekhnicheskaya Str. 26, 194021 St. Petersburg, Russia
3 Institute of Chemistry, Saratov State University, Astrakhanskaya Str. 83, 410012 Saratov, Russia
4 Research Park, RC XRD, St. Petersburg State University, Universitetskaya nab. 7–9, 199034 St. Petersburg, Russia
5 RnD Center IFTE, Politekhnicheskaya Str. 26, 194021 St. Petersburg, Russia
6 Department "Chemistry and Chemical Technology", Petrochemical Institute, Omsk State Technical University, Mira Ave. 11, 644050 Omsk, Russia
7 Department of Electronics, St. Petersburg State Electrotechnical Univeristy, ul. Professora Popova 5, 197022 St. Petersburg, Russia
* Correspondence: bobyl@theory.ioffe.ru

Abstract: Phase-homogeneous $LiFePO_4$ powders have been synthesized. The content of impurity crystalline phases was less than 0.1%, according to synchrotron diffractometry (SXRD) data. Anisotropic crystallite sizes $\overline{L}_{V[hkl]}$ were determined by XRD. A low resistance covering layer of mechanically strong ferric-graphite-graphene composite with impregnated ferric (Fe^{3+}) particles < 10 nm in size increases the cycleability compared to industrial cathodes. In accordance with the corrosion model, the destruction of the Fe^{3+}-containing protective layer of crystallites predominates at the first stage, and at the second stage Fe escapes into the electrolyte and to the anode. The crystallite size decreases due to amorphization that starts from the surface. The rate capability, $Q(t)$, has been studied as a function of $\overline{L}_{V[hkl]}$, of the correlation coefficients r_{ik} between crystallite sizes, of the Li diffusion coefficient, D, and of the electrical relaxation time, τ_{el}. For the test cathode with a thickness of 8 μm, the values of $D = 0.12\ nm^2/s$, $\tau_{el} = 8$ s were obtained. To predict the dependence $Q(t)$, it is theoretically studied in ranges closest to experimental values: $D = 0.5 \div 0.03\ nm^2/s$, $\tau_{el} = 8/1$ s, average sizes along [010] $\overline{L}_1 = 90/30$ nm, averaged $\overline{r} = 0/1$.

Keywords: energy storage; electrochemical battery; Mössbauer spectroscopy; synchrotron XRD; energy technology; lattice structure; storage degradation; anisotropic crystallite; electrode powder

Citation: Agafonov, D.; Bobyl, A.; Kamzin, A.; Nashchekin, A.; Ershenko, E.; Ushakov, A.; Kasatkin, I.; Levitskii, V.; Trenikhin, M.; Terukov, E. Phase-Homogeneous $LiFePO_4$ Powders with Crystallites Protected by Ferric-Graphite-Graphene Composite. *Energies* **2023**, *16*, 1551. https://doi.org/10.3390/en16031551

Academic Editor: Roman Vadimovich Davydov

Received: 27 December 2022
Revised: 26 January 2023
Accepted: 1 February 2023
Published: 3 February 2023

1. Introduction

The values of the battery capacity, Q_0, the rate capability, $Q(t)$, and the maximum number of discharge-charge cycles are important target battery parameters. As shown in [1], the phase homogeneity of $LiFePO_4$ suffers from the impurity phases which appear as synthesis residues at low temperatures (<550 °C) and ferric (Fe^{3+}) compounds at high temperatures in the presence of residual oxygen. The first-principal modelling of Li-Fe-P-O_2 phase diagrams [2] supported that.

Several attempts have been made to develop a technology for producing highly efficient and phase-homogeneous electrode powders using: various raw materials and processing methods [3,4], variations in the composition and proportions of the loaded raw materials [5–7], growth duration and additional multi-stage post-growth annealing [8,9], extra pure initial chemicals of so-called "battery quality" [10], one-pot synthesis [11,12], etc. Coating of crystallites with various functional layers was also found useful [13]: metal oxides, glasses, etc. and the most popular carbon-based materials [14–31] were tried as the coating substances. The latter include various carbon phases [14], nanocarbons of different shapes (spheres, tubes and pores) [15], graphene [16–18]. Sucrose [19–21], glucose [22,23], adipic acid [24–26], polyvinyl alcohol [27,28], polymeric additives [29,30], and ferrocene [31] were used as catalysts for graphitization.

Our approach is based on (i) one-pot $LiFePO_4$ liquid-phase synthesis from high purity lithium and iron acetates taken as the starting materials, (ii) multi-stage thermal processing including acetic acid evaporation at its boiling temperature and protective coating formation from adipic acid and polyvinyl alcohol, (iii) elimination of potential impurity sources, (iv) stage-by-stage control of the effect of the regime parameters on crystallite sizes and coating quality.

X-ray diffractometry (XRD) is commonly used to characterize the presence and size of crystallites in electrode powders. Its synchrotron version (SXRD) [32–34] is the most sensitive to the impurities in crystallites. Various surface-sensitive methods are used to analyze the $LiFePO_4$ particle surface [19,35–38]: Auger electron spectroscopy (AES), electron diffraction (EDS) [39], X-ray photoelectron spectroscopies (XPS) [40–42], inductively coupled plasma (ICP) [39].

We used both synchrotron and laboratory X-ray sources SXRD and XRD, respectively, to increase the sensitivity towards the presence of impurity phases and to determine anisotropic size distributions of $LiFePO_4$ crystallites, as well as to study degradation during cycling of test cathodes.

Mössbauer spectroscopy (MS) makes it possible to detect iron-containing compounds and study their properties [43–49]. In particular, dependence of hyperfine spectral parameters on the charge of Fe ions [45–48] have been used to study reversible delithiation processes in $LiFePO_4$ crystallites. Line broadening associated with an increase in the disorder around Fe^{2+} towards the crystallite surface was observed; oxidation suppression by surface carbon was also found [45]. In this work, the MS results were used for determination of the proportion of the Fe^{3+} ferric impurity compounds in $LiFePO_4$ (which is itself a Fe^{2+} compound).

To reveal the role of ferric compounds, the following results are important: (1) up to 5% ferric states are typically present at regular lattice sites with reduced symmetry [50], which is not detected by XRD; (2) lowering the content of X-ray inactive ferric compounds from 17% to 9% improves the electrochemical parameters, even compared with those materials where the content was zero [51]; (3) presence of significant amounts of Fe^{3+} cations at broad interfaces improves electronic and ionic conductivity [52]; (4) a significant number of ferric states was detected on the surface of crystallites with MS and XPS [53]; (5) MS and ICP studies showed that an increase in the synthesis time led to an increase in the fraction of iron atoms with a long-range order compared to the fraction of iron atoms in the Li/Fe-PO_4 amorphous phase [54]; (6) initial powders contained ferric compounds $Li_3Fe_2(PO_4)_3$ and α-Fe_2O_3, which became X-ray inactive after 6 h annealing at 700 °C [55]; (7) after low-temperature precipitation at 106 °C, the powder contained an amorphous ferric phase $LiFePO_4(OH)$, which transformed into crystalline $LiFePO_4$ after annealing under reducing conditions; after annealing at 500 °C, the crystallite size increased from 17 nm to 35 nm [56]; (8) with differential scanning calorimetry (DSC) and derivative thermogravimetry (DTG), small particles of $Li_3Fe_2(PO_4)_3$ and presumably of Fe_2O_3 were detected on the surface of $LiFePO_4$ crystallites after their oxidation below 470–475 °C [57].

We used Mössbauer spectroscopy, EDX and XRD studies to elucidate the non-trivial role of ferric compounds, in particular, in the protective layer formation and its degradation.

The combined use of transmission electron microscopy (TEM) and Raman spectroscopy (RS) is important in study of the various carbon phases in the cathode materials initial powders [58–80]. Using RS, it was found that some carbon materials mixtures have a unique combination of properties: mechanical characteristics, strength, chemical inertness and biocompatibility [61–63]. Among a large number of different grades of carbon black [65], the most famous is Ketjen black [70], which is widely used in electrode powder technologies [71–76] and is added especially when it is necessary to sharply increase the electronic conductivity [75,76].

The novelty and importance of Raman spectroscopies line intensity analysis lies in the study of the graphene phase's order degree, the graphite phase mechanical strength, and the multilayer graphene phase conductivity compared to industrial $LiFePO_4$ powders. Thus, the development of phase-homogeneous electrode powders may further increase battery capacity and cyclability, and optimize the rate capability.

In this work we synthesized $LiFePO_4$ powders and coatings on the surface of their crystallites. The phase homogeneity was tested with SXRD, and its effect on the test cathodes cyclability was shown. Composition and properties of the layers covering the crystallites was studied using TEM and RS. MS was used to discover the role of ferric compounds, and XRD and EDX were used to study the mechanisms of degradation of the $LiFePO_4$ test cathodes. $LiFePO_4$ test cathodes were subjected to galvanostatic measurements. The hierarchy of relaxation times of electrochemical recharging of the cathodes was determined using the data on anisotropic size distribution of crystallites. We studied the dependence of $Q(t)$ of the test cathodes on the crystallite parameters, the lithium diffusion coefficient D along [010] and the quality of their coating in terms of electric relaxation time τ_{el}. $Q(t)$ calculations allowed us to assess the possibility of improving the technology.

2. Synthesis of $LiFePO_4$ Powders and Surface Coatings, and Characterization Techniques

2.1. Synthesis of $LiFePO_4$ Powder and Coatings

We used a modified version of the liquid-phase synthesis of $LiFePO_4$ based on lithium and iron acetates as initial reagents [7,81]. They have good solubility in water and low thermal stability at the synthesis temperatures; acetic acid has a low boiling point which ensures its distillation during synthesis, and volatile components can be easily removed from the reaction zone during annealing [82–89]. Initial acetates were prepared from metallic iron and lithium carbonate by interaction with acetic acid (Snabtechmet), grade A.C.S. These materials are available and have a high degree of purity for large-scale production. Chemicals were used as received from the manufacturers; additional reduction of atmospheric impact was provided at the stages of pre-drying and annealing. Mechanochemical activation of the starting materials in a liquid medium promoted formation of an intermediate synthesis product, which was converted into $LiFePO_4$ at a temperature lower by 100 °C than in the standard procedures. This created the prerequisites for obtaining $LiFePO_4$ with a reduced content of impurity phases [87]. Mechanochemical activation was carried out in a saturated solution of ammonium dihydroorthophosphate in distilled water. Table 1 shows the following synthesis steps:

Table 1. Sequence of $LiFePO_4$ liquid-phase synthesis.

№	Steps	Reagents
1	Preparation of acetates	$Fe + LiCO_3 + CH_3COOH \rightarrow Fe(CH_3COO)_2$, $LiCH_3COO$, H_2O
2	Organic Additives	AA, PA
3	Phosphoric acid	H_3PO_4
4	Pre-drying at 100 °C	Evaporation of CH_3COOH, H_2O
5	Annealing 1.5 h at 400 °C in an Ar atmosphere	Evaporation of CH_3COOH, H_2O, CO_2, $((CH_2))_4CO$
6	Annealing 1.5 h at 670 °C in an Ar atmosphere	Crystallization of $LiFePO_4$

1. Lithium acetate is obtained by direct acetic acid action on lithium carbonate. The Fe^{2+} ion is easily oxidized by atmospheric oxygen to Fe^{3+}, therefore, iron acetate is prepared by placing a calculated amount of iron in a flask containing an excess of acetic acid. The flask is stoppered and kept under vacuum until dissolved with stirring with a magnetic stirrer and heated at the end of the process.

2. Two organic additives, adipic acid (AA) [88–91] and polyvinyl alcohol (PA) [92–94], are used to obtain and condition the synthesized composite. AA crystallizes upon reaction mixture evaporation and decomposes upon heat treatment, resulting in a precipitate with a loosened structure. PA is adsorbed on the precipitate surface and contributes to the small crystals formed during the interaction of phosphoric acid and lithium and iron acetates.

3. Phosphoric acid is used as a source of phosphorus ions.

4. Pre-drying to evaporate CH_3COOH and H_2O in a stream of hot air, which prevented agglomeration of the starting materials.

5–6. Annealing for evaporation at different temperatures and crystallization. After drying in a stream of hot air, the sample was pelletized and placed in a sealed muffle furnace, which was constantly purged with especially pure nitrogen. The furnace was heated to 400 °C, the sample was held at this temperature for 1.5 h, then the heating was turned off and the furnace was expected to cool in a nitrogen flow to a temperature of 25 °C. The resulting intermediate was subjected to repeated grinding and tableting. This approach does not lead to the formation of the final product due to the low temperature, but it makes it possible to obtain a dense mixture of initial substances with a large interfacial surface and a fixed contact between the phases of the initial substances. Then the tableted sample is placed in a sealed muffle furnace, which is constantly purged with especially pure nitrogen; the temperature in the furnace rises to 670 °C. This approach is tested in a series of experiments near this temperature. The selected temperature regime provides the maximum $LiFePO_4$ capacity of 65 mAh/g at 20 C rate. At all stages of preparation and synthesis, the above measures are taken, in particular, to exclude the transition of Fe^{2+} to Fe^{3+}.

Preliminary technological experiments were also carried out, in particular using: iron acetate $(CH_3COO)_2Fe$, iron oxalate $FeC_2O_4 \times 2H_2O$, lithium carbonate Li_2CO_3, ammonium dihydroorthophosphate ($NH_4H_2PO_4$) from VEKTON, grade A.C.S. Depending on the composition and the annealing modes, the following impurity crystalline phases were observed in the samples: Fe_3O_4 (ICDD 00-019-0629), Fe_2O_3 (ICDD 00-039-1346), $FeCO_3$ (ICDD 00-029-0696), $FePO_4$ (ICDD 00-050-1635). To compare the quality and target parameters of the developed sample N1, SPbTU, the following industrial $LiFePO_4$ powders were used: N2, Phostech Lithium [95]; N3 OCELL Technologies N4, Golden Light Energy. The powders had specific capacities ranging from 145 mAh/g to 167 mAh/g at 0.1 C rate [7,81], given in the Table 2 at 20 C rate.

Table 2. SXRD composition of impurity crystallites in powders [96], XRD anisotropic sizes of $LiFePO_4$ crystallites $\overline{L}_{V[hkl]}$ averaged over the length of their columns [97] (from [98]). MS fractions of Fe^{2+} and Fe^{3+} compounds (description below). Forecast cycling of a test cathode after 100 times cycling at 1 C rate. The errors of $\overline{L}_{V[hkl]}$ reported in parentheses characterize the reproducibility.

	Impurity Phase, %	Q, mAh/g	$\overline{L}_{V[100]}$, nm	$\overline{L}_{V[010]}$, nm	$\overline{L}_{V[001]}$, nm	Forecast Cycling	Mössbauer	
		20 C Rate					Fe^{2+}, %	Fe^{3+}, %
N1	not detected	63	66 (5)	82 (5)	89 (7)	3500	96	4
N2	Li_3PO_4, 1.01 (2)	58	145 (26)	131 (13)	185 (17)	1000	95	5
N3	Li_3PO_4, 2.39 (3) Fe_2P, 2.34 (3) Fe_3P, 2.02 (3)	57	141 (5)	146 (15)	165 (7)	5000	92	8
N4	not detected	40	230 (20)	261 (8)	242 (30)	800	98	2

The novelty of our synthesis in comparison with [87–89] is the following set:

1. One-pot $LiFePO_4$ liquid-phase synthesis using chemically pure lithium and iron acetates as starting materials.
2. Using raw materials of organic nature.

3. Using the tableting operation after the pyrolysis of organic matter, at 400 °C. This operation facilitates the course of the final synthesis of the topochemical reaction.
4. Drying, before heat treatment, was carried out in a stream of hot air, which prevented the agglomeration of the starting materials.
5. The synthesis intermediate is tableted, heat-treated at 400 °C, then re-milled and re-tableted before the final heat treatment at 670 °C. This made it possible to obtain a material of high phase purity.
6. Before repeated tableting, adipic acid was introduced, the pyrolysis of which in an inert medium (high-purity nitrogen) led to encapsulation of $LiFePO_4$ in a carbon shell.
7. At all stages of the preparation, measures were taken to exclude the transition of Fe^{2+} to Fe^{3+}, in particular, by isolation from atmospheric moisture.

2.2. Characterization Techniques

Note that conventional XRD is less sensitive than SXRD due to the lower intensity of laboratory radiation sources [32–34,96]. The SXRD experiments were done at the Structural Materials Science station of the Siberia-2 synchrotron radiation source of Kurchatov Institute Research Center [96]. Measurements were made at a wavelength of λ = 0.68886 Å in the transmission (Debye–Scherrer) mode with a Fujifilm Imaging Plate memory layer as a 2D detector, sample-to-detector distance of 200 mm, and exposure time of 15 min.

The anisotropic crystallite sizes $\overline{L}_{V[hkl]}$ averaged over the length of the columns [97] were determined by XRD [98]. Bruker D8 Discover diffractometer was used in a parallel-beam linear-focus mode at 2θ = 15–125 deg, and MAUD software was utilized for profile fitting [94]. The primary beam was conditioned with a double-bounce channel-cut Ge220 monochromator to provide $CuKa_1$ radiation with a wavelength of 1.54056 Å. The Cagliotti coefficients of the instrumental profile function were refined by fitting the data for a LaB_6 powder specimen (NIST SRM 660c).

The powders were studied with a JEOL JEM 2100 high-resolution TEM at an accelerating voltage of 200 kV and crystal lattice resolution of 0.14 nm. The instrument was equipped with an INKA 250 Xray spectrometer. An image of the single-crystal gold lattice with the (111) interplanar spacing of 0.235 nm was used as reference for linear scale calibration.

Raman spectra were measured at room temperature in the "backscattering" geometry on a LabRam HR 800 spectrometer equipped with a confocal microscope. The measurements used the exciting light wavelengths of 532 nm and 633 nm, focused on the surface of the sample into a spot with a diameter of ~1 μm. In this case, the laser radiation power on the sample was maintained at a level of 2.0 mW. The use of a 600 pcs/mm diffraction grating made it possible to obtain a spectral resolution no worse than 2.5 cm^{-1}.

The Mössbauer effect was measured at room temperature on 57Fe nuclei in the γ-ray transmission geometry through powders sputtered onto aluminum foil with a spot diameter of 20 mm. The movement of the 57Co(Rd) γ-radiation source in the spectrometer was carried out with a constant acceleration of the reference signal in the form of a triangle. Velocity calibration was performed using α-iron foil for two Doppler shift velocities of the gamma-ray source.

3. $LiFePO_4$ Powders Phase Homogeneity Studies Using SXRD and Its Effect on the Test Cathodes Cyclability

3.1. Phase Homogeneity Using SXRD

The results are shown in Figure 1 and the quantitative composition of impurity crystalline phases are given in Table 2.

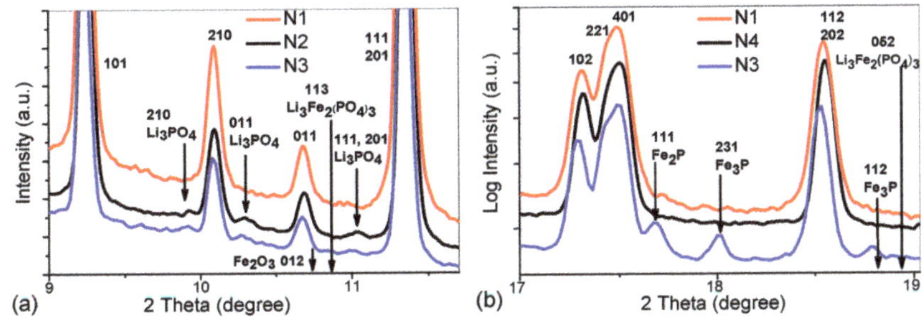

Figure 1. SXRD measurements of the developed and industrial $LiFePO_4$ powders, N1 and N2–4, respectively, for 2 Theta intervals 9.0–11.7 (**a**) and 17.0–19.1 (**b**). Arrows near the abscissa indicate the expected peak positions.

3.2. Cycling Test Cathode Cells

A test electrode was prepared of 80 wt.% powder, 10 wt.% acetylene black and 10 wt.% polyvinylidene fluoride (PVDF). It was applied as a sample homogenized suspension of acetylene black in a 5 wt.% solution of PVDF in N- methylpyrrolidone (analytical grade) on 1 cm^2 aluminum plate, 0.4 mm thick, after which it was dried at 120 °C in air for 12 h [99]. To reduce the errors of the galvanostatic measurements of rate capability at small times, the cathode thickness was minimal, about 8 μm. The measurements included charge-discharge cycles with constant current loads: sequentially from 0.1 to 20 C rates in one cycle, then 15 cycles each, completing 10 cycles with a load of 1 rate. The current density of 1 C rate corresponded to 170 μA per 1 mg of sample and was analyzed at the potentials in the range of 2.6–4.3 V relative to the lithium reference electrode. Figure 2 shows the galvanostatic measurements results. Cycling predictions were obtained by making 150 charge-discharge cycles and by recalculating to the point of the capacitance reduction to 80 % of its initial value at a current of 1 C rate indicated in Table 2.

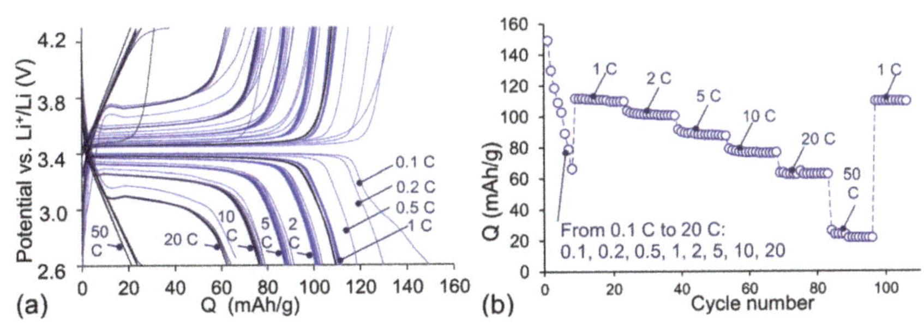

Figure 2. (**a**) Galvanostatic charge-discharge curves, (**b**) variation of capacity with the cycle number. The values of the charge and discharge currents (C rate -units) are on the diagrams.

The following conclusions can be drawn from the results in Table 2:

1. No impurity crystallites were found in the samples N1 and N4. However, the cyclability of the latter was significantly worse than that of the developed sample N1, despite the trend towards an increase in the role of the [100] surface of larger crystallites in their cyclability. Consequently, the absence of impurity crystallites is a necessary but obviously not a sufficient condition.
2. Sample N1 had a high cyclability, but lower than that in the sample N3, which contained several impurity phases. In addition to lowering the resistance, these phases probably catch the degradation products, which slows down the formation of harmful impurity phases. It can be concluded that it is promising to search for such

compositions or isovalent doping, including various mixed solid solutions, starting from our pure technology.

4. Crystallite Coating Composition and Properties Studied with TEM and RS

4.1. TEM Study

TEM images of $LiFePO_4$ powders are shown in Figure 3. Several types of particles are observed in the powders:

- Large 40–150 nm particles of $LiFePO_4$ are observed in all TEM images; an example is shown in Figure 3a for the developed sample N1.
- In almost all samples, nanocrystalline 5–10 nm $Li_3Fe_2(PO_4)_3$ particles are observed (Figure 3b,e,f) with lattice spacing of 0.428 nm, which corresponds to (200) or (−121) planes. It should be noted, however, that the (011) planes of $LiFePO_4$ have a similar interplanar distance. The ferric compound $Li_3Fe_2(PO_4)_3$ on the surface of $LiFePO_4$ crystallites appears as a result of insufficient oxygen content to complete the oxidation reaction.
- Figure 3f shows crystallites with interplanar spacing of 0.220 nm that corresponds presumably to the (321) planes of Fe_3P in the N3 sample, even though a similar distance can be found in the structures of other phases. According to [91,101,102], at T > 850 °C and in the presence of carbon, $LiFePO_4$ is reduced to form Fe_3P. As seen from Table 2, a significant amount of that phase is reliably detected in the sample N3 using SXRD; the plate-like shape of the particles was described in detail in [103].
- Various structures of the carbon layers encapsulating the $LiFePO_4$ particles can be seen in the samples. More ordered carbon shells are up to 5 nm thick and the amorphous shells are up to 20 nm thick. In some cases, particles without a carbon shell are observed. The properties of the carbon coatings will be discussed below in the RS section.

In none of the samples have particles of Fe2O3 been found. Perhaps they can be observed in other technologies, considering also some identification uncertainty with $Li_3Fe_2(PO_4)_3$, so the coating we observed was called a ferric-graphite-graphene composite.

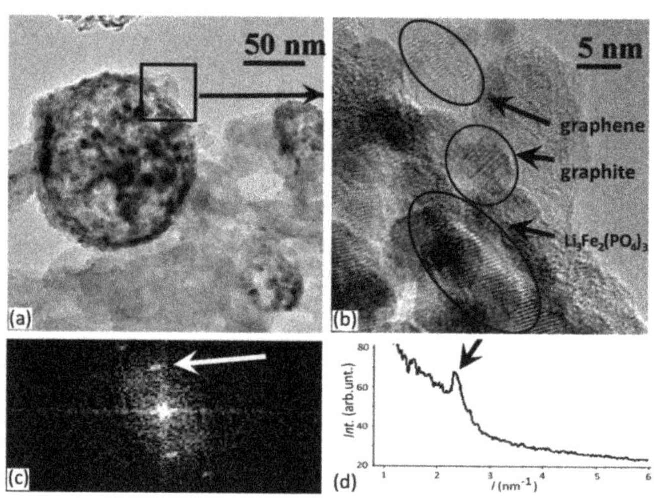

Figure 3. *Cont.*

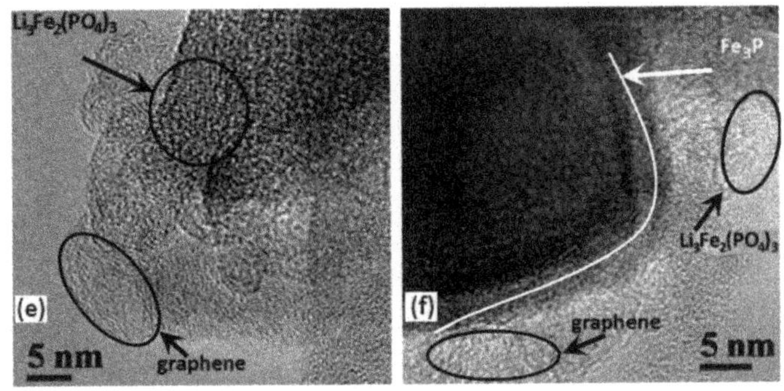

Figure 3. (**a**–**d**)—TEM images of the developed N1 powder of 3 crystallites (**a**); (**b**) is an enlarged part of (**a**). Areas of the ferric-graphite-graphene composite coating layer are marked (**b**); FT snapshot obtained from this TEM image (**c**), a maximum close to the interplanar spacing of 0.428 nm (**d**). (**e**,**f**)—TEM images of industrial powders: N4 (**e**) and N3 (**f**) containing the impurity crystallite phase of Fe_3P. Regions characteristic of multilayer graphene and the most probable $Li_3Fe_2(PO_4)_3$ are also marked.

4.2. Raman Spectroscopy Studies of Carbon Phases

Figure 4 shows the Raman spectra of the Sample N1 taken in the ranges from 800 to 3700 cm^{-1} [104].

Table 3. The Raman spectra parameters: E1, I1 and W1—position of the maximum, area and width of one Lorentz approximation for the 1st peak and so on for the others shown in Figure 4 for the excitation wavelengths of 532 nm and 633 nm.

	E1, I1, W1	**E2, I2, W2**	**E3, I3, W3**	**E4, I4, W4**	**E5, I5, W5**	**E6, I6, W6**	**E7, I7, W7**	**ID/IG (I2/I4)**	**Isp2/Isp3 ((I2 + I4)/(I1 + I3))**	**(I1 + … AI)/ (I5 + … I7)**
					Excitation 532 nm					
N1	1203 64,354 130	1346 464,462 191	1513 109,518 128	1596 183,271 64	2681 84,869 391	2915 114,677 376	3176 7613 112	2.53	3.73	3.96
N4	1187 45,591 94	1341 733,854 190	1527 130,196 119	1599 244,016 58	2668 146,229 337	2919 231,601 344	3180 17,221 113	3.01	5.56	2.92
N2	1190 22,138 110	1342 288,779 188	1522 53,844 120	1600 98,169 59	2651 37,209 307	2909 88,666 380	3185 5240 117	2.94	5.09	3.53
					Excitation 633 nm					
N1	1201 63,280 131	1346 475,058 195	1514 106,974 126	1596 182,032 64	2716 65,290 328	2949 38,599 234	3178 9359 128	2.61	3.86	7.3
N4	1193 11,389 112	1332 133,742 180	1528 21,170 123	1602 37,709 56	2630 16,408 293	2883 20,492 318	3178 692 78	3.55	5.26	5.42
N2	1198 7329 117	1331 69,319 179	1516 11,644 127	1601 19,955 57	2594 9304 340	2870 9744 334	3178 350 130	3.47	4.70	5.58

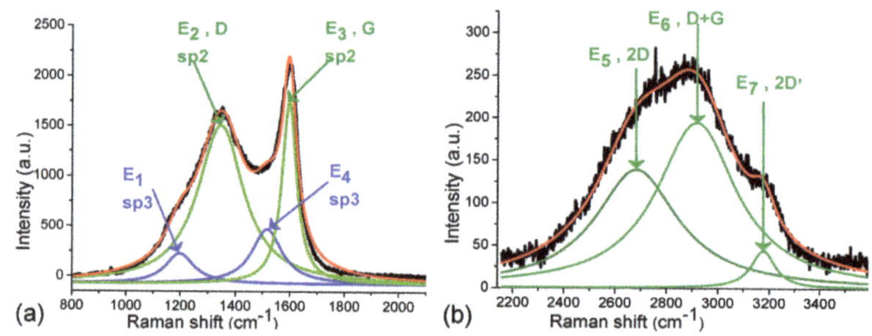

Figure 4. Sample N1 Raman spectra at 532 nm in the range 800–2100 cm^{-1} (**a**) and second ordered RS in the range 1900–3700 cm^{-1} (**b**). Decomposition into the Lorentz components with energies in their maxima E1–E7, given in Table 3, is shown.

The RS spectrum analysis is performed by the component separation method using a Lorentz line shape [105,106]. To improve reliability, two close excitation lines 532 nm and 633 nm are also used, since it is expected that the main conclusions from the results of comparing the line amplitudes should coincide in two measurement series. Table 3 includes all decomposition parameters useful for comparison between this and other studies, especially the second order linewidths.

According to [58–60], two lines at 1518 cm^{-1} and 1201 cm^{-1} have already been observed in disordered carbon black and diamond-like carbons. That could imply that the short-range vibrations of the sp3-coordinated carbons contribute to the disordered spectra. Unique sp2/sp3 nanohybrids as bulky nanodiamonds (NDs) and sp2 concentric onion-like carbons (OLC) [61,62] with outstanding mechanical performance, sufficient chemical inertness, excellent biocompatibility, high mechanical strength [61–63] are possible.

To interpret the lines in Figure 4, we use the sequence proposed in [64]. The spectra show lines arising from light scattering in spectral regions close to their position in the spectra of multilayer graphene or graphene-like layers: lines D (E2), G (E4), 2D (E5), D + G (E6), 2D′ (E7) [64–66]. Line G in the spectrum of graphene corresponds to nonresonant light scattering involving an optical phonon of E2g symmetry with a small wave vector. This phonon is caused by vibrations of carbon atoms in the layer plane. The appearance of the D line in the spectra is explained by resonant scattering involving electronic states from two nonequivalent K- and K′-points of the Brillouin zone and an optical phonon with a large wave vector. This process is forbidden by the quasi-momentum selection rule, but the condition for its conservation can be satisfied if the crystal lattice defect also participates in the scattering process. In structurally perfect graphene samples, line D should not be observed. The second order spectra in Figure 4b are markedly broadened. The nature of the 2D line, an overtone of the D line, is also associated with resonant light scattering involving electronic states. The quasi-momentum conservation conditions for such a process are always satisfied, so the 2D line will be present in the graphene spectrum even if it does not contain the D line. The combination D + G corresponds to a defect-induced double resonance "inter-valley" scattering process which is allowed through a defect-induced triple resonance process. The second order spectra lines are observed in graphene oxide, GO, [67,68], which is produced through graphite chemical oxidation and subsequent exfoliation via sonication, and in various modifications of carbon black (CB) powders [69].

It should be noted that the positions and widths of the observed second ordered three lines in the range 1900–3700 cm^{-1} correspond to those described in [70–72,77–79]. In particular, the papers [74,79] describe the dependence of the position of the 2D line, with a maximum of E5, on the magnitude of the deformation. At the same time, its width in Figure 4b is almost 10 times larger than that described in [64–66] and is comparable to its displacement at deformations of about 1%, which indicates the presence of a significant deformation disorder in the studied samples compared to multilayer graphene.

The carbon fraction diagnostics using Raman spectroscopy data is based on an analysis of the position, the width, and the intensity ratio of the observed lines. In our case, the integral intensity is obtained from the results of decomposition into components. The best known are the intensity ratios ID/IG and Isp^2/Isp^3 [64–66], in the used designations I2/I4 and (I2 + I4)/(I1 + I3), respectively. Obviously, the ratios of the total line intensities of the first and second order can also be useful, i.e., (I1 + ... I4)/(I5 + ... I7) [68]. Comparing the ratios of the line intensities, we can draw the following conclusions:

For the developed sample, the ID/IG (I2/I4) ratio was minimal compared to the industrial control samples. This indicates a smaller amount of disorder in the multilayer graphene subsystem and smaller sizes of clusters in their amorphous part [64–66].

The Isp^2/Isp^3 ((I2 + I4)/(I1 + I3)) ratio was also minimal in comparison with the control industrial samples. This means a larger role of graphite short-range sp3 bonds in the mechanically stronger amorphous part compared to the graphene multilayer lobes and the bridges between crystallites [58–60].

The ratio (I1 + ... AI)/(I5 + ... I7) was at a maximum suggesting a higher conductivity of the multilayer graphene due to a higher degree of screening of overtone photon-phonon interactions by carriers [64–66,68].

5. The Role of Ferric Compounds Studied with MS; Degradation Mechanisms of $LiFePO_4$ Test Cathodes with XRD

5.1. Mössbauer Spectroscopy Studies

The obtained Mössbauer spectra are processed by the least squares method using the Lamb–Mössbauer factors [107]. Table 4 shows the MS calculations results obtained with a high Doppler shift rate of the gamma source. Figure 5 shows the Mössbauer spectra of sample N1, and Table 4 shows the calculation results. From Figure 5 and Table 4, it can be seen that the Mössbauer spectra consist of two doublets superimposed on each other, and no additional lines indicating the presence of another phase are observed. The values of the Fe^{2+} and Fe^{3+} absorption lines relative intensities are determined from the experimental spectra; the line half-widths are given in Table 4.

Table 4. Parameters of hyperfine interactions obtained by mathematical processing of Mössbauer spectra.

Line Marking	IS, mm/s	QS, mm/s	G, mm/s	Int, (%)	Charge State Fe
Fe^{2+}	0.981 ± 0.001	2.928 ± 0.001	0.281 ± 0.001	94.8 ± 0.2	Fe^{2+}
1	1.156 ± 0.040	1.818 ± 0.028	0.345 ± 0.025	1.4. ± 0.3	Fe^{2+}
2	0.172 ± 0.080	0.830 ± 0.080	0.345 ± 0.025	2.6 ± 0.4	Fe^{3+}
3	0.404 ± 0.400	0.452 ± 0.060	0.345 ± 0.025	1.2. ± 0.3	Fe^{3+}

As can be seen in Figure 5, MS at room temperature (295 K) does not show any magnetic ordering lines traces, and the spectrum consists of quadrupole doublets. This means that the sample is in a paramagnetic state and no magnetic ordering traces or relaxation processes are observed. High intensity doublet lines are symmetrical. The spectra show a doublet with a small linewidth, maximum intensity, and hyperfine interaction (HFI) parameters: IS = 0.981(1) mm/s, QS = 2.926(2) mm/s. A doublet with similar HFI parameters IS = 1.23 mm/s and QS = 2.96 mm/s is observed for iron ions in the olivine structure, which corresponds to the high-spin iron Fe^{2+} in an octahedral M2 environment [48,80,108]. In the case of $LiFePO_4$ with the olivine structure, a doublet with IS = 1.22 mm/s and QS = 2.80 mm/s, which is attributed to Fe^{2+} ions, is also observed [45,46]. However, the IS values obtained from experimental MS (Figure 5) are somewhat lower (0.981 mm/s). The small linewidth of the dominant doublet (0.281 mm/s) means that the Fe^{2+} ions occupy positions in the well-ordered $LiFePO_4$ phase structure. The IS and QS values are close to those obtained using the density functional theory (DFT) calculations, taking into account both the spin polarization and the correlation of Fe 3d electrons [109]. This can be explained by the high-spin configuration of Fe^{2+} ions in a distorted octahedral environment formed by oxygen ions.

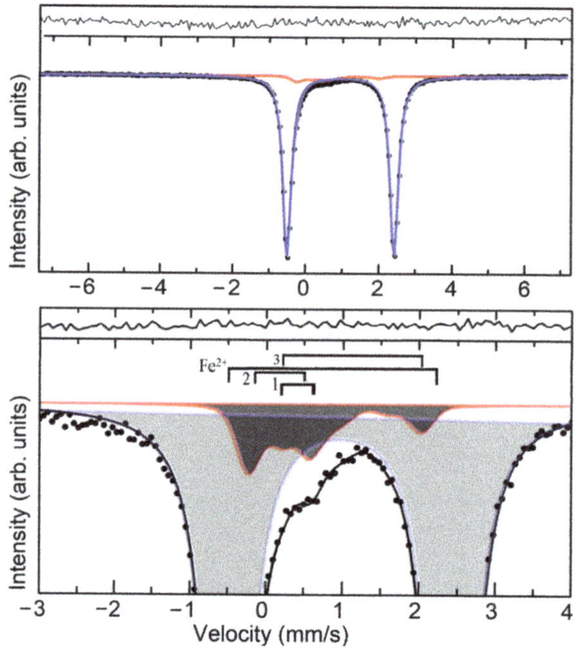

Figure 5. Experimental MS of the developed $LiFePO_4$ (N1) powder. Upper and lower parts—spectra are recorded with high and low speed of Doppler shift of the gamma-ray source, respectively. The best fitting results of the model spectrum are shown as a solid line.

In addition to the dominant doublet on the MS (Figure 5), in the range of velocities from −0.25 to +0.7 mm/s, low intensity broad lines (G = 0.565 mm/s) were also observed. The nature of these lines formation is a subject of discussion in the literature [110–113]. This small contribution is often ascribed to a lithium-deficient phase in $Li_{1-x}Fe^{2+}{}_{1-x}Fe^{3+}{}_xPO_4$ or to partial reduction of $LiFePO_4$ in an Ar/H_2 atmosphere, leading to the formation of amorphous impurity phases such as $FePO_4$ and/or Fe_2P obtained by high-temperature annealing in a partially reducing atmosphere of Ar/H_2 [111,112]. Based on the results of [113], it can be argued that the Fe^{2+} and No. 3 lines observed on the $LiFePO_4$ MS (Figure 5) with the HFI parameters IS = 0.981 (1) and QS =2.928 (1) mm/s IS= 1.156 (0.04) mm/s and QS = 1.818 (0.028) mm/s, belong to $LiFePO_4$ of the olivine type and $FePO_4$ in the amorphous state, respectively. Comparison of the [114–116] results indicates the absence of any ferromagnetic or ferrimagnetic impurities, such as Fe_2O_3, in the samples, at least in the samples with carbon-coated crystallites. The similarity of valence and local coordination states of Fe^{2+} ions in glasses and in $LiFePO_4$ crystals may be the reasons for the easy formation of $LiFePO_4$ crystals during the crystallization of lithium iron phosphate glasses (see [117] and references therein). Therefore, the following mechanism of crystallization in glasses was considered (see [118] and references therein): first, $LiFePO_4$ crystals are formed in glasses with a high content of Fe^{2+} ions, after which $Li_3Fe_2(PO_4)_3$ crystals appear in the remaining glass phase enriched with Fe^{3+} ions.

Thus, the following conclusions can be drawn:

In the initial equilibrium compositions of $LiFePO_4$, the Fe^{3+} content in the samples is much higher than the value that the electrochemical decrease in the Li content could provide. In the developed sample, it was at least 6–8%. The amount of Fe^{3+} in the literature varies from 2% to 30% [18,51,54–56].

A ferric compound, most likely in the form of $Li_3Fe_2(PO_4)_3$, is found in TEM images (Figure 3) in the form of nanocrystallites on the surface of $LiFePO_4$ particles in both sets of—those synthesized and industrial ones. This agrees with the results obtained in [52,53,57].

For high cyclability and low sample resistance, it is necessary to have an optimal amount of Fe^{3+} ferric compounds, which appear as by-products of $LiFePO_4$ synthesis. As can be seen from Table 2, the Fe^{3+} content of about 2% will be insufficient. The values of 5–8% will be optimal for the capacity value [18,56,119] and cycling; significantly larger contents will be excessive [18,54–56] due to comparability of Fe^{3+} concentration with the total content of Fe ions.

These results, together with the results of the RS studies, demonstrate that the surface of the developed powder is a high-quality, low-resistance and mechanically strong ferric-graphite-graphene composite with inclusions of the $Li_3Fe_2(PO_4)_3$ ferric (Fe^{3+}) compound with crystallite sizes < 10 nm, which increases cyclability compared to industrial cathodes.

5.2. *Ageing Mechanisms of $LiFePO_4$ and Test Cathode Study Using XRD and EDX*

Degradation of $LiFePO_4$ crystallites is only a part of the degradation of the cathode and of the battery as a whole [120–123]. The batteries were studied in the post-mortem state [120]; the possibility of using characterization techniques with spatial resolution from Å to mm-cm was discussed in [121]. The effect of stress factors (time, temperature and state-of-charge) on battery degradation during long-term testing up to 44 months was shown in [122]; the degradation mechanisms were classified into three levels—atomic, interface and electrode scale [123]. In [124] a review of manufacturer-provided characteristics of Li-ion batteries was made. A Radon–Nikodym based approach, where probability density is built first and then used to average observable dynamic characteristic was developed and applied to determination of relaxation rate distribution from experimental measurements.

Degradation of electrode powders can include deterioration of the conductive carbon network near the interfaces [125] and its amorphization [126], appearance of cracks in crystallites [125] and amorphization of their surface [127], impurity atoms introduction into the working crystallites [128]. To reduce the Fe diffusion into the electrolyte, nano-carbon coatings are used [129]. A pyrrole (PPy) coating suppresses Fe dissolution and allows for extended retention of the olivine structure [127]. Modification of carbon by using ZnO [130], and Poly(styrene sulfonic acid) membranes by polymerization of aniline improves the coating and reduces its resistance [131,132].

Separately, degradation of crystallites may result from chemical and mechanical attacks by stress–corrosion and erosion–corrosion [133]. In the case of acids present in the electrolyte, such as HF [134], impurities catalyze these attacks: iron-rich phases have a lower corrosion potential relative to $LiFePO_4$, and phosphorus-rich impurity has a higher value [135]. Corrosion proceeds especially actively at the points of concentration of mechanical stresses, in places where cracks appear on the surface of crystallites [133]. Another example of such attacks is the formation of amorphous layers of $LiFePO_4(OH)$ on the surface of crystallites when powders are kept in a humid atmosphere, or due to the moisture and OH groups residual presence in batteries [136].

Thus, based on the literature analysis, we can conclude that, in general, the growth mechanism and $LiFePO_4$ crystallites degradation is a complex chemical and electrochemical process. To describe the first irreversible phase of high-temperature degradation, their explanation is combined corrosion, stress– or erosion–induced. A significant part of the degradation in temperature ranges from growth to 100 °C can be described by the Avrami–Erofe'ev reversible mechanisms [137], Ostwald ripening reaction [138] and Ostwald's rule of stages [139], provided that the cathode powder volume is preserved, for example, by excluding its components' diffusion into the electrolyte.

5.3. *Test Cathode Aging Study Using XRD and EDX*

XRD Bruker D8 Discover diffractometer was used to determine also the unit cell volume, *V* (Table 5 and Figure S2).

Table 5. Results of degradation of the test $LiFePO_4$ cathodes after 100-fold cycling at 1C discharge rate; the extrapolated prediction of their cycling was given above in Table 2. ΔQ—capacity reduction, Fe atomic % obtained using EDX (Section S1), cell volumes before and after cycling, Vb-Va—their changes.

	ΔQ, mAh/g	EDX, Fe, %		$\bar{L}_{VXRD}$, nm		$V = a \times b \times c$, nm^3		*Vb-Va*, nm^3
		Before	After	Before	After	Before	After	Before–After
N1	0.7	13.7	5.47	190 (20)	160 (50)	291.183	291.37	−0.09
N2	4.0	6.28	2.98	220 (50)	210 (50)	291.376	290.069	1.303
N3	0.4	7.45	2.92	119 (11)	164 (17)	290.723	290.633	0.09
N4	6.2	11.17	12.5	800 (200)	150 (20)	291.184	290.88	0.304

Table 5 lists the results which correspond to the first aging phase characterized with a partial destruction (or stabilization) of the protective ferric-graphite-graphene layer on the surface of the crystallites. As can be seen from Figure S1, significant deviations from stoichiometry within the $LiFePO_4$ olivine structure are possible when the equality $2\ V_{Li} \approx Fe_{Li}$ is satisfied. With these deviations, the cell volume increases and, according to [54,140], for every two lithium vacancies $2\ V_{Li}$, a Fe_{Li} defect arises (iron in lithium position).

Based on Table 5 and Figures S1 and S2 the following conclusions can be drawn:

According to EDX measurements, during the first aging phase, the Fe content in the N1-3 test cathodes decreases mainly due to its diffusion from the intercrystallite space into the electrolyte; an increase in the Fe content in the lowest quality sample N4 indicates the beginning of its crystallite destruction.

For the samples N1–4, a decrease in the size of crystallites is observed, and for the samples N2–4, a decrease in the volume of unit cells Vb-Va is observed, which is proportional to a decrease in the number of cycles. The latter also means a decrease in the Fe_{Li} and V_{Li} defect concentrations, i.e., the crystallites approach the stoichiometric composition. However, an increase in deviation from stoichiometry is observed for the developed sample N1 in the first phase of degradation.

In the experiments with incompletely discharged test cathodes, an increase in crystallite size up to 60 nm is observed (significantly smaller than in the initial powder) as the concentration of the $FePO_4$ phase increases (see Figure S2). In this case, the unit cell dimensions of the remaining crystallites decrease, i.e., they approach the stoichiometric composition. This means that smaller crystallites have a greater deviation from stoichiometry, by 5% V_{Li} relative to the average value. Note that the XRD measurement of the discharged test cathodes is lengthy; therefore, the process of relaxation of partially discharged crystallites already ends as a result of the Li redistribution relaxation between crystallites in these samples [141,142].

Thus, the $LiFePO_4$ cathode degradation occurs in two stages: at the first stage, the layer on the surface of the crystallites is destroyed; at the second stage, Fe escapes into the electrolyte and onto the anode with a decrease in the size of the crystallites due to thickening of the amorphized near-surface layer. In Figure 6, this two-stage process scheme is shown; it is close to that previously proposed for describing the Ostwald ripening reaction and Ostwald's rule of $LiFePO_4$ crystallization stages [137–139]. According to these works, at high temperatures, the growth part consists of Ostwald ripening stages from the metastable state of the feedstock with free energy $\Delta G > 0$ up to the equilibrium state with zero energies. The presence of a local electrochemical potential must also be included in the height of the barriers.

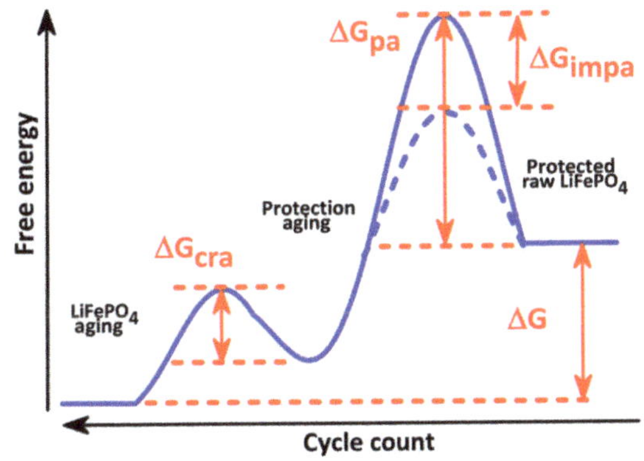

Figure 6. Scheme of two-stage $LiFePO_4$ cathode degradation. Here, the free energy in the metastable state is ΔG at the battery operating temperatures. With an increased number of cycles, the protective ferric-graphite-graphene layer is destroyed with the corrosion activation energy ΔG_{pa}; a decrease in this activation energy by ΔG_{impa} occurs in the presence of impurities that catalyze the corrosion processes. The second stage is degradation with activation energy of ΔG_{cra} of unprotected crystallite destruction into an amorphous phase.

6. The Galvanostatic Measurements of $LiFePO_4$ Test Cathodes

The main target quantitative parameters of the electrodes are: rate capability $Q(t)$ and capacity Q_0, limit value at charging time $t \rightarrow \infty$. These parameters are actively used in the development of electrodes [143,144], batteries [145,146] and supercapacitors [147,148] to assess the quality of crystallites [149] and their carbon coatings [150,151] in studies of degradation of powders and batteries in general [120–123]. In addition to these parameters, three characteristic discharge/charge times associated with RC electrical (τ_{el}), diffusion relaxation (τ_d) and electrochemical reaction at the electrode/electrolyte interface (τ_c) are also important [152]. In turn, the first two consist of 3 components each, so we have 7 characteristic times in total. It was shown in [99] that the crystallite shape engineering task aiming to optimize the rate capability and increase the cathode capacity can be divided into two subtasks: 1. Achieving a large rate capability (and capacity) at big times or increasing the rate capability at small times. 2. Decreasing characteristic discharge/charge times to increase the rate capability at small times, which can be partially solved by improving the quality of their coating.

To develop the analytical dependence $Q(t)$ and use it to describe the results of galvanostatic measurements of test cathodes, it is necessary to establish a hierarchy among these 7 relaxation times of electrochemical charge exchange. To do this, it is necessary to determine the value of the specific interfacial area in the electrodes [153], taking into account the anisotropic size distribution of crystallites.

6.1. *Test Cathode Aging Study Using XRD and EDX*

Let us define the term "specific interfacial area" as $a_s = S/V$, where S is the total area of projections of the test cathode crystallites onto the (010) plane, and V is their total volume. It was shown in [98,99] that the combined use of the results of TEM and XRD measurements makes it possible to determine these parameters using the anisotropic size distribution of $LIFePO_4$ powder crystallites, which is described by a 3-dimensional lognormal function:

$$f(\overline{L}) = \frac{1}{L_1 L_2 L_3 \sqrt{(2\pi)^3 \det K}} \exp\left[-\frac{1}{2}\left(\ln\overline{L} - \ln\overline{\overline{L}}\right)^T \overline{\overline{\Lambda}}^{-1}\left(\ln\overline{L} - \ln\overline{\overline{L}}\right)\right], \quad (1)$$

where $\overline{L} = \begin{bmatrix} L_1 \\ L_2 \\ L_3 \end{bmatrix}$—crystallite sizes, $\overline{\overline{L}} = \begin{bmatrix} \overline{L_1} \\ \overline{L_2} \\ \overline{L_3} \end{bmatrix}$—their means, $\overline{\Lambda} = \begin{bmatrix} \sigma_1^2 & r_{12}\sigma_1\sigma_2 & r_{13}\sigma_1\sigma_3 \\ r_{21}\sigma_2\sigma_1 & \sigma_2^2 & r_{23}\sigma_2\sigma_3 \\ r_{31}\sigma_3\sigma_1 & r_{32}\sigma_3\sigma_2 & \sigma_3^2 \end{bmatrix}$—correlation moment matrix, and the product, $Cov_{ik} = r_{ik}\sigma_i\sigma_k$—covariances, r_{ik}—correlation coefficients between the i-th and k-th anisotropic distributions with possible values from 0 to 1 [154], excluding negative values. Table 6 shows the parameters of the developed samples N1 and N2, described earlier in [99].

Table 6. Parameters of N1 and N2 samples. Column 1—parameters of crystallites, average size $\overline{L}_1$ (nm) and variance σ_1 of Lognormal distribution along the [010] axis, etc. and columns 2, 3—parameters for [100], [001] axes. Columns 4–6 are the correlation coefficients, while 7 are their average values. Column 8 shows the total area S of the cross sections of crystallites on the (010) plane, 9 shows the diffusion coefficients, 10 is the electrical relaxation time (see Section 6.2 below).

	1	2	3	4	5	6	7	8	9	10
	$\overline{L}_1, \sigma_1$	$\overline{L}_2, \sigma_2$	$\overline{L}_3, \sigma_3$	r_{12}	r_{13}	r_{23}	$\overline{r}$	a_s, m^2	D, nm^2/s	τ_{el}, s
N1	60, 0.41	49, 0.40	72, 0.38	0.87	0.64	0.73	0.75	2.0×10^7	0.16 (0.4)	8
N2	92, 0.43	108, 0.41	160, 0.35	0.72	0.56	0.53	0.60	3.1×10^7	0.3–2.1 (0.4)	20

The calculation of a_s is performed through the following steps:

- using weight (0.015 g) of the initial amount in the $LiFePO_4$ powder sample and its pycnometric density (3.6 g/cm^3), the total volume of all crystallites in the cathode is calculated (V = 4.2×10^{18} nm^3),
- the average crystallite volume is calculated (Mathematica 12 notation):

$$\overline{v_{pr}} = \textbf{Total}\left[\left(\overline{f} \circ \overline{v}\right), 3\right], \quad (2)$$

where $\overline{v}$—3-dimensional N-bit matrix of particle volumes, each element of which for ellipsoid particles has a volume $\frac{\pi}{6} L_{in} * L_{jn} * L_{kn}$. Index n runs over values from 1 to N, while L_{in}, L_{jn} and L_{kn} are the sizes of crystallites along the [010], [100] and [001] axes, respectively; $\overline{f}$ is the discretization of function (1) normalized to 1 in the form of a 3-dimensional N-bit matrix, each element of which means the probability of occurrence of the crystallites with the corresponding sizes. The **Total** operator means the matrix elements product and all products summation:

- dividing V by $\overline{v_{pr}}$ we obtain the number of particles N_{ct} in the cathode and the S value by calculating an equation similar to (2). Instead of $\overline{v}$ it uses the matrix $\overline{s}$—particle area projections onto the (010) plane, and the program line is as follows:

$$S = N_{ct}\, \textbf{Total}\left[\left(\overline{f} \circ \overline{s}\right), 3\right], \quad (3)$$

- as a result, for the developed powder, we obtain the number of crystallites in the cathode $N_{ct} = 4.1 \times 10^{12}$; the total areas $S = 8.3 \times 10^{16}$ nm^2 and the specific interfacial area $a_s \approx 2 \times 10^7$ m^{-1} can be calculated. The results for 2 samples are shown in Table 6.

Since we have limited ourselves to powder improvement technology, electrochemical tests were carried out by fabricating thin test cathodes using a three-electrode cell [99]. In this case, only 3 out of 7 characteristic times will remain: diffusion relaxation τ_d along the crystallite [010] axis columns, RC electric τ_{el} associated with the coating of crystallites, and the response time to the electrode/electrolyte interface t_c. To estimate the latter, we use the value of specific interfacial area a_s obtained above, as well as the analyses given in [152–154], quantitative calculations and the following expression:

$$t_c = \frac{F\varepsilon c_e}{(1 - t_+^0)\left|a_s i_0 exp\left(\frac{\alpha_c F}{RT}\eta\right)\right|}, \tag{4}$$

where Faraday's constant F = 96,487 C·mol^{-1}, porosity ε = 0.3, electrolyte concentration c_e = 1000 mol·m^{-3}, transference number t_+^0 = 0.4, cathodic transfer coefficient α_c = 0.5 taken from [155–157], and the values $a_s = 2 \times 10^7$ m^{-1}, surface overpotential η = 0.1 V and reference exchange current density = 1.5 A/m^2 are obtained from the developed sample measurements. Substituting numerical values into (4), we obtain t_c = 0.5 s. Thus, comparing the value of t_c with those obtained in [155], we are convinced that in our sample with 8 µm cathode thickness, its value is indeed the minimum in the hierarchy of electrochemical charge exchange relaxation times, which allows simplifying the $Q(t)$ analytical dependence, leaving only the parameters of $f(\overline{L})$ and τ_d, τ_{el}.

6.2. *Q(t) Dependence on Crystallite Parameters, Lithium Diffusion Coefficient D along [010] and the Quality of Their Coating (Electrical Relaxation Time τ_{el})*

To develop an analytical model $Q(t)$, we make the following assumptions:

1. In [152,158], for some current source, for which $Q(t)$ asymptotically approaches the limit value Q_M at $t\to\infty$, and at $t\to 0$ it approaches the dependence $\frac{Q_M}{2}\left(\frac{\tau}{t}\right)^{-n}$, the following empirical equation was proposed:

$$Q(t) = Q_M\left[1 - \left(\frac{\tau}{t}\right)^n\left(1 - e^{-\left(\frac{\tau}{t}\right)^{-n}}\right)\right], \tag{5}$$

   where τ is the time constant, and the exponent values n are defined in [152] for batteries and supercapacitors, as 0.5 and 1.0, respectively. That is, the exponent n is equal to the slope tangent of the dependence $Q(t)$ in double logarithmic coordinates at small t. Equation (5) can be interpreted as follows: over time, $Q(t)$ reaches its limit value Q_M with probability $[1 - P]$, where P is equal to the subtract in the square bracket of Equation (5) and has the meaning of the process probability not being implemented due to the limited rate [159].
2. According to [99], the crystallite is divided into columns with a cross-sectional area $dx_3 * dx_2$ along the Li diffusion direction—axis [010], along which the coordinate axis x_1 is directed. The crystallite rate capability $q_{cr}(t, L_1, L_2, L_3)$ is determined by integrating over the plane (010) the rate capability $q_{se}(t)$ of length M column:

$$q_{cr}(t,\ L_1, L_2, L_3) = \int_{-\frac{L_3}{2}}^{\frac{L_3}{2}} dx_3 \int_{-\frac{L_2}{2}}^{\frac{L_2}{2}} q_{se}(t)dx_2, \tag{6}$$

   where L_1, L_2 and L_3 are the crystallite dimensions along the [010], [100] and [001] axes, respectively.
3. Sequential charge carriers flow in a crystallite column through a capacitor (the model of a dense electric double layer) and an element with distributed parameters (the model of Warburg element diffusion of the stage limiting the rate of the Faradaic process). The model can be considered similar to the electrical circuit in which the capacitor and the Warburg element are series-connected [42,160].
4. The chain Figure 7 corresponds to the probabilistic equation of the sequence of events [154]:

$$q_{se}(t) = q_M(1 - P_C)(1 - P_W) = q_M\left[1 - \left(\frac{\tau_{el}}{t}\right)^1\left(1 - e^{-\left(\frac{\tau_{el}}{t}\right)^{-1}}\right)\right]\left[1 - \left(\frac{\tau_d}{t}\right)^{0.5}\left(1 - e^{-\left(\frac{\tau_d}{t}\right)^{-0.5}}\right)\right], \tag{7}$$

in which the rate capability $q_{se}(t)$ is not realized with probability P_C and P_W in time t. The dependence $\tau_d = \frac{M^2}{\pi^2 D}$ is used, where D = Const, which describes the process of diffusion (desorption) from a finite size M with associated boundary conditions [161].

5. Next, similarly to Equations (2) and (3), the desired dependences are calculated:

$$Q(t, D, \tau_{el}) = Total\left[\left(\overline{f} \circ \overline{q_{cr}}\right), 3\right], \tag{8}$$

6. Figure 8 shows the fitting of the dependence of the calculated sum results (7) on the discharge time with reference to the experimental normalization value of the rate capability at t_{nr}= 80 s (10 C rate), which is intermediate between the dominant contributions.

Figure 7. Cathode equivalent circuit of a crystallite column consisting of a capacitor C1 and Warburg element W1 series connection.

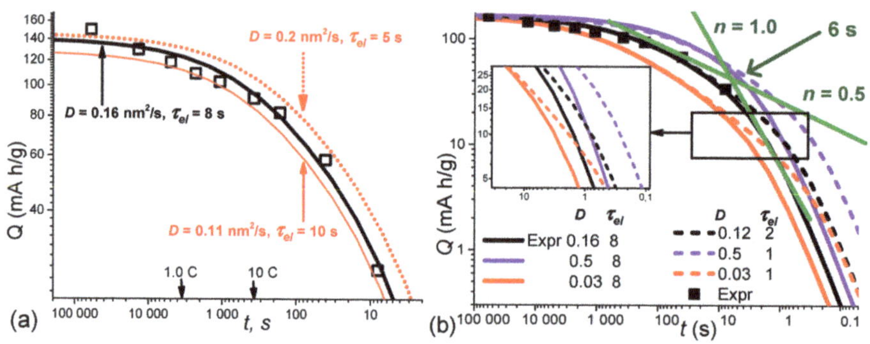

Figure 8. (**a**) Theoretical $Q(t, D, \tau_{el})$ dependence, Equation (8), on the discharge time for the developed powder with the steps of fitting the most optimal black curve to the experimental points. (**b**) The same optimal curve and experimental points on large scales along the axes and large values of D and τ_{el} to demonstrate tilt angles. Straight lines with slopes n corresponding to the Warburg element and capacitor are shown, with the intersection at the point t = 6 s close to the obtained value τ_{el} = 8 s.

The theoretical dependence $Q(t)$ can use two relaxation times τ_d and τ_{el}, which are the most important in the hierarchy of relaxation times. The procedure for fitting $Q(t, D, \tau_{el})$, expression (7), to the experimental $Q(t)$ includes the use of the distribution parameters of anisotropic crystallite sizes, as well as the normalization value $Q(t_{nr})$ at some t_{nr}. For the obtained value $\tau_{el} = 8$ s, the above estimate of the need to fulfill the inequality $\tau_{el} < t_c = 0.5$ s is performed with a large margin. As can be seen from Tables 2 and 6, the crystallite average sizes along the [010] axis of the developed sample N 1 are reduced by 2–3 times relative to the rest. At the same time, it is significant that the crystallite average volume is even more reduced, in particular, by a factor of 30 compared to sample N4. This fact also indicates the high quality of the developed powder, since a decrease in the crystallite volume should obviously reduce the crystallite lifetime (cycling).

6.3. Q(t) Calculation in the Ranges of D, τ_{el} u $\overline{r}$, Close to the Values of the Developed $LiFePO_4$ Powder to Assess the Possibility of Improving Technologies

As can be seen from Figure 8b, a 25% decrease in D and a decrease in τ_{el} by a factor of 4 have practically no effect on the value of $Q(t)$ in the practically important range of discharge rates up to 50 C rate; the latter corresponds to the rightmost experimental point. From Figure 8b, it is also seen (purple solid and dotted curves), that an obvious way to improve the powders would be technologies aimed at increasing D by a factor of three and reducing τ_{el} to extremely small values equal to $t_c = 0.5$ s.

In [99 + SI] a significant number of different calculations are presented in wide ranges of anisotropic distribution parameters of crystallite sizes, which are in agreement with known experimental results. Below, the dependence of $Q(t)$ on the values of the correlation coefficients of Equation (1) (the correlation between the crystallite linear sizes) will be described.

It should be noted that experimental determination of the correlation coefficients is quite reliable for crystallites with plate- and bar-like shapes, using SEM or TEM microscopy. The largest face of the crystallites is located in the object microscope plane and, in principle, no XRD measurement is required. However, there frequently are more complicated situations, with less anisotropy (<5-fold). Here, only complementary XRD and TEM measurements are possible [98]. The essence of the technique is that the difference between the column averaged $\overline{L}_{i\ XRD}$ and the volume averaged $\overline{L}_{i\ TEM}$ can be related to the average sizes of real measurements $\overline{L}_i$ for log-normal distributions. In this case, the TEM measurement results produce a correlation cloud between the longitudinal and transverse sizes of crystallites and the corresponding marginal distribution functions [159]. Figure 9a shows such a cloud obtained by digitizing TEM images of the developed sample. Luckily, they are described by a lognormal function (normal in the ln coordinate). This allows us to divide them into 3 components along the crystallographic axes and then, using a fitting procedure, find the correlators r_{ik} which are also shown in Table 6. The procedure is described in [99] and implies using the correlation between the longitudinal and transverse particle sizes (see Figure 9a) r_{bs} as a trial to obtain r_{ik}. The program fits the row-by-row sum of the 3-D matrix $\overline{f}$ to the corresponding 3-D marginal size distributions along the crystallographic axes. Figure 9a shows the ordering of points, which is due to the discreteness of the particle size digitization, as well as the coincidence of some particles in sizes.

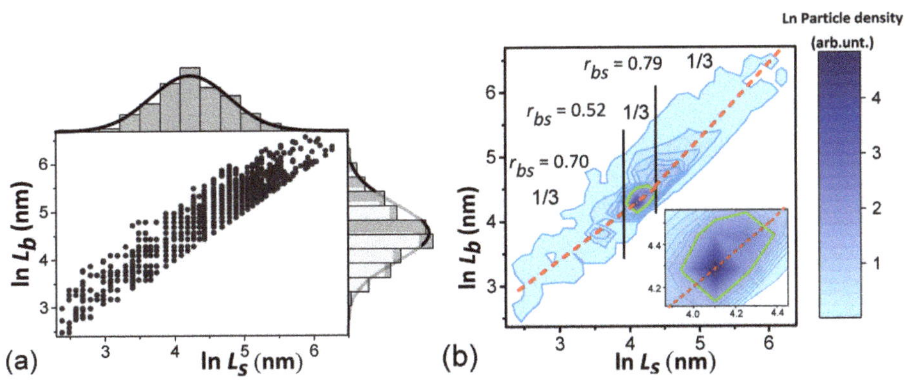

Figure 9. (**a**) Correlation cloud of transverse and longitudinal sizes L_s and L_b of $LiFePO_4$ for 4326 particles and the corresponding marginal functions of their distributions fitted with Lognormal functions. (**b**) Particle-density isolines map divided into parts by vertical straight lines with small, intermediate and large particles 1/3 of their total; red dotted lines—smoothed average values. The correlation coefficients of these parts are indicated; the inset shows an enlarged part near the particle density maximum.

Figure 9b shows a crystallite density distribution contour map. To take the correlations into account, the cloud can be discretized into sections and correlations calculated for each section. An example for 3 sections is shown in Figure 9b. However, to reduce the error in determining r_{ik}, it is necessary to increase the number of digitized particles, which will significantly increase the processing time, considering the overlap of the particles in TEM images [98]. On the other hand, the difference between the r_{ik} values the studied samples are relatively small, no more than 3 tenths; therefore, to determine the dependence of $Q(t)$ on this parameter, we use the $\bar{r}$ average value in a wider range from 0 to 1. The calculation results shown in Figure 10 are carried out using the average value $\bar{r}$ indicated in Table 6.

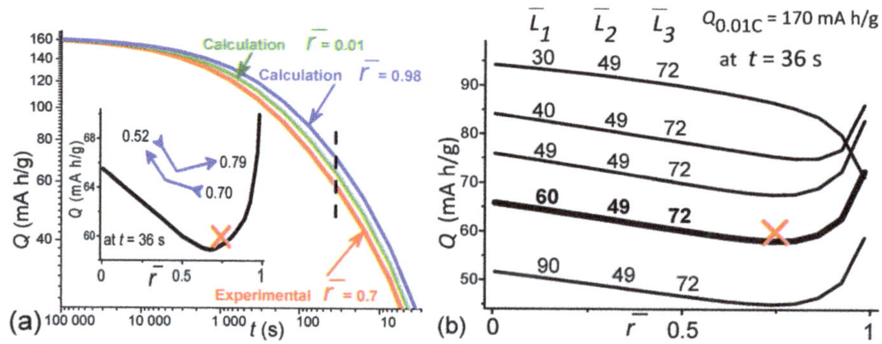

Figure 10. (**a**) $Q(t)$ dependences on the mean values of the correlation coefficients $\bar{r}$ In the inset, $Q(t)$ dependence on $\bar{r}$ at $t = 36$ s (marked by a vertical dotted line); the red cross marks the sample N1 $\bar{r}$ value. A scheme of $\bar{r}$ changes during measurements from small to large t is shown. (**b**) $Q(t)$ dependences on $\bar{r}$ normalized to $Q_{0.01}$ at $t = 36$ s for the indicated average crystallite sizes; the thick curve and the red cross are for N1 sample.

As can be seen from Figure 10b, the dependence of $Q(t)$ on $\bar{r}$ in the range of short discharge times (high power) is nonmonotonic for the developed sample. Note that it satisfies the inequality $\bar{L}_1 > \bar{L}_2$, and $\bar{L}_1 \sim \bar{L}_3$. To increase $Q(t)$ in the region of small t, when these relations are satisfied, it is necessary to increase $\bar{r}$, in other words, to reduce the cloud width in Figure 9b. As can be seen in the inset to Figure 10a, this can be expected to be quite realistic, since the $\bar{r}$ achieved value is on the threshold of its sharp rise. However, when developing a powder with particle sizes $\bar{L}_1 < \bar{L}_2$, $\bar{L}_3$, the $Q(t)$ dependence decreases as $\bar{r}$ decreases; i.e., it is necessary to strive for an increase in the cloud width.

To discuss the mechanisms of the discovered $Q(t)$ dependence on $\bar{r}$, it is necessary to pay attention to the nonmonotonic change in the r_{bs} value with increasing particle sizes observed in Figure 9b. The inset to Figure 10a shows a scheme according to which Li leaves small particles with $r_{bs} = 0.70$ at small discharge t (high currents), and then, as the time t increases, Li additionally leaves medium-sized particles with $r_{bs} = 0.52$, and finally, Li leaves large particles with $r_{bs} = 0.79$. Since $Q(t)$ is strongly dependent on the particle size, all these factors together may possibly lead to the observed dependence of $Q(t)$ on $\bar{r}$.

Thus, as shown in Figure 8b, to improve the developed powder technology, it is necessary to increase the lithium diffusion coefficient D in $LiFePO_4$ crystallites by up to three times. In this case, it is necessary to improve their coatings quality from reducing the electrical relaxation time τ_{el} to the electrochemical reaction duration at the electrode/electrolyte interface $\tau_c = 0.5$ s. For an additional increase in $Q(t)$ (and power) in the region of short recharge times, it is necessary to optimize the correlation coefficients between the crystallites anisotropic sizes. They can be controlled by the cloud width of the correlations between the transverse and longitudinal dimensions of TEM particle images (see Figure 9b).

7. Conclusions

1. One-pot synthesis has been developed for $LiFePO_4$ powders with impurity phase content of less than 0.1%, with 2–3 times smaller crystallites along the [010] axis, with 2–3 times greater cycling compared to the industrial samples, and with the particles covered by a mechanically strong, low-resistance ferric-graphite-graphene composite protective layer with inclusions of ferric (Fe^{3+}) compound particles 5–10 nm

in size. The ordered carbon shell thickness reaches 5 nm, and the amorphous shell is up to 20 nm.

2. To detect impurity crystallite phases, SXRD was used, since conventional XRD is less sensitive due to the lower intensity of laboratory sources compared to the synchrotron.
3. Control of adipic acid and polyvinyl alcohol concentrations and use of multistage annealing modes makes it possible to control the coating quality. The composite layer improves cyclability compared to industrial cathodes.
4. The role of ferric Fe^{3+} compounds:
    - the content of ferric Fe^{3+} compounds is much higher (at least 6–8%) than the value expected from the electrochemical decrease in the Li content. The amount of Fe^{3+} reported in the literature varies from 2% to 30% (Table 7).
    - in a controlled way, Fe^{3+} compounds can be formed on the surface when the volatile components are not completely removed during $LiFePO_4$ synthesis from an intermediate low-temperature amorphous phase.
    - To obtain highly cyclable and low resistant samples, it is necessary to have some optimal amount of the Fe^{3+} ferric compounds, which appear as $LiFePO_4$ synthesis by-products. EDX studies of the tested cathode show that the total number of Fe atoms is reduced compared to the original samples. We have not detected Fe_2O_3, but it was observed in other technologies.
    - According to the corrosion degradation model, an increase in the cycle number leads to a decrease in the ferric Fe^{3+} compounds content on the surface of crystallites. These compounds play a certain sacrificial role [162,163], disappearing as the cathode resource is exhausted, and impurity phases can play the role of a catalyst for this breakdown. However, some of them, such as iron phosphide, weaken the catalytic activity. The degradation occurs in two stages: at the first stage, the layer on the crystallite surface is destroyed, and at the second stage, Fe escapes into the electrolyte and onto the anode with a decrease in the crystallite size due to increasing amorphization of the near-surface layer of crystallites.

Table 7. Parameters of LiFePO4 powders: developed, studied industrial and described in the references, as well as comments on the comparison procedure.

	Developed	Industrial	From References
Growth T, °C	Two steps 400 °C, 670 °C	unknown	[1] 400–800; [3] 700; [5,18] 650; [6] 550; [8] 810; [19,22] 650–700; [24] 670; [33] 550–800; [37] 550; [49,55] 700; [51] 600.
Technology	one-pot liquid-phase	unknown	[143] More than 10 types, main: Solvothermal, Hydrothermal, Stripping synthesis, Sol-gel.
Protected layer	ferric-graphite-graphene	mostly ferric-graphite	[143] More than 30 types, main: [14] different carbon; [15] nanocarbons; [16–18] graphene; [19–21] sucrose; [22,23] glucose; [24–26] adipic acid; [27,28] polyvinyl alcohol; [29,30] polymeric additive; [31] ferrocene.
Q, mAh/g	0.1 C 15110 C 82	128–163 72–80	[151] commercial 121–160, best MWCNT, essentially mixed; [143] capacity growth of commercial powder from 160 to 208 mAh/g, more than theoretical 170 mAh/g (*)
Particle sizes, nm	$\overline{L}_{V[100]}$ 66 $\overline{L}_{V[010]}$ 82 $\overline{L}_{V[001]}$ 89	141–230 131–261 165–242	[1] 300–7000; [22] 240–3000; [90] 180–300; [118] 500–30,200; [136] 120; [138] 95–280; [141] 60–1000; [144] 40–500; [149] 20–140; [151] 90–300; [164] 30–158. (**)
Cell volumes, ^{0}A			[22] 290.63–290.94; [36] 291.08–292.07; [88] 289.7–291.2; [90] 291.3–292.3; [114] 291.33–291.63; [118] 289.8–291.9.

Table 7. *Cont.*

	Developed	Industrial	From References
Cycling	3500	800–1000, 5000 with ferric impurity	[121] 50–600; [143] 50–1000 with different capacity retentions 92–100% (**)
BET, m^2/g	12.5	9.4–13.4	[1] 1–20; [6] 32–66; [16] 49.3–59.4; [19] 49.6; [121] 15.5; [132] 50; [136] 19–25; [164] 35.
D, nm^2/s	0.12	0.25–0.45	[3] 2; [6] 2–3; [11,16] 0.01–1; [12] 4.9–7.2; [16] 3.4–8.8, 1.8–1000; [22] 109; [36] 1.2–8.2; [51] 900–2400; [119] 7400–42,000; [128] 0.83–27.3; [149] 1; [151] 0.67–11.7; [165] 1–100; [166] 0.01–10. (***)
τ_{el}, s	8	3–30	[152] 3–200
t_c, s	0.5	0.5	[155] 0.1 to >100
Fe^{3+}, %	4	2–8	[36] 0.28–0.39; [51] 9–17; [52] 5.16; [54] 2–12; [56] 5–26; [111] 7; [117] 7–25; [119] 5.4–17; [140] 1.13–17

(*) The identification of commercial trends also requires an analysis of market feasibility information, or its public accumulation. An example would be NREL chart www.nrel.gov/pv/cell-efficiency.html (accessed on 14 November 2022) about laboratory solar cells and separately about industrial modules. (**) Difficult to compare because there are no universal certification requirements for measurements. (***) According to [165] using geometrical area of electrode, BET area, particle spherical surface etc. might distort the D values. The used dimension nm^2/s is more descriptive in relation to particle sizes [99].

5. Galvanostatic studies of the N1 sample test cathodes were carried out in 3 stages with an assessment of the possibility to further improve the technology.
   5.1 To develop a theoretical dependence $Q(t)$ that takes into account the 3D lognormal crystallite size distribution $f(\overline{L})$, the response time of the electrode/electrolyte interface t_c is estimated using the specific interfacial area in the electrodes $a_s = S/V$, where S is the total the projected area of the test cathode crystallites on the (010) plane, and V is their total volume. The value $t_c = 0.5$ s is obtained.
   5.2 Comparing the t_c value with those obtained in [155], it appears to be a minimal one in the hierarchy of relaxation times of electrochemical charge exchange. This makes it possible to simplify the theoretical equation for the $Q(t)$ dependence on the $f(\overline{L})$ parameters. Fitting the theoretical dependence to the experimental data gives the value of the Li diffusion coefficient, $D = 0.12\ nm^2/s$. The value of $\tau_{el} = 8$ s satisfies the inequality $\tau_{el} > t_c = 0.5$ s
   5.3 $Q(t)$ calculations in the ranges of diffusion coefficients D, electrical relaxation times τ_{el}, and correlation coefficients $\bar{r}$ close to the values characteristic of the developed $LiFePO_4$ powder show that a decrease in D by 25% and a decrease in τ_{el} by a factor of 4 has practically no effect on the $Q(t)$ value in the practically important range of discharge rates up to 50 C. Improving the powder technology should be aimed at increasing D three times and reducing τ_{el} to extremely small values closer to $t_c = 0.5$ s. For an additional increase in $Q(t)$ (and power) in the short recharge time region, it is necessary to optimize the values of the correlation coefficients between the anisotropic crystallite sizes.
6. Table 7 summarizes the obtained parameters and compares them with the known ones, taking into account comments on them.

Supplementary Materials: The following supporting information can be downloaded at: https://www.mdpi.com/article/10.3390/en16031551/s1, Figure S1: The dependence of the $LiFePO_4$ orthorhombic olivine structure unit cell volume on the values of Li deficiency and Fe excess; Figure S2: The results of XRD measurements of N1 test cathode depending on its discharge degree; Section S1: SEM and EDX studies of samples before and after test cathodes 100-fold cycling made from developed and commercial powders.

Author Contributions: D.A., powder synthesis, methodology; A.B., conceptualization, writing, software; A.K., MS investigation, analysis; A.N., SEM investigation, analysis; E.E., data curation, original draft preparation; A.U., electrochemistry investigation, analysis; I.K., XRD investigation, analysis; V.L., RS investigation, analysis; M.T., TEM investigation, analysis; E.T., data analysis, validation. All authors have read and agreed to the published version of the manuscript.

Funding: This research received no external funding.

Data Availability Statement: Not applicable.

Acknowledgments: Sincere thanks to V.G. Malyshkin for discussing the aging mechanisms. XRD studies were performed in Resource Center of St. Petersburg State University, SXRD studies—in Synchrotron radiation source of Kurchatov Institute Research Center, SEM studies—in Federal Joint Research Center "Material science and characterization in advanced technology", TEM studies—in Omsk Regional Center of Collective Usage SB RAS.

Conflicts of Interest: The authors declare no conflict of interest.

References

1. Yamada, A.; Chung, S.C.; Hinokuma, K.J. Optimized $LiFePO_4$ for Lithium Battery Cathodes. *Electrochem. Soc.* **2001**, *148*, A224–A229. [CrossRef]
2. Ong, S.P.; Wang, L.; Kang, B.; Ceder, G. Li–Fe–P–O_2 Phase Diagram from First Priciples Calculations. *Chem. Mater.* **2008**, *20*, 1798–1807. [CrossRef]
3. Bezerra, C.A.G.; Davoglio, R.A.; Biaggio, S.R.; Bocchi, N.; Rocha-Filho, R.C. High-purity $LiFePO_4$ prepared by a rapid one-step microwave-assisted hydrothermal synthesis. *J. Mater. Sci.* **2021**, *56*, 10018–10029. [CrossRef]
4. Henriksen, C.; Mathiesen, J.K.; Ravnsbæk, D.B. Improving capacity and rate capability of Li-ion cathode materials through ball milling and carbon coating—Best practice for research purposes. *Solid State Ion.* **2020**, *344*, 115–152. [CrossRef]
5. Liu, Y.; Luo, G.Y.; Gu, Y.J.; Wu, F.Z.; Mai, Y.; Dai, X.Y. Study on the Preparation of $LiFPO_4$ by Hydrothermal Method. *IOP Conf. Ser. Mater. Sci. Eng.* **2020**, *761*, 012004. [CrossRef]
6. Nan, C.; Lu, J.; Li, L.; Li, L.; Peng, Q.; Li, Y. Size and shape control of $LiFePO_4$ nanocrystals for better lithium ion battery cathode materials. *Nano Res.* **2013**, *6*, 469–477. [CrossRef]
7. Kudryavtsev, E.N.; Sibiryakov, R.V.; Agafonov, D.V.; Naraev, V.N.; Bobyl, A.V. Modification of liquid-phase synthesis of lithium-iron phosphate, a cathode material for lithium-ion battery. *Russ. J. Appl. Chem.* **2012**, *85*, 879–882. [CrossRef]
8. Janssen, Y.; Santhanagopalan, D.; Qian, D.; Chi, M.; Wang, X.; Hoffmann, C.S.; Meng, Y.; Khalifah, P.G. Reciprocal Salt Flux Growth of $LiFePO_4$ Single Crystals with Controlled Defect Concentrations. *Chem. Mater.* **2013**, *25*, 4574–4584. [CrossRef]
9. Abakumov, A.M.; Fedotov, S.S.; Antipov, E.V.; Tarascon, J.-M. Solid state chemistry for developing better metal-ion batteries. *Nat. Commun.* **2020**, *11*, 4976. [CrossRef]
10. Ming, X.-L.; Wang, R.; Li, T.; Wu, X.; Yuan, L.-J.; Zhao, Y. Preparation of Micro-Nano-Structured $FePO_4{\cdot}2H_2O$ for $LiFePO_4$ Cathode Materials by the Turbulent Flow Cycle Method. *ACS Omega* **2021**, *6*, 18957–18963. [CrossRef]
11. Jittmonkong, K.; Roddecha, S.; Sriariyanun, M. One-pot Synthesis of $LiFePO_4$ Nano-particles Entrapped in Mesoporous Melamine-Formaldehyde Matrix as the Promising Cathode Materials for the Next Generation Lithium Ion Batteries. *Mater. Today Proc.* **2019**, *17*, 1284–1292. [CrossRef]
12. Zhang, B.; Wang, S.; Liu, L.; Li, Y.; Yang, J. One-Pot Synthesis of $LiFePO_4$/N-Doped C Composite Cathodes for Li-ion Batteries. *Materials* **2022**, *15*, 4738. [CrossRef]
13. Kaurz, G.; Gates, B.D. Review–Surface Coatings for Cathodes in Lithium Ion Batteries: From Crystal Structures to Electrochemical Performance. *J. Electrochem. Soc.* **2022**, *169*, 043504. [CrossRef]
14. Khodabakhshi, S.; Fulvio, P.F.; Andreoli, E. Carbon black reborn: Structure and chemistry for renewable energy harnessing. *Carbon* **2020**, *162*, 604–649. [CrossRef]
15. Ren, W.; Wang, K.; Yang, J.; Tan, R.; Hu, J.; Guo, H.; Duan, Y.; Zheng, J.; Lin, Y.; Pan, F. Soft-contact conductive carbon enabling depolarization of $LiFePO_4$ cathodes to enhance both capacity and rate performances of lithium ion batteries. *J. Power Sources* **2016**, *331*, 232–239. [CrossRef]
16. Wang, X.; Feng, Z.; Huang, J.; Deng, W.; Li, X.; Zhang, H.; Wen, Z. Graphene-decorated carbon-coated $LiFePO_4$ nanospheres as a high-performance cathode material for lithium-ion batteries. *Carbon* **2018**, *127*, 149–157. [CrossRef]

17. Li, Y.; Qi, F.; Guo, H.; Guo, Z.; Li, M.; Wu, W. Characteristic investigation of an electrochemical-thermal coupled model for a $LiFePO_4$/Graphene hybrid cathode lithium-ion battery. *Case Stud. Therm. Eng.* **2019**, *13*, 100387. [CrossRef]
18. Zhang, B.; Xu, Y.; Wang, J.; Ma, X.; Hou, W.; Xue, X. Electrochemical performance of $LiFePO_4$/graphene composites at low temperature affected by preparation technology. *Electrochim. Acta* **2021**, *368*, 137575. [CrossRef]
19. Xi, Y.; Lu, Y. Toward Uniform In Situ Carbon Coating on Nano-$LiFePO_4$ via a Solid-State Reaction. *Ind. Eng. Chem. Res.* **2020**, *59*, 13549–13555. [CrossRef]
20. Zhang, H.; Wang, L.; Chen, Y.; Wen, X. Regenerated $LiFePO_4$/C for scrapped lithium iron phosphate powder batteries by pre-oxidation and reduction method. *Ionics* **2022**, *28*, 2125–2133. [CrossRef]
21. Ravet, N.; Chouinard, Y.; Magnan, J.F.; Besner, S.; Gauthier, M.; Armand, M. Electroactivity of natural and synthetic triphylite. *J. Power Sources* **2001**, *97–98*, 503–507. [CrossRef]
22. Huynh, L.T.N.; Nguyen, H.H.A.; Tran, T.T.D.; Nguyen, T.T.T.; Nguyen, T.M.A.; La, T.H.; Tran, V.M.; Le, M.L.P. Electrode composite $LiFePO_4$@ Carbon: Structure and electrochemical performances. *J. Nanomater.* **2019**, *2019*, 2464920. [CrossRef]
23. Altin, E.; Altundag, S.; Gultek, E.; Altin, S. $Li_{1+x}FePO_4$ (x = 0–0.5) production from Fe^{3+} sources by glass-ceramic technique with different carbon sources and investigation of structural, thermal and electrochemical performance. *J. Non-Cryst. Solids* **2022**, *586*, 121546. [CrossRef]
24. Lim, H.H.; Jang, I.C.; Lee, S.B.; Karthikeyan, K.; Aravindan, V.; Lee, Y.S. The important role of adipic acid on the synthesis of nanocrystalline lithium iron phosphate with high rate performance. *J. Alloys Compd.* **2010**, *495*, 181–184. [CrossRef]
25. Xu, H.; Jing, M.; Huang, Z.; Li, J.; Wang, T.; Yuan, W.; Ju, B.; Shen, X. Cross-Linked Polypropylene Oxide Solid Electrolyte Film with Enhanced Mechanical, Thermal, and Electrochemical Properties via Additive Modification. *ACS Appl. Polym. Mater.* **2021**, *3*, 6539–6547. [CrossRef]
26. Ramkumar, V.; Gardas, R.L. Structural Arrangements of Guanidinium-Based Dicarboxylic Acid Ionic Liquids and Insights into Carbon Dioxide Uptake through Structural Voids. *Cryst. Growth Des.* **2022**, *22*, 3646–3655. [CrossRef]
27. Guan, L.; Liu, M.; Yu, F.; Qiu, T.; Zhou, T.; Lin, X. A $LiFePO_4$ regeneration method based on PVAc alcoholysis reaction. *Renew. Energy* **2021**, *175*, 559–567. [CrossRef]
28. Wang, Y.; Liu, Z.; Zhou, S. An effective method for preparing uniform carbon coated nano-sized $LiFePO_4$ particles. *Electrochim. Acta* **2011**, *58*, 359–363. [CrossRef]
29. Ravet, N.; Gauthier, M.; Zaghib, K.; Goodenough, J.B.; Mauger, A.; Gendron, F.; Julien, C.M. Mechanism of the Fe^{3+} Reduction at Low Temperature for $LiFePO_4$ Synthesis from a Polymeric Additive. *Chem. Mater.* **2007**, *19*, 2595–2602. [CrossRef]
30. Liao, Y.; Li, G.; Xu, N.; Chen, T.; Wang, X.; Li, W. Synergistic effect of electrolyte additives on the improvement in interfacial stability between ionic liquid based gel electrolyte and $LiFePO_4$ cathode. *Solid State Ion.* **2019**, *329*, 31–39. [CrossRef]
31. Doeff, M.M.; Wilcox, J.D.; Kostecki, R.; Lau, G. Optimization of carbon coatings on $LiFePO_4$. *J. Power Sources* **2006**, *163*, 180–184. [CrossRef]
32. Kim, S.; Mathew, V.; Kang, J.; Gim, J.; Song, J.; Jo, J.; Kim, J. High rate capability of LiFePO cathodes doped with an excess amount of Ti. *Ceram. Int.* **2016**, *42*, 7230–7236. [CrossRef]
33. Chang, H.-H.; Chang, C.-C.; Wu, H.-C.; Guo, Z.-Z.; Yang, M.-H.; Chiang, Y.-P.; Sheu, H.-S.; Wu, N.-L. Kinetic study on low-temperature synthesis of $LiFePO_4$ via solid-state reaction. *J. Power Sources* **2006**, *158*, 550–556. [CrossRef]
34. Lin, F.; Liu, Y.; Yu, X.; Cheng, L.; Singer, A.; Shpyrko, O.G.; Xin, H.L.; Tamura, N.; Tian, C.; Weng, T.-C.; et al. Synchrotron X-ray Analytical Techniques for Studying Materials Electrochemistry in Rechargeable Batteries. *Chem. Rev.* **2017**, *117*, 13123–13186. [CrossRef] [PubMed]
35. Sun, X.; Xu, Y. Fe excess in hydrothermally synthesized $LiFePO_4$. *Mater. Lett.* **2012**, *84*, 139–142. [CrossRef]
36. Zhang, Y.; Alarco, J.A.; Nerkar, J.Y.; Best, A.S.; Snook, G.A.; Talbot, P.C.; Cowie, B.C.C. Observation of Preferential Cation Doping on the Surface of $LiFePO_4$ Particles and Its Effect on Properties. *ACS Appl. Energy Mater.* **2020**, *3*, 9158–9167. [CrossRef]
37. Kapaev, R.R.; Novikova, S.A.; Kulova, T.L.; Skundin, A.M.; Yaroslavtsev, A.B. Synthesis of $LiFePO_4$ nanoplatelets as cathode materials for Li-ion batteries. *Nanotechnol. Russ.* **2016**, *11*, 757–760. [CrossRef]
38. Kim, J.-K.; Jeong, S.M. Physico-electrochemical properties of carbon coated $LiFePO_4$ nanoparticles prepared by different preparation method. *Appl. Surf. Sci.* **2020**, *505*, 144630. [CrossRef]
39. Watts, J.F.; Wolstenholme, J. *An Introduction to Surface Analysis by XPS and AES*; John Wiley & Sons Ltd: Hoboken, NJ, USA, 2020; pp. 1–288.
40. Grissa, R.; Abramova, A.; Tambio, S.-J.; Lecuyer, M.; Deschamps, M.; Fernandez, V.; Greneche, J.-M.; Guyomard, D.; Lestriez, B.; Moreau, P. Thermomechanical Polymer Binder Reactivity with Positive Active Materials for Li Metal Polymer and Li-Ion Batteries: An XPS and XPS Imaging Study. *ACS Appl. Mater. Interfaces* **2019**, *11*, 18368–18376. [CrossRef]
41. Kumar, Y.A.; Kim, H.-J. Effect of Time on a Hierarchical Corn Skeleton-Like Composite of CoO@ZnO as Capacitive Electrode Material for High Specific Performance Supercapacitors. *Energies* **2018**, *11*, 3285. [CrossRef]
42. Yoon, J.-H.; Kumar, Y.A.; Sambasivam, S.; Hira, S.A.; Krishna, T.N.V.; Zeb, K.; Uddin, W.; Kumar, K.D.; Obaidat, I.M.; Kim, S.; et al. Highly efficient copper-cobalt sulfide nano-reeds array with simplistic fabrication strategy for battery-type supercapacitors. *J. Energy Storage* **2020**, *32*, 101988. [CrossRef]
43. Eds Long, G.J.; Grandjean, F. *Mossbauer Spectroscopy Applied to Magnetism and Materials Science*; Springer Science and Business Media: New York, NY, USA, 1996; Volume 1, pp. 1–479.

44. Wu, Y.; Holze, R. Surface Science in Batteries. In *Surface and Interface Science: Applications of Surface Science I*, 1st ed.; Wandelt, K., Ed.; Wiley-VCH Verlag GmbH & Co. KGaA: Hoboken, NJ, USA, 2020; pp. 381–427. [CrossRef]
45. Andersson, A.S.; Kalska, B.; Häggström, L.; Thomas, J.O. Lithium extraction/insertion in $LiFePO_4$: An X-ray diffraction and Mössbauer spectroscopy study. *Solid State Ion.* **2000**, *130*, 41–52. [CrossRef]
46. Jalkanen, K.; Lindén, J.; Karppinen, M. Local structures in mixed LixFe1–yMyPO4 (M=Co, Ni) electrode materials. *J. Solid State Chem.* **2015**, *230*, 404–410. [CrossRef]
47. Khalifi, M.E.; Lippens, P.-E. First-Principles Investigation of the 57Fe Mössbauer Parameters of $LiFePO_4$ and FePO. *J. Phys. Chem. C* **2016**, *120*, 28375–28389. [CrossRef]
48. Yaroslavtsev, S.; Novikova, S.; Rusakov, V.; Vostrov, N.; Kulova, T.; Skundin, A.; Yaroslavtsev, A. LiFe1-XMgXPO4/C as cathode materials for lithium-ion batteries. *Solid State Ion.* **2018**, *317*, 149–155. [CrossRef]
49. Hosono, E.; Wang, Y.; Kida, N.; Enomoto, M.; Kojima, N.; Okubo, M.; Matsuda, H.; Saito, Y.; Kudo, T.; Honma, I.; et al. Synthesis of Triaxial $LiFePO_4$ Nanowire with a VGCF Core Column and a Carbon Shell through the Electrospinning. *ACS Appl. Mater. Interfaces* **2010**, *2*, 212–218. [CrossRef] [PubMed]
50. Bini, M.; Ferrari, S.; Capsoni, D.; Mustarelli, P.; Spina, G.; Giallo, F.D.; Lantieri, M.; Leonelli, C.; Rizzutie, A.; Massarotti, V. Pair distribution function analysis and Mo¨ssbauer study of defects in microwave-hydrothermal $LiFePO_4$. *RSC Adv.* **2012**, *2*, 250–258. [CrossRef]
51. Bazzi, K.; Mandal, B.P.; Nazri, M.; Naik, V.M.; Garg, V.K.; Oliveira, A.C.; Vaishnava, P.P.; Nazri, G.A.; Naik, R. Effect of surfactants on the electrochemical behavior of $LiFePO_4$ cathode material for lithium ion batteries. *J. Power Sources* **2014**, *265*, 67–74. [CrossRef]
52. Yaroslavtsev, S.A.; Vostrov, N.I.; Novikova, S.A.; Kulova, T.L.; Yaroslavtsev, A.B.; Rusakov, V.S. Study of Delithiation Process Features in $LixFe_{0.8}M_{0.2}PO_4$ (M = Mg, Mn, Co, Ni) by Mössbauer Spectroscopy. *J. Phys. Chem. C* **2020**, *124*, 13026–13035. [CrossRef]
53. Reklaitis, J.; Davidonis, R.; Dindune, A.; Valdniece, D.; Jasulaitiene, V.; Baltrunas, D. Characterization of $LiFePO_4$/C composite and its thermal stability by Mossbauer and XPS spectroscopy. *Phys. Status Solidi B* **2016**, *253*, 2283–2288. [CrossRef]
54. Jensen, K.M.Ø.; Christensen, M.; Gunnlaugsson, H.P.; Lock, N.; Bøjesen, E.D.; Proffen, T.; Iversen, B.B. Defects in Hydrothermally Synthesized $LiFePO_4$ and LiFe1-xMnxPO4 Cathode Materials. *Chem. Mater.* **2013**, *25*, 2282–2290. [CrossRef]
55. Lazarević, Z.Ž.; Križan, G.; Križan, J.; Milutinović, A.; Ivanovski, V.N.; Mitrić, M.; Gilić, M.; Umićević, A.; Kuryliszyn-Kudelska, I.; Romčević, N.Ž. Characterization of $LiFePO_4$ samples obtained by pulse combustion under various conditions of synthesis. *J. Appl. Phys.* **2019**, *126*, 085109. [CrossRef]
56. Sun, S.; An, Q.; Tian, Z.; Zhao, X.; Shen, X. Low-Temperature Synthesis of $LiFePO_4$ Nanoplates/C Composite for Lithium Ion Batteries. *Energy Fuels* **2020**, *34*, 11597–11605. [CrossRef]
57. Jie, Y.; Yang, S.; Hu, F.; Li, Y.; Ye, L.; Zhao, D.; Jin, W.; Chang, C.; Lai, Y.; Chen, Y. Gas evolution characterization and phase transformation during thermal treatment of cathode plates from spent $LiFePO_4$ batteries. *Thermochim. Acta* **2020**, *684*, 178483. [CrossRef]
58. Wilcox, J.D.; Doeff, M.M.; Marcinek, M.; Kostecki, R. Factors influencing the quality of carbon coatings on $LiFePO_4$. *J. Electrochem. Soc.* **2007**, *154*, A389–A395. [CrossRef]
59. Rouzaud, J.N.; Oberlin, A.; Beny-Bassez, C. Carbon films: Structure and microtexture (optical and electron microscopy, Raman spectroscopy). *Thin Solid Film.* **1983**, *105*, 75–96. [CrossRef]
60. Bonhomme, F.; Lassegues, J.C.; Servant, L. Raman spectroelectrochemistry of a carbon supercapacitor. *J. Electrochem. Soc.* **2001**, *148*, E450–E458. [CrossRef]
61. Duan, X.; Tian, W.; Zhang, H.; Sun, H.; Ao, Z.; Shao, Z.; Wang, S. sp^2/sp^3 Framework from Diamond Nanocrystals: A Key Bridge of Carbonaceous Structure to Carbocatalysis. *ACS Catal.* **2019**, *9*, 7494–7519. [CrossRef]
62. Osswald, S.; Yushin, G.; Mochalin, V.; Kucheyev, S.O.; Gogotsi, Y. Control of sp^2/sp^3 Carbon Ratio and Surface Chemistry of Nanodiamond Powders by Selective Oxidation in Air. *J. Am. Chem. Soc.* **2006**, *128*, 11635–11642. [CrossRef]
63. Vejpravová, J. Mixed sp2–sp3 Nanocarbon Materials: A Status Quo Review. *Nanomaterials* **2021**, *11*, 2469. [CrossRef]
64. Ferrari, A.C.; Basko, D.M. Raman spectroscopy as a versatile tool for studying the properties of graphene. *Nat. Nanotechnol.* **2013**, *8*, 235–246. [CrossRef]
65. Wu, J.-B.; Lin, M.-L.; Cong, X.; Liu, H.-N.; Tan, P.-H. Raman spectroscopy of graphene-based materials and its applications in related devices. *Chem. Soc. Rev.* **2018**, *47*, 1822–1873. [CrossRef]
66. Wu, J.-B.; Lin, M.-L.; Tan, P.-H. Raman Spectroscopy of Monolayer and Multilayer Graphenes Chapter 1. In *Raman Spectroscopy of Two-Dimensional Materials*; Tan, P.-H., Ed.; Springer Series in Materials Science 276; Springer Nature Singapore Pte Ltd.: Singapore, 2019.
67. Bokobza, L.; Bruneel, J.-L.; Couzi, M. Raman spectra of carbon-based materials (from graphite to carbon black) and of some silicone composites. *Carbon* **2015**, *1*, 77–94. [CrossRef]
68. Casiraghi, C. Doping dependence of the Raman peaks intensity of graphene close to the Dirac point. *Phys. Rev. B* **2009**, *80*, 233407. [CrossRef]
69. Sun, L.; Deng, Q.; Fang, B.; Li, Y.; Deng, L.; Yang, B.; Ren, X.; Zhang, P. Carbon-coated LiFePO4 synthesized by a simple solvothermal method. *CrystEngComm* **2016**, *18*, 7537–7543. [CrossRef]
70. Public Spectra. Raman Spectrum of Ketjen Black. CAS Registry Number 1333-86-4.

71. Vivo-Vilches, J.F.; Celzard, A.; Fierro, V.; Devin-Ziegler, I.; Brosse, N.; Dufour, A.; Etienne, M. Lignin-based carbon nanofibers as electrodes for vanadium redox couple electrochemistry. *Nanomaterials* **2019**, *9*, 106. [CrossRef]
72. Vadahanambia, S.; Chunb, H.-H.; Jung, K.H.; Park, H. Nitrogen doped holey carbon nano-sheets as anodes in sodium ion battery. *RSC Adv.* **2016**, *6*, 38112–38116. [CrossRef]
73. Tan, H.; Xu, L.; Geng, H.; Rui, X.; Li, C.; Huang, S. Nanostructured $Li_3V_2(PO_4)_3$ Cathodes. *Small* **2018**, *14*, 1800567. [CrossRef]
74. Feng, C.; Chen, Y.; Liu, D.; Zhang, P. Conductivity and electrochemical performance of $LiFePO_4$ slurry in the lithium slurry battery. *IOP Conf. Ser. Mater. Sci. Eng.* **2017**, *207*, 012076. [CrossRef]
75. Lion Specialty Chemicals Co. *KETJENBLACK Highly Electro-Conductive Carbon Black*; Lion: Tokyo, Japan, 2020.
76. Folaranmi, G.; Bechelany, M.; Sistat, P.; Cretin, M.; Zaviska, F. Activated Carbon Blended with Reduced Graphene Oxide Nanoflakes for Capacitive Deionization. *Nanomaterials* **2021**, *11*, 1090. [CrossRef]
77. Knight, D.S.; White, W.B. Characterization of diamond films by Raman spectroscopy. *J. Mater. Res.* **1989**, *4*, 385–393. [CrossRef]
78. Cong, C.; Wang, Y.; Yu, T. Raman Spectroscopy Study of Two-Dimensional Materials Under Strain. Chapter 6. In *Raman Spectroscopy of Two-Dimensional Materials*; Tan, P.-H., Ed.; Springer Series in Materials Science: Berlin/Heidelberg, Germany, 2019; Volume 276, pp. 111–129. [CrossRef]
79. Yoon, D.; Son, Y.-W.; Cheong, H. Strain-dependent splitting of the double-resonance Raman scattering band in graphene. *Phys. Rev. Lett.* **2011**, *106*, 155502. [CrossRef] [PubMed]
80. Davydov, V.V.; Grebenikova, N.M.; Smirnov, K.Y. Method of Monitoring the State of Flowing Media with Low Transparency That Contain Large Inclusions. *Meas. Tech.* **2019**, *62*, 519–526. [CrossRef]
81. Kamzin, A.S.; Bobyl, A.V.; Ershenko, E.M.; Terukov, E.I.; Agafonov, D.V.; Kudryavtsev, E.N. Structure and Electrochemical Characteristics of $LiFePO_4$ Cathode Materials for Rechargeable LiIon Batteries. *Phys. Solid State* **2013**, *55*, 1385–1394. [CrossRef]
82. Doeff, M.; Hu, M.Y.; McLarnon, F.; Kostecki, R. Effect of surface carbon structure on the electrochemical performance of $LiFePO_4$. *Electrochem. Solid State Lett.* **2003**, *6*, A207–A209. [CrossRef]
83. Higuchi, M.; Katayama, K.; Azuma, Y.; Yukawa, M.; Suhara, M. Synthesis of $LiFePO_4$ cathode material by microwave processing. *J. Power Sources* **2003**, *119–121*, 258–261. [CrossRef]
84. Yang, J.; Xu, J.J. Nonaqueous sol-gel synthesis of high-performance $LiFePO_4$. *Electrochem. Solid State Lett.* **2004**, *7*, A515–A518. [CrossRef]
85. Deb, A.; Bergmann, U.; Cramer, S.P.; Cairns, E.J. Structural investigations of $LiFePO_4$ electrodes and in situ studies by Fe X-ray absorption spectroscopy. *Electrochim. Acta* **2005**, *50*, 5200–5207. [CrossRef]
86. Fu, Q.; Sarapulova, A.; Trouillet, V.; Zhu, L.; Fauth, F.; Mangold, S.; Welter, E.; Indris, S.; Knapp, M.; Dsoke, S.; et al. In Operando Synchrotron Diffraction and in Operando X-ray Absorption Spectroscopy Investigations of Orthorhombic V2O5 Nanowires as Cathode Materials for Mg-Ion Batteries. *J. Am. Chem. Soc.* **2019**, *141*, 2305–2315. [CrossRef]
87. Kim, D.-H.; Kim, J. Synthesis of lithium manganese phosphate nanoparticle and its properties. *J. Phys. Chem. Solids* **2007**, *68*, 734–737. [CrossRef]
88. Lee, S.B.; Jang, I.C.; Lim, H.H.; Aravindan, V.; Kim, H.S.; Lee, Y.S. Preparation and electrochemical characterization of $LiFePO_4$ nanoparticles with high rate capability by a sol–gel method. *J. Alloys Compd.* **2010**, *491*, 668–672. [CrossRef]
89. Son, C.G.; Chang, D.R.; Kim, H.S.; Lee, Y.S. Synthesis and Electrochemical Properties of Nanocrystalline $LiFePO_4$ Obtained by Different Methods. *J. Electrochem. Sci. Technol.* **2011**, *2*, 103–109. [CrossRef]
90. Son, C.G.; Yang, H.M.; Lee, G.W.; Cho, A.R.; Aravindan, V.; Kim, H.S.; Kim, W.S.; Lee, Y.S. Manipulation of adipic acid application on the electrochemical properties of $LiFePO_4$ at high rate performance. *J. Alloys Compd.* **2011**, *509*, 1279–1284. [CrossRef]
91. Li, M.; Mu, B. Effect of different dimensional carbon materials on the properties and application of phase change materials: A review. *Appl. Energy* **2019**, *242*, 695–715. [CrossRef]
92. Jugović, D.; Mitrić, M.; Cvjetićanin, N.; Jančar, B.; Mentus, S.; Uskoković, D. Synthesis and characterization of $LiFePO_4$/C composite obtained by sonochemical method. *Solid State Ion.* **2008**, *179*, 415–419. [CrossRef]
93. Zhao, B.; Jiang, Y.; Zhang, H.; Tao, H.; Zhong, M.; Jiao, Z. Morphology and electrical properties of carbon coated $LiFePO_4$ cathode materials. *J. Power Sources* **2009**, *189*, 462–466. [CrossRef]
94. Chen, H.; Chen, Y.; Gong, W.; Xiang, K.; Sun, B.; Liu, J. Preparation and electrochemical performance of $LiFePO_4$/C composite with network connections of nano-carbon wires. *Mater. Lett.* **2011**, *65*, 559–561. [CrossRef]
95. Gauthier, M. Phosphate materials for lithium batteries and energy storage. *Procedia Eng.* **2012**, *46*, 234–254. [CrossRef]
96. Chernyshov, A.A.; Veligzhanin, A.A.; Zubavichus, Y.V. Structural materials science end-station at the Kurchatov synchrotron radiation source: Recent instrumentation upgrades and experimental results. *Nucl. Instr. Meth. Phys. Res. A* **2009**, *603*, 95–98. [CrossRef]
97. Ectors, D.; Goetz-Neunhoeffer, F.; Neubauer, J. A generalized geometric approach to anisotropic peak broadening due to domain morphology. *J. Appl. Cryst.* **2015**, *48*, 189–194. [CrossRef]
98. Bobyl, A.; Kasatkin, I. Anisotropic crystallite size distributions in $LiFePO_4$ powders. *RSC Adv.* **2021**, *11*, 1379–1385. [CrossRef] [PubMed]
99. Bobyl, A.; Nam, S.-C.; Song, J.-H.; Ivanishchev, A.; Ushakov, A. Rate Capability of $LiFePO_4$ Cathodes and the Shape Engineering of Their Anisotropic Crystallites. *J. Electrochem. Sci. Technol.* **2022**, *13*, 438–452. [CrossRef]
100. Julien, C.M.; Mauger, A.; Zaghib, K. Surface effects on electrochemical properties of nano-sized $LiFePO_4$. *J. Mater. Chem.* **2011**, *21*, 9955–9968. [CrossRef]

101. Herle, P.S.; Ellis, B.; Coombs, N.; Nazar, L.F. Nano-network electronic conduction in iron and nickel olivine phosphates. *Nat. Mater.* **2004**, *3*, 147–152. [CrossRef] [PubMed]
102. Kosova, N.V.; Devyatkina, E.T. Lithium Iron Phosphate Synthesis Using Mechanical Activation. *Chem. Sustain. Dev.* **2012**, *20*, 61–68.
103. Ershenko, E.; Bobyl, A.; Boiko, M.; Zubavichus, Y.; Runov, V.; Trenikhin, M.; Sharkov, M. Fe3P impurity phase in high-quality $LiFePO_4$: X-ray diffraction and neutron-graphical studies. *Ionics* **2017**, *23*, 2293–2300. [CrossRef]
104. Babaev, A.A.; Zobov, M.E.; Kornilov, D.Y.; Tkachev, S.V.; Terukov, E.I.; Levitskii, V.S. Optical and electrical properties of graphene oxide. *Opt. Spectrosc.* **2018**, *125*, 1014–1018. [CrossRef]
105. Mallet-Ladeira, P.; Puech, P.; Toulouse, C.; Cazayous, M.; Ratel-Ramond, N.; Weisbecker, P.; Vignoles, G.L.; Monthioux, M.A. Raman study to obtain crystallite size of carbon materials: A better alternative to the Tuinstra–Koenig law. *Carbon* **2014**, *80*, 629–639. [CrossRef]
106. Puech, P.; Kandara, M.; Paredes, G.; Moulin, L.; Weiss-Hortala, E.; Kundu, A.; Ratel-Ramond, N.; Plewa, J.-M.; Pellenq, R.; Monthioux, M. Analyzing the Raman spectra of graphenic carbon materials from kerogens to nanotubes: What type of information can be extracted from defect bands? *C* **2019**, *5*, 69. [CrossRef]
107. Aldon, L.; Perea, A.; Womes, M.; Ionica-Bousquet, C.M.; Jumas, J.C. Determination of the Lamb-Mössbauer factors of $LiFePO_4$ and $FePO_4$ for electrochemical in situ and operando measurements in Li-ion batteries. *J. Solid State Chem.* **2010**, *183*, 218–222. [CrossRef]
108. Li, Z.; Shinno, I. Next nearest neighbor effects in triphylite and related phosphate minerals. *Mineral. J.* **1997**, *19*, 99–107. [CrossRef]
109. Lippens, P.-E.; Khalifi, M.E.; Chamas, M.; Perea, A.; Sougrati, M.-T.; Ionica-Bousquet, C.; Aldon, L.; Olivier-Fourcade, J.; Jumas, J.-C. How Mössbauer spectroscopy can improve Li-ion batteries. *Hyperfine Interact* **2012**, *206*, 35–46. [CrossRef]
110. Xia, X.; Wang, Z.; Chen, L. Regeneration and characterization of air-oxidized $LiFePO_4$. *Electrochem. Commun.* **2008**, *10*, 1442–1444. [CrossRef]
111. Yu, D.Y.W.; Donoue, K.; Kadohata, T.; Murata, T.; Matsuta, S.; Fujitani, S. Impurities in $LiFePO_4$ and their influence on material characteristics. *J. Electrochem. Soc.* **2008**, *155*, A526–A530. [CrossRef]
112. Dhindsa, K.S.; Mandal, B.P.; Bazzi, K.; Lin, M.W.; Nazri, M.; Nazri, G.A.; Naik, V.M.; Garg, V.K.; Oliveira, A.C.; Vaishnava, P.; et al. Enhanced electrochemical performance of graphene modified $LiFePO_4$ cathode material for lithium ion batteries. *Solid State Ion.* **2013**, *253*, 94–100. [CrossRef]
113. Shiratsuchi, T.; Okada, S.; Yamaki, J.; Yamashit, S.; Nishid, T. Cathode performance of olivine-type $LiFePO_4$ synthesized by chemical lithiation. *J. Power Sources* **2007**, *173*, 979–984. [CrossRef]
114. Maccario, M.; Croguennec, L.; Wattiaux, A.; Suard, E.; Le Cras, F.; Delmas, C. C-containing $LiFePO_4$ materials-Part I: Mechano-chemical synthesis and structural characterization. *Solid State Ion.* **2008**, *179*, 2020–2026. [CrossRef]
115. Salah, A.; Zaghib, K.; Mauger, A.; Gendron, F.; Julien, C. Magnetic studies of the carbothermal effect on $LiFePO_4$. *Phys. Status Solidi A Appl. Mater. Sci.* **2006**, *203*, R1–R3. [CrossRef]
116. Salah, A.A.; Mauger, A.; Zaghib, K.; Goodenough, J.B.; Ravet, N.; Gauthier, M.; Gendron, F.; Julien, C.M. Reduction Fe^{3+} of impurities in $LiFePO_4$ from pyrolysis of organic precursor used for carbon deposition. *J. Electrochem. Soc.* **2006**, *153*, A1692–A1701. [CrossRef]
117. Hirose, K.; Honma, T.; Doi, Y.; Hinatsu, Y.; Komatsu, T. Mössbauer analysis of Fe ion state in lithium iron phosphate glasses and their glass-ceramics with olivine-type $LiFePO_4$ crystals. *Solid State Commun.* **2008**, *146*, 273–277. [CrossRef]
118. Wang, Z.; Su, S.; Yu, C.; Chen, Y.; Xia, D. Synthesises, characterizations and electrochemical properties of spherical-like $LiFePO_4$ by hydrothermal method. *J. Power Sources* **2008**, *184*, 633–636. [CrossRef]
119. Bazzi, K.; Nazri, M.; Naik, V.; Garg, V.; Oliveira, A.; Vaishnava, P.; Nazri, G.; Naik, R. Enhancement of electrochemical behavior of nanostructured $LiFePO_4$/Carbon cathode material with excess Li. *J. Power Sources* **2016**, *306*, 17–23. [CrossRef]
120. Waldmann, T.; Iturrondobeitia, A.; Kasper, M.; Ghanbari, N.; Aguesse, F.; Bekaert, E.; Daniel, L.; Genies, S.; Gordon, I.J.; Loble, M.W.; et al. Post-mortem analysis of aged lithium-ion batteries: Disassembly methodology and physico-chemical analysis techniques. *J. Electrochem. Soc.* **2016**, *163*, A2149–A2164. [CrossRef]
121. Pender, J.P.; Jha, G.; Youn, D.H.; Ziegler, J.M.; Andoni, I.; Choi, E.J.; Heller, A.; Dunn, B.S.; Weiss, P.S.; Penner, R.M.; et al. Electrode degradation in lithium-ion batteries. *ACS Nano* **2020**, *14*, 1243–1295. [CrossRef]
122. Sui, X.; Swierczynski, M.; Teodorescu, R.; Stroe, D.-I. The Degradation Behavior of $LiFePO_4$/C Batteries during Long-Term Calendar Aging. *Energies* **2021**, *14*, 1732. [CrossRef]
123. Wang, L.; Qiu, J.; Wang, X.; Chen, L.; Cao, G.; Wang, J.; Zhang, H.; He, X. Insights for understanding multiscale degradation of $LiFePO_4$ cathodes. *eScience* **2022**, *2*, 125–137. [CrossRef]
124. Bobyl, A.V.; Zabrodskii, A.G.; Malyshkin, V.G.; Novikova, O.V.; Terukov, E.I.; Agafonov, D.V. Degradation of Li-ion batteries. Application of Generalized Radon–Nikodym Approach to Direct Estimation of Degradation Rate Distribution. *Izvestiya RAN Energetika* **2018**, *2*, 46–58. Available online: https://www.researchgate.net/publication/309175377 (accessed on 23 November 2016).
125. Scipioni, R.; Jørgensen, P.S.; Stroe, D.I.; Younesi, R.; Simonsen, S.B.; Norby, P.; Hjelm, J.; Jensen, S.H. Complementary analyses of aging in a commercial $LiFePO_4$/graphite 26650 cell. *Electrochim. Acta* **2018**, *284*, 454–468. [CrossRef]
126. Scipioni, R.; Jørgensen, P.S.; Ngo, D.-T.; Simonsen, S.B.; Liu, Z.; Yakal-Kremski, K.J.; Wang, H.; Hjelm, J.; Norby, P.; Barnett, S.A.; et al. Electron microscopy investigations of changes in morphology and conductivity of $LiFePO_4$/C electrodes. *J. Power Sources* **2016**, *307*, 259–269. [CrossRef]

127. Lia, X.; Jiang, F.; Qu, K.; Wang, Y.; Pan, Y.; Wang, M.; Liu, Y.; Xu, H.; Chen, J.; Huang, Y.; et al. First Atomic-Scale Insight on Degradation in Lithium Iron Phosphate Cathodes by Transmission Electron Microscopy. *J. Phys. Chem. Lett.* **2020**, *11*, 4608–4617. [CrossRef]
128. Zhao, N.; Li, Y.; Zhao, X.; Zhi, X.; Liang, G. Effect of particle size and purity on the low temperature electrochemical performance of $LiFePO_4$/C cathode material. *J. Alloys Compd.* **2016**, *683*, 123–132. [CrossRef]
129. Wang, J.; Yang, J.; Tang, Y.; Li, R.; Liang, G.; Sham, T.-K.; Sun, X. Surface aging at olivine $LiFePO_4$: A direct visual observation of iron dissolution and the protection role of nano-carbon coating. *J. Mater. Chem. A* **2013**, *1*, 1579–1586. [CrossRef]
130. Chen, X.; Li, Y.; Wang, J. Enhanced Electrochemical Performance of $LiFePO_4$ Originating from the Synergistic Effect of ZnO and C Co-Modification. *Nanomaterials* **2021**, *11*, 12. [CrossRef]
131. Tan, S.; Tieu, J.H.; Bélanger, D. Chemical Polymerization of Aniline on a Poly(styrene sulfonic acid) Membrane: Controlling the Polymerization Site Using Different Oxidants. *J. Phys. Chem. B* **2005**, *109*, 14085–14092. [CrossRef] [PubMed]
132. Wang, Y.; Wang, Y.; Hosono, E.; Wang, K.; Zhou, H. The Design of a $LiFePO_4$/Carbon Nanocomposite with a Core–Shell Structure and Its Synthesis by an In Situ Polymerization Restriction Method. *Angew. Chem. Int. Ed.* **2008**, *47*, 7461–7465. [CrossRef]
133. Callister, W.D.; Rethwisch, D.G. *Materials Science and Engineering*, 9th ed.; John Wiley & Sons, Inc.: Hoboken, NJ, USA, 2014; 990p.
134. Blyr, A.; Sigala, C.; Amatucci, G.; Guyomard, D.; Chabre, Y.; Tarascon, J.-M. Self-Discharge of $LiMn_2O_4$/C Li-Ion Cells in Their Discharged State: Understanding by Means of Three-Electrode Measurements. *J. Electrochem. Soc.* **1998**, *145*, 194–209. [CrossRef]
135. Wang, J.; Tang, Y.; Yang, J.; Li, R.; Liang, G.; Sun, X. Nature at $LiFePO_4$ aging process: Roles of impurity phases. *J. Power Sources* **2013**, *238*, 454–463. [CrossRef]
136. Cuisinier, M.; Dupré, N.; Moreau, P.; Guyomard, D. NMR Monitoring of Electrode/Electrolyte Interphase in the Case of Air-exposed and Carbon Coated $LiFePO_4$. *J. Power Sources* **2013**, *243*, 682–690. [CrossRef]
137. Chen, J.; Bai, J.; Chen, H.; Graetz, J. In Situ Hydrothermal Synthesis of $LiFePO_4$ Studied by Synchrotron X-ray Diffraction. *J. Phys. Chem. Lett.* **2011**, *2*, 1874–1878. [CrossRef]
138. Zhu, J.; Fiore, J.; Li, D.; Kinsinger, N.M.; Wang, Q.; DiMasi, E.; Guo, J.; Kisailus, D. Solvothermal Synthesis, Development, and Performance of $LiFePO_4$ Nanostructures. *Cryst. Growth Des.* **2013**, *13*, 4659–4666. [CrossRef]
139. Chung, S.-Y.; Kim, Y.-M.; Kim, J.-G.; Kim, Y.-J. Multiphase transformation and Ostwald's rule of stages during crystallization of a metal phosphate. *Nat. Phys.* **2009**, *5*, 68–73. [CrossRef]
140. Malik, R.; Burch, D.; Bazant, M.; Ceder, G. Particle Size Dependence of the Ionic Diffusivity. *Nano Lett.* **2010**, *10*, 4123–4127. [CrossRef]
141. Orikasa, Y.; Maeda, T.; Koyama, Y.; Murayama, H.; Fukuda, K.; Tanida, H.; Arai, H.; Matsubara, E.; Uchimoto, Y.; Ogumi, Z. Transient Phase Change in Two Phase Reaction between $LiFePO_4$ and $FePO_4$ under Battery Operation. *Chem. Mater.* **2013**, *25*, 1032–1039. [CrossRef]
142. Raj, H.; Rani, S.; Sil, A. Two-Phase Composition ($LiFePO_4$/$FePO_4$) and Phase Transformation Dependence on Charging Current: In Situ and Ex Situ Studies. *Energy Fuels* **2020**, *34*, 14874–14881. [CrossRef]
143. Ahsan, Z.; Ding, B.; Cai, Z.; Wen, C.; Yang, W.; Ma, Y.; Zhang, S.; Song, G.; Javed, M.S. Recent Progress in Capacity Enhancement of $LiFePO_4$ Cathode for Li-Ion Batteries. *J. Electrochem. Energy Convers. Storage* **2021**, *18*, 010801. [CrossRef]
144. Tolganbek, N.; Yerkinbekova, Y.; Kalybekkyzy, S.; Bakenov, Z.; Mentbayeva, A. Current state of high voltage olivine structured $LiMPO_4$ cathode materials for energy storage applications: A review. *J. Alloys Compd.* **2021**, *882*, 160774. [CrossRef]
145. Masias, A.; Marcicki, J.; Paxton, W.A. Opportunities and challenges of lithium ion batteries in automotive applications. *ACS Energy Lett.* **2021**, *6*, 621–630. [CrossRef]
146. Mahmud, S.; Rahman, M.; Kamruzzaman, M.; Ali, M.O.; Emon, M.S.A.; Khatun, H.; Ali, M.R. Recent advances in lithium-ion battery materials for improved electrochemical performance: A review. *Results Eng.* **2022**, *15*, 100472. [CrossRef]
147. Moniruzzaman, M.; Kumar, Y.A.; Pallavolu, M.R.; Arbi, H.M.; Alzahmi, S.; Obaidat, I.M. Two-Dimensional Core-Shell Structure of Cobalt-Doped@MnO_2 Nanosheets Grown on Nickel Foam as a Binder-Free Battery-Type Electrode for Supercapacitor Application. *Nanomaterials* **2022**, *12*, 3187. [CrossRef]
148. Kumar, Y.A.; Kim, H.-J. Enhanced electrochemical performance of nanoplate nickel cobaltite (NiCo2O4) supercapacitor applications. *RSC Adv.* **2019**, *9*, 1115–1122. [CrossRef]
149. Yang, Z.; Dai, Y.; Wang, S.; Yub, J. How to make lithium iron phosphate better: A review exploring classical modification approaches in-depth and proposing future optimization methods. *J. Mater. Chem. A* **2016**, *4*, 18210–18222. [CrossRef]
150. Wen, L.; Guan, Z.; Wang, L.; Hu, S.; Lv, D.; Liu, X.; Duan, T.; Liang, G. Effect of Carbon-Coating on Internal Resistance and Performance of Lithium Iron Phosphate Batteries. *J. Electrochem. Soc.* **2022**, *169*, 050536. [CrossRef]
151. Ramasubramanian, B.; Sundarrajan, S.; Chellappan, V.; Reddy, M.V.; Ramakrishna, S.; Zaghib, K. Recent Development in Carbon-$LiFePO_4$ Cathodes for Lithium-Ion Batteries: A Mini Review. *Batteries* **2022**, *8*, 133. [CrossRef]
152. Tian, R.; Park, S.; King, P.; Cunningham, G.; Coelho, J.; Nicolosi, V.; Coleman, J. Quantifying the factors limiting rate performance in battery electrodes. *Nat. Commun.* **2019**, *10*, 1933. [CrossRef] [PubMed]
153. Newman, J.; Balsara, N.P. *Electrochemical Systems*, 4th ed.; John Wiley & Sons: Hoboken, NJ, USA, 2021; 608p.
154. Rice, J. *Mathematical Statistics and Data Analysis*; Brooks/Cole Cengage Learning: Belmont, CA, USA, 2007; p. 138.
155. Jiang, F.; Peng, P. Elucidating the performance limitations of lithium-ion batteries due to species and charge transport through five characteristic parameters. *Sci. Rep.* **2016**, *6*, 32639. [CrossRef]

156. Fang, W.; Ramadass, P.; Zhang, Z.J. Study of internal short in a Li-ion cell-II. Numerical investigation using a 3D electrochemical-thermal model. *J. Power Sources* **2014**, *248*, 1090–1098. [CrossRef]
157. Davydov, V.V.; Myazin, N.S.; Velichko, E.N. Characteristics of spectrum registration of condensed medium by the method of nuclear-magnetic resonance in a weak field. *Tech. Phys. Lett.* **2017**, *43*, 607–610. [CrossRef]
158. Higgins, T.M.; Coleman, J.N. Avoiding resistance limitations in high-performance transparent supercapacitor electrodes based on large-area, high-conductivity PEDOT: PSS films. *ACS Appl. Mater. Interfaces* **2015**, *7*, 16495–16506. [CrossRef]
159. Triola, M.F. *Elementary Statistics Technology Update*, 11th ed.; Pearson Education: London, UK, 2012; 888p.
160. Lvovich, V.F. *Distributed Impedance Models. Impedance Spectroscopy: Applications to Electrochemical and Dielectric Phenomena*; John Wiley & Sons, Inc.: Hoboken, NJ, USA, 2012; 368p.
161. Beckman, I. *Higher Mathematics: The Mathematical Apparatus of Diffusion*; Yurayt Publishing House: Moscow, Russia, 2020; 459p.
162. Zhang, Y.; Zhang, W.; Shen, S.; Yan, X.; Wu, R.; Wu, A.; Lastoskie, C.; Zhang, J. Sacrificial Template Strategy toward a Hollow LiNi1/3Co1/3Mn1/3O2 Nanosphere Cathode for Advanced Lithium-Ion Batteries. *ACS Omega* **2017**, *2*, 7593–7599. [CrossRef]
163. Wang, D.; Zhang, Z.; Hong, B.; Lai, Y. Self-sacrificial organic lithium salt enhanced initial columbic efficiency for safer and greener lithium-ion batteries. *Chem. Commun.* **2019**, *55*, 10737–10739. [CrossRef]
164. Yoo, J.W.; Zhang, K.; Patil, V.; Lee, J.T.; Jung, D.-W.; Pu, L.S.; Oh, W.; Yoon, W.-S.; Park, J.H.; Yi, G.-R. Porous supraparticles of LiFePO4 nanorods with carbon for high rate Li-ion batteries. *Mater. Express* **2018**, *8*, 316–324. [CrossRef]
165. Churikov, A.V.; Ivanishchev, A.V.; Ushakov, A.V.; Romanova, V.O. Diffusion aspects of lithium intercalation as applied to the development of electrode materials for lithium-ion batteries. *J. Solid State Electrochem.* **2014**, *18*, 1425–1441. [CrossRef]
166. Zhang, W.-J. Structure and performance of $LiFePO_4$ cathode materials: A review. *J. Power Sources* **2011**, *196*, 2962–2970. [CrossRef]

energies

MDPI

Article

Simulation of Electromagnetic Implosion of Metal Shells to Obtain Supercritical Fluids

Valery Antonov [1,*] and Nicolay Kalinin [2]

1 Institute of Physics and Mathematics, Peter the Great St. Petersburg Polytechnic University, 195251 St. Petersburg, Russia
2 Centre of Nanoheterostructure Physics, Ioffe Institute, 194021 St. Petersburg, Russia
* Correspondence: antonovvi@mail.ru; Tel.: +79-(21)-9499887

Abstract: This study analyzes the conditions for creating the energy density necessary to obtain supercritical fluids of substances with parameters (temperature T > 1 eV, density N > 1022 cm^{-3}, specific energy density ε > 100 kJ/g). The calculations are carried out on the basis of the one-dimensional (1D) two-temperature (2T) magneto hydrodynamic radiation model, which takes into account the physical processes occurring in the energy storage, switching system and the pulsed plasma load-a cylindrical compressible conductive shell. Developing a mathematical model, we assumed that physical processes were self-consistent. The simulation results were presented as time dependences of the main process parameters. Calculations showed that it becomes possible to sharpen the radiation pulse and pressure in the shock wave. As a result, we formulated the requirements for a laboratory energy source to establish the characteristics of a current pulse flowing through a conductive cylindrical shell and its dimensions (radius and thickness) necessary to achieve the goal.

Keywords: supercritical fluids; magneto hydrodynamic radiation model; shock wave

check for updates

Citation: Antonov, V.; Kalinin, N. Simulation of Electromagnetic Implosion of Metal Shells to Obtain Supercritical Fluids. *Energies* **2022**, *15*, 8953. https://doi.org/10.3390/en15238953

Academic Editor: Roger Gläser

Received: 31 October 2022
Accepted: 23 November 2022
Published: 26 November 2022

1. Introduction

This work is a part of fundamental research aimed at obtaining materials with programmable features, which can be used for various purposes, including medical ones. In this article, we consider a numerical analysis of the conditions for creating supercritical fluids by the pulse method [1–3]. In this case, the compression and heating of substances to temperatures and densities exceeding their critical values occurs by a thin-walled metal shell in the process of its electromagnetic implosion. The study conducts full-scale experiments and mathematical modeling of the under-study processes. At present, there has been no single theory provided for the accurate assessment of the matter's thermodynamic properties in a sufficiently wide region of the phase space. The properties of matter in most mathematical models are described by the equations of state and formulas for transport coefficients. The correct choice of equations, therefore, is an important condition for the correct simulation of the conductor's electrical explosion.

There has been considerable interest from researchers on the state of the supercritical fluid of substances with high critical point parameters. This is due to the substance's unique properties in the field of supercritical parameters: phase transitions, compressibility, transfer coefficients, etc. [4–24]. A large group of related physical phenomena are determined by the fact that the possibility of these states means abrupt decay appears. As a result, an aerosol can form with a characteristic particle size of the order of 10–100 nm. To obtain supercritical fluids, it is necessary to create a high energy density in the substance, which is achievable using the methods of pulsed energy injection into a substance with a maximum peak power.

There are several possibilities for obtaining supercritical fluid. The simulation of electrical breakdown in a liquid is given in our previous works [25–27].

One of the methods for creating the required energy density in metals is an electric explosion capable of providing a high current density of 108 A/cm^2 and a high specific power of Joule heating of a substance. However, in this case, the requirements for the power system electrical circuit are extremely high. Requirements may be reduced by using mechanisms of spatial power concentration in the substance. In particular, this can be the magnetic implosion of cylindrical metal shells or multi-wire assemblies on the axis of the considered configuration.

2. Materials and Methods

The electromagnetic implosion of plasma shells makes it possible to obtain large magnetic field pressures with a short-rise front (~10 ns), which can be effectively converted into thermal plasma energy. Moreover, the conversion of magnetic pressure pulses into thermal pressure pulses can be carried out in the exacerbation mode [7].

In this research, we analyze the characteristics of a high-current Z-discharge through the plasma shell (liner), which allows powerful pressure pulses with amplitudes exceeding the critical pressure of many substances to be obtained. Figure 1 shows the configuration of the high-current Z-discharge considered in this work and its power supply system:

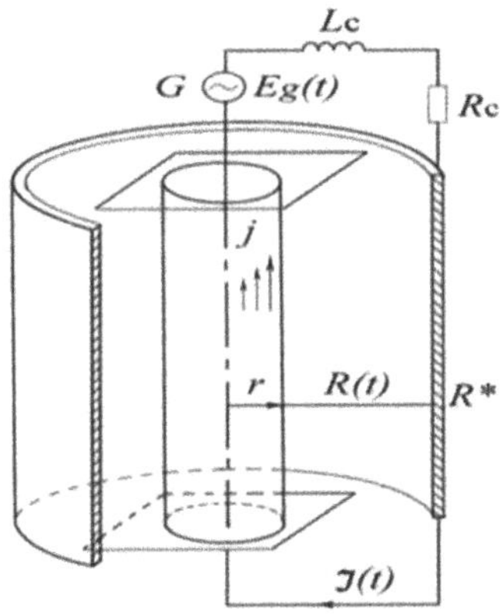

Figure 1. The electrical circuit of the discharge.

The main element of the plasma load is a cylindrical plasma shell consisting of heavy, highly-emitting ions. The plasma envelope is covered by a massive cylindrical shell coaxial, which acts as a "reverse conductor". In the center, along the axis of the plasma shell, a cylindrical plasma bunch with an initial diameter approximately ten times less than the inner initial diameter of the shell can be located. In such a "liner" system, the kinetic energy of the shell when it reaches the axis turns into the energy of a converging cylindrical shock wave. Moreover, before the onset of the shock wave, the plasma of the shell becomes denser; as a result of the compression of the shell, the possibility of sharpening the radiation pulses and pressure in the shock wave appears.

We consider a discharge power supply system based on lines with distributed parameters, which is more promising for use in comparison with systems with lumped parameters. The generator voltage is considered to be a given function of time. The wave impedance of the forming lines is also set as a parameter of the problem (Figure 1).

To describe the dynamics of the plasma shell, we use the approximation of single-fluid two-temperature radiation magnetic hydrodynamics (RMHD). We assume that electrons and plasma ions move as a single medium with one hydrodynamic velocity u and that the electrostatic field is small. In this case, each component has its own temperature T_i and T_e of ions and electrons, respectively. The system of equations was solved in the axisymmetric region, considering the azimuthal and longitudinal components of the electromagnetic field. Ions are heavy particles, considered as one subsystem with internal energy per unit mass E_i and pressure P_i. Electrons, light particles, are another subsystem of the medium with internal energy per unit mass E_e and pressure P_e, respectively. Both subsystems have the same hydrodynamic velocity.

To calculate the plasma composition, we used a simplified version of the impact-radiation model, which takes into account electron impact ionization, dielectronic recombination, triple collision recombination, and photo recombination The main processes that determine the ionization composition of plasma are ionization by electron impact, photo recombination at low density (in the "corona"), and recombination in triple collisions, which plays a role at a sufficiently high density. At a high density of heavy particles, the multiplicity of ionization can be determined using the Saha formulas [8].

The kinetics of ionization can be taken into account using the average ion model, in which the ionization state of the plasma is calculated from the average ion Z [14]. The change in Z in time is described by the equation:

$$\frac{dZ}{dt} = Z\cdot(v_i - v_r - v_\omega), \tag{1}$$

in which the rates of ionization by electron impact v_i, photo recombination v_r, and triple recombination v_ω are taken in the interpolation form.

$$v_i = 10^{-7}\cdot N_i\left(\frac{I_H}{I_Z}\right)^{3/2}\exp\left(-\frac{I_Z}{T_e}\right)\frac{(I_Z/Te)^{1/2}}{1+(I_Z/T_e)},$$

$$v_r = 3.310^{-31}Z\cdot N_i^2\left(\frac{I_H}{I_Z}\right)^{3/2}\left(\frac{I_H}{T_e}\right)^{3/2}\frac{(I_Z/T_e)^{1/2}}{1+(I_Z/T_e)},$$

$$v_\omega = 6\cdot 10^{-14}\cdot\lambda\cdot N_i\left(\frac{I_Z}{I_H}\right)^{1/2}\left(\frac{I_Z}{T_e}\right)^{3/2}\frac{1}{1+(I_Z/Te)}$$

At high density, in the equilibrium case $dZ/dt = 0$, we have the system of Saha equations, and in the case of low density and high temperature, the coronal approximation. In this case, the ionization processes are not in equilibrium with increasing temperature.

From a comparison of the reaction rate constants, it follows that photo recombination plays a predominant role under the condition:

$$N_e = ZN_i << 10^{17}\left(\frac{T_e}{I_H}\right)$$

Otherwise, photo processes can be ignored. Then, in the case of equilibrium, we arrive at the Saha model in the Raiser approximation.

$$Z = 3.1610^2 m_i \rho^{-1} T_e^{3/2} e^{-\frac{I(Z)}{T}}$$

To simulate plasma dynamics using the RMHD method, we use Lagrangian variables. The system of equations describing the main features of the process under consideration has the following form:

$$\frac{du}{dt} = -r\frac{\partial P}{\partial m} + F;\ \frac{dr}{dt} = u;\ P = P_i + P_e + P_\omega;$$

$$P_\omega = -\rho(\nu+\nu_a)\frac{\partial u}{\partial m} + \mu_a\rho\left(\frac{\partial u}{\partial m}\right)^2; F = -\frac{\rho}{\mu_0 r}\frac{\partial}{\partial m}(rB)^2;$$

$$\frac{d\varepsilon_i}{dt} = -P_i\frac{d}{dt}\left(\frac{1}{\rho}\right) - Q_{ei} - P_\omega\frac{d}{dt}\left(\frac{1}{\rho}\right); \tag{2}$$

$$W_e = -\chi_e\rho^2\frac{\partial T_e}{\partial m};\ Q_j = \frac{1}{\mu_0\sigma}E_z\frac{\partial}{\partial m}(rB_\varphi);$$

$$\frac{d}{dt}\left(\frac{B_\varphi}{r\rho}\right) = -\frac{\partial E_z}{\partial m};\ E_z = \frac{\rho}{\mu_0\sigma}\frac{\partial}{\partial m}(rB_\varphi).$$

Here:

ε_i, ε_g—specific internal energy of the ionic and electronic components;
Q_{ei}—rate of energy transfer from ions to electrons;
W_e—heat flow in the electronic component;
Q_j—power of joule heat;
χ_e¯coefficient of electronic thermal conductivity;,
σ—plasma electrical conductivity;
E_z—longitudinal component of the electrical field;
B_ϕ—azimuthal ccomponent of the magnetic field
S—radiation energy flux.

The contribution of radiation to the energy equation for the electronic component is determined by the expression:

$$\frac{\partial S}{\partial m} = \int_0^\infty k_\nu \cdot (U_{\nu p} - U_\nu) \cdot d\nu, U_\nu = 2\int_{-1}^{1} d\gamma \int_{-1}^{1} \frac{I_\nu}{\sqrt{1-\mu^2}} d\mu. \quad (3)$$

In the numerical solution, the frequency integral is replaced by the sum over the spectral intervals, in which the absorption coefficient is assumed to be independent of the frequency:

$$\frac{\partial S}{\partial m} = \sum_{i=1}^{k} k_i (U_{ip} - U_i). \quad (4)$$

The magnetic field at the shell boundary is determined by the current flowing through the shell, which depends on the characteristics of the electrical circuit. For the stationary case, the equation is written as:

$$Ug(t) - R_\rho I - L_C\frac{dI}{dt} - \frac{d}{dt}(L_\Delta I) = 0. \quad (5)$$

Here, L_C is the inductance of the electrical circuit; L_Δ is the gap inductance "reverse" conductor, the outer surface of the plasma envelope.

As equations of the state of the shell plasma and the plasma bunch, we applied the generalized equations of state of the "average ion" type [10], in which we used the results of calculating the ionization according to the simplified impact-radiation model described above. The calculation of plasma transfer coefficients is based on a wide-range semi-empirical model described in [13,14].

The equations of plasma dynamics, together with the equations of the electric circuit, radiation transfer, and the ionic composition of plasma, are solved by a completely conservative implicit difference scheme implemented by the combined sweep method.

3. Results

We considered the process of the electromagnetic implosion of a cylindrical plasma shell with an outer radius of 0.02 m and an inner radius of 0.01 cm, obtained as a result of an electric explosion of a cylindrical aluminum shell. The simulation results are shown in Figures 2–5. The initial values of the plasma temperature and density are T_0 = 0.7 eV and ρ_0 = 0.05 g/cm^3, respectively. The characteristics of the initial state of the plasma are chosen so that the initial temperature of the plasma is approximately equal to the critical temperature of the metal, and the density is less than the density of the metal at the critical point by more than an order of magnitude. The values of the critical parameters for aluminum and copper are presented in Table 1.

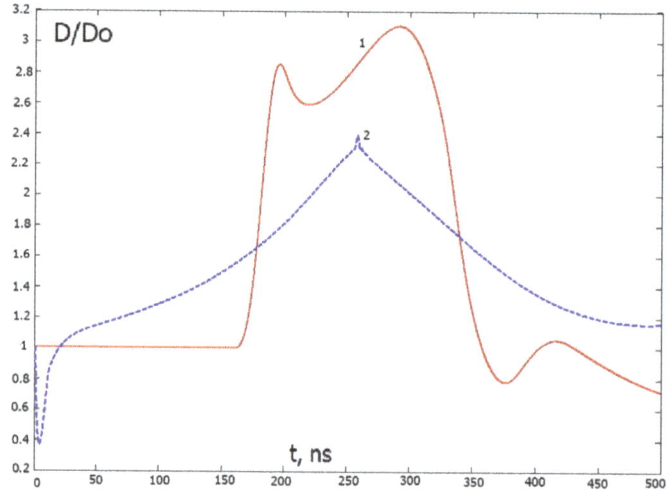

Figure 2. Change in density on the axis (1) and at the plasma boundary (2). The plasma density on the axis exceeds the critical value for 15 ns.

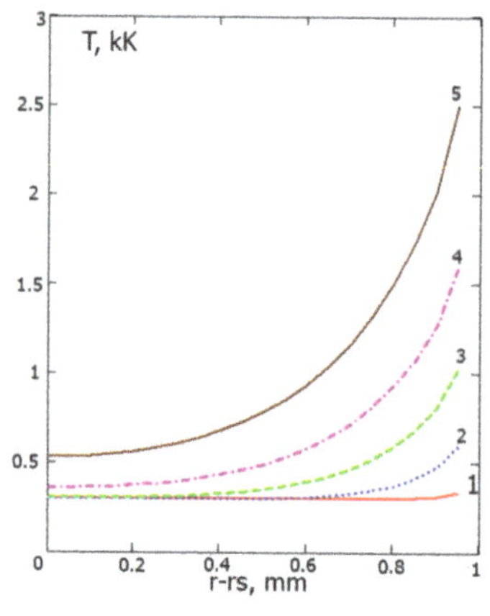

Figure 3. Radial temperature distributions in a copper cladding 1 mm thick at different times; *t*, ns: 1–5, 2–20, 3–50, 4–100, 5–200.

Table 1. Critical parameters of metals.

Temperature, K	Density, g/cm^3	Pressure, GPa
Aluminum, 8000	0.64	0.45
Cuprum, 8400	0.89	0.75

The temporary change in the discharge current is shown in Figure 2. The amplitude of the current pulse is ~90 kA; the front of the current rise is about 20 ns. The magnetic fields pressure of the magnetic field on the shell's surface PB = 6 GPa.

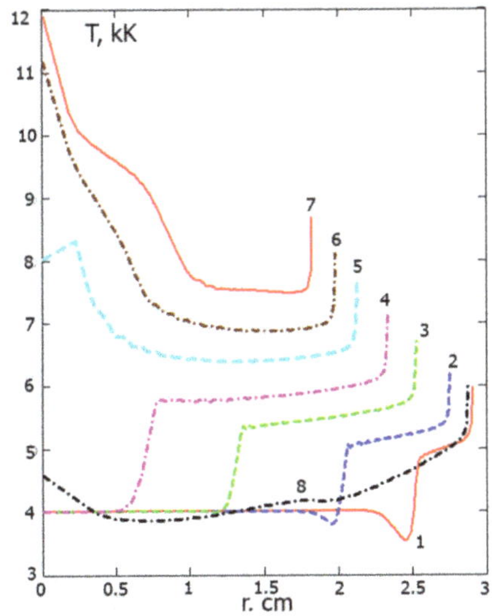

Figure 4. Temperature distributions of plasma compressed by a copper sheath; t, ns: 1–50, 2–80, 3–120, 4–150, 5–180, 6–200, 7–300, 8–360.

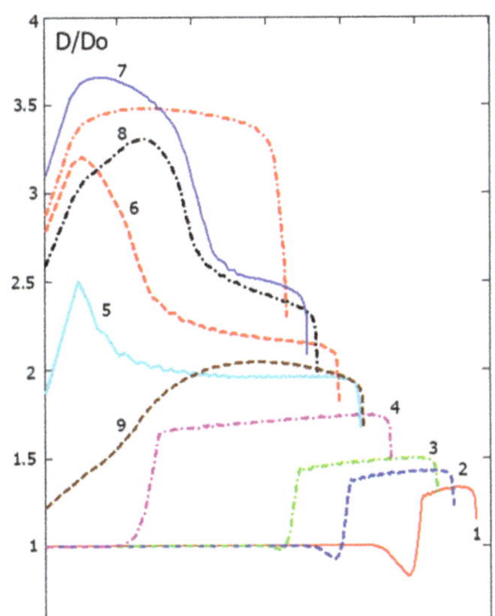

Figure 5. Density distributions of the plasma compressed by the shell at different times; t, ns: 1–50, 2–80, 3–100, 4–150, 5–180, 6–200, 7–220, 8–260, 9–290.

4. Discussion

The study revealed the following features of the process.

In order for the plasma compression to be effective, it is necessary to form a magnetic piston that compresses the plasma towards the axis. Therefore, the current rise time t_0 must be less than the plasma compression time $t_{imp} \sim R_0/U_A$, where R_0 is the initial plasma radius and U_A is the Alfen velocity.

The relationship $D_s << R_0$ must also be satisfied; where D_s is the thickness of the skin layer. These conditions determine the requirements for matching the characteristics of the current pulse and the plasma load, as well as the initial states of the plasma (temperature and density).

The efficiency of gas (plasma) compression decreases if during the compression time the magnetic field of the current flowing through the shell penetrates into the shell. Therefore, the thickness Δr of the shell is chosen from the condition that, during the compression time, the magnetic field of the discharge current does not penetrate inside the shell, i.e., skin layer thickness $\delta S < \Delta r$. No less important is the condition that, during the compression, the shell substance remains in a condensed (solid or liquid) state, i.e., will not heat up above the temperature at which intensive vaporization begins. These conditions determine the shell thickness Δr. We find this quantity by considering the nonlinear diffusion of a pulsed magnetic field into a cylindrical heating shell.

Under the action of an increasing magnetic field, a shock wave propagates through the plasma of the liner. As a result, the ions are heated more strongly than electrons. Electrons continuously lose energy mainly through radiation. The ion-electron energy exchange due to elastic collisions is not effective due to the small value of the mass ratio of electrons and ions.

The magnetic field of the discharge current acting on the outer boundary of the plasma leads to the formation of a relatively narrow layer of dense high-temperature, highly-ionized plasma near the outer boundary of the shell. In the process of accelerating the shell to the axis, the plasma density in this layer increases to values greater than the initial plasma density in the shell by more than an order of magnitude. The temperature of electrons and the average charge of ions increase several times.

It should be noted that, at the stage of acceleration of the shell toward the axis, the inclusion of radiation does not have a strong effect on the plasma dynamics. When the radiation is taken into account, the maximum electron temperature and the thickness of the dense high-temperature layer turn out to be no more than 20% less than without taking it into account.

At the final stage, the plasma envelope accelerated to high speeds accumulates at the axis. As a result, the kinetic energy is converted into the thermal energy of the plasma, and a clot of dense high-temperature, highly ionized plasma is formed, in which a pulse pressure is formed that exceeds the maximum pressure of the magnetic field created by the discharge current. The pressure pulse duration is $\Delta t = \Delta r / U_r$. In this study, there is a maximum pressure that is several times greater than the pressure at the critical point.

A plasma bunch formed on the axis effectively emits in the short-wave spectral range. At this stage, radiation plays an essential role. Plasma parameters determined with and without radiation transfer can differ by more than two times. As the results of calculations of the photon path lengths show, averaged "over the bar" and "over the Rosseland" are many times larger than the characteristic size of the cross-section of the plasma bunch formed on the axis. Therefore, plasma can be considered transparent in the continuum.

5. Conclusions

An idealized version of the plasma shell electromagnetic implosion, which does not take into account important factors such as line radiation and the development of instabilities, was considered. According to preliminary estimates, these factors can reduce the plasma parameters, which are formed approximately 1.5–2 times in the process of shell accumulation at the axis. This limits, but does not exclude, the possibility of obtaining the considered method of supercritical states of substances using moderately high currents ($I \sim 100$ kA).

Thus, the calculations show the fundamental possibility of obtaining a supercritical fluid by the method of electromagnetic implosion.

It is possible to continue this research work in the following directions. Firstly, the current pulses considered in the calculations can be obtained, but technologically, this is not an easy task, due to the stringent requirements for the total inductance of the electro-physical system. Secondly, the paper considers rather high initial parameters of the plasma state.

In this regard, it is expedient to consider implosion at longer fronts of the current pulse and at lower initial plasma parameters.

Author Contributions: Conceptualization. V.A. and N.K.; methodology V.A. and N.K.; software, N.K.; validation, V.A. and N.K.; formal analysis V.A.; writing—original draft preparation, V.A. All authors have read and agreed to the published version of the manuscript.

Funding: This research received no external funding.

Conflicts of Interest: The authors declare no conflict of interest.

References

1. Al'tshuler, L.V.; Zhuchenko, V.S.; Levin, A.D. *Shock Waves and Extreme States of Matter*; Nauka: Moscow, Russia, 2000.
2. Fortov, V.E. *Extreme States of Matter*; FIZMATLIT: Moscow, Russia, 2009.
3. Bakulin, Y.D.; Luchinsky, A.V. Estimates of the possibility of obtaining high densities energy in the electric explosion of cylindrical shells. *J. Appl. Mech. Tech. Phys.* **1980**, *1*, 115–122.
4. Bolmatov, D.; Brazhkin, V.; Trachenko, K. Thermodynamic behavior of supercritical matter. *Nat. Commun.* **2013**, *4*, 2331. [CrossRef] [PubMed]
5. Bolmatov, D.; Zhernenkov, M.; Zav'yalov, D.; Tkachev, S.N.; Cunsolo, A.; Cai, Y.Q. The Frenkel Line: A direct experimental evidence for the new thermodynamic boundary. *Sci. Rep.* **2015**, *5*, 15850. [CrossRef] [PubMed]
6. Bolmatov, V.; Brazhkin, V.; Trachenko, K. Helium at elevated pressures: Quantum liquid with non-static shear rigidity. *J. Appl. Phys.* **2013**, *113*, 103514. [CrossRef]
7. Gasilov, V.A.; Zakharov, S.V.; Krukovsky, A.Y.; Skorovarov, K.V. Generation of intense radiation fluxes and megabar shock waves by compressing liners. *Plasma Phys.* **1995**, *21*, 399–406.
8. Frank-Kamenetsky, D.A. *Lectures on Plasma Physics*; Atomizdat: Moscow, Russia, 1968.
9. Burtsev, V.A.; Ermolaev, Y.L.; Kalinin, N.V.; Petrov, I.B. Imploding plasma dynamics and Radiation. *Plasma Devices Oper.* **1994**, *2*, 239–262. [CrossRef]
10. Afanasyev, Y.V.; Gamaliy, E.G.; Rozanov, V.B. Basic equations of dynamics and kinetics of laser plasma. Theory of heating and compression of low-entropy thermonuclear targets. *Trudy FIAN* **1982**, *134*, 10–31.
11. Basko, M.M. Metallic equation of state in the mean ion approximation. *High Temp. Thermophys.* **1985**, *23*, 388–396.
12. Basko, M.M. Generalized van der Waals equation of state for in line use in hydrodynamic code. *Keldysh Inst. Prepr.* **2018**, *112*, 28. [CrossRef]
13. Prut, V.V. Semiempirical Model of Equation of State for Condensed Media. *Therm. Phys. High Temp.* **2005**, *43*, 711–726. [CrossRef]
14. Bespalov, I.M.; Polishchuk, A.Y. Method for calculating the degree of ionization and the thermal and electrical conductivity over a wide range of density and temperature. *Sov. Tech. Phys. Lett* **1989**, *15*, 39–41.
15. Polishchuk, A.Y. Optical properties of plasma in extreme states. *Therm. Phys. High Temp.* **1990**, *28*, 656–663.
16. Zeldovich, Y.B.; Raizer, Y.P. *Physics of Shock Waves and High-Temperature Hydrodynamic Phenomena*; FIZMATLIT: Moscow, Russia, 2008.
17. Velikovich, V.L.; Golberg, S.M.; Lieberman, M.A.; Felberg, F.S. Hydrodynamics of plasma compression with a frozen-in magnetic field by a thin cylindrical wall. *JETF* **1985**, *88*, 445–460.
18. Lieberman, M.A.; Lichtenberg, A.J. *Principles of Plasma Discharges and Materials Processing*; John Wiley & Sons, Inc.: New York, NY, USA, 1994.
19. Samukawa, S.; Hori, M.; Rauf, S.; Tachibana, K.; Bruggeman, P.; Kroesen, G.; Whitehead, J.C.; Murphy, A.B.; Gutsol, A.F.; Starikovskaia, S.; et al. The 2012 Plasma Roadmap. *J. Phys. D Appl. Phys.* **2012**, *45*, 253001. [CrossRef]
20. Li, J.; Hao, L.; Li, J. Theoretical modeling and numerical simulations of plasmas generated by shock waves. *Sci. China Technol. Sci.* **2019**, *62*, 2204–2212. [CrossRef]
21. Hu, S.X.; Karasiev, V.V.; Recoules, V.; Nilson, P.M.; Brouwer, N.; Torrent, M. Interspecies radiative transition in warm and superdense plasma mixtures. *Nat. Commun.* **2020**, *11*, 1989. [CrossRef]
22. Weidl, M.S.; Winske, D.; Jenko, F.; Niemann, C. Hybrid simulations of a parallel collisionless shock in the large plasma device. *Phys. Plasmas* **2016**, *12*, 122102. [CrossRef]
23. Heuer, P.V.; Weidl, M.S.; Dorst, R.S.; Schaeffer, D.B.; Bondarenko, A.S.; Tripathi, S.K.P.; Compernolle, B.V.; Vincena, S.; Constantin, C.G.; Niemann, C.; et al. Observations of a Field-Aligned Ion/Ion-Beam Instability in a Magnetized Laboratory Plasma. *Phys. Plasmas* **2018**, *25*, 032104. [CrossRef]
24. Chahal, B.S.; Ghai, Y.; Saini, N.S. Low-frequency shock waves in a magnetized superthermal dusty plasma. *J. Theor. Appl. Phys.* **2017**, *11*, 181–189. [CrossRef]
25. Antonov, V.; Kalinin, N.; Kovalenko, A. Dynamic generation of supercritical water fluid in a strong electrical discharge in a liquid. *IOP Conf. Ser. Mater. Sci. Eng.* **2016**, *158*, 012007. [CrossRef]

26. Ryakhovskiy, A.I.; Antonov, V.I.; Kalinin, N.V. The EOS choice effect on the simulated results obtained for an underwater electrical explosion of conductors. *St. Petersburg Polytech. State Univ. J. Phys. Math.* **2017**, *10*, 26–37. [CrossRef]
27. Kalinin, N.; Antono, V.; Vaganov, S.; Kovalenko, A. The use of inhomogeneous forming lines in the laser power supply system on a high-current Z-discharge plasma. In *X-ray Lasers 2018: Proceedings of the 16th International Conference on X-ray Lasers*; Springer Nature: Berlin, Germany, 2020; Volume 241.

energies

MDPI

Article

Online Multiphase Flow Measurement of Crude Oil Properties Using Nuclear (Proton) Magnetic Resonance Automated Measurement Complex for Energy Safety at Smart Oil Deposits

Rustem Kashaev [1], Nguyen Duc Ahn [1], Valeriya Kozelkova [1], Oleg Kozelkov [1] and Valentin Dudkin [2,*]

1 Department of Instrumentation and Mechatronics, Kazan State Power Engineering University, 420066 Kazan, Russia
2 Department of Photonics and Communication Lines, The Bonch-Bruevich Saint Petersburg State University of Telecommunication, 193232 St. Petersburg, Russia
* Correspondence: vidoodkin@mail.ru

Abstract: The necessity of a flow express control of oil dispersed system (ODS) properties, such as crude oil, oil products, water–oil emulsions, and polluted waters, is substantiated. This control is necessary for the production and preparation of oil for transportation through the pipeline and oil refining, oil products, and wastewater treatment systems. A developed automatic measuring complex (AMC) is used to implement the concept of digital oil deposits. The primary measuring device is a relaxometer developed by us based on nuclear (proton) magnetic resonance (PMR). The design and operation algorithm of the AMC and the relaxometer are described. Equations have been developed to determine the ODS characteristics using the measured PMR parameters. This makes it possible to determine the flow rates of crude oil, the concentration of water in the oil, the concentration of asphaltene, resins, and paraffins in the oil, as well as the density, viscosity, and molecular weight of the oil. Additionally, it is possible to determine the dispersed distribution of water droplets in emulsions in oil production and treatment units. Data on this distribution will improve the management of separation processes. It has been established that the implemented control of multiphase ODS using PMR parameters (relaxation times, populations of proton phases, and amplitudes of spin-echo signals) makes it possible, using AMC, to assess the consumption of electricity in technological processes at the digital oil deposits, as well as during the transportation of oil and oil products through pipelines. AMC makes it possible to reduce electrical energy consumption in technological installations and reduce pollution emissions into wastewater. The advantages of using the developed AMC are shown in examples of its application. Such as an assessment of the influence of the gas factor on electricity consumption during oil transportation through pipelines or compensation for the additional moment of resistance on the shaft of the submersible motor, which is caused by surface tension forces at the interface of water droplets in the emulsion.

Keywords: express-control; smart digital oil deposit; nuclear magnetic resonance; energy save; oil disperse systems; fuel emulsions; correlations; relaxation times; ODS properties

Citation: Kashaev, R.; Ahn, N.D.; Kozelkova, V.; Kozelkov, O.; Dudkin, V. Online Multiphase Flow Measurement of Crude Oil Properties Using Nuclear (Proton) Magnetic Resonance Automated Measurement Complex for Energy Safety at Smart Oil Deposits. *Energies* **2023**, *16*, 1080. https://10.3390/en16031080

Academic Editor: Dan Gabriel Cacuci

Received: 17 December 2022
Revised: 10 January 2023
Accepted: 16 January 2023
Published: 18 January 2023

1. Introduction

One of the critical factors in oil production and refining is cost reduction [1–6]. Another important area in the oil and gas sector is to reduce the negative impact on the environment at various stages of oil production, processing, and transportation [7–11]. Implementing projects based on the formation of digital smart fields (DSFs) fits into managing oil production, preparation, and pumping through pipelines [12–14]. The main elements of DSFs in the oil and gas complex are automatic measuring complexes (AMC). Their use on various DSFs increases production by up to 10–25% and reduces electrical energy consumption by up to 8% [15,16]. The latter is extremely important, especially when using autonomous power stations [17–19]. Digital automation of old oil and gas fields provides an opportunity to convert them into new stages of exploitation [20,21]. It is necessary since the last cycles of field operation are characterized by an increase in high-viscosity, high-water (up to >95%) oil with high concentrations of asphaltene, resins, and paraffins. In this situation, it is necessary to constantly monitor the state of the media and control technological processes and the operation of various installations. For example, in the Romashkinskoye field of the Republic of Tatarstan (Russian Federation), there are only 15,000 wells in the final stages, producing oil with an average water content of 87%. Electric energy consumption for the maintenance of these wells has increased by 48% over the past three years. In addition, there is an increase in the concentration of asphaltene–resins–paraffins (ARP) deposits in the oil produced at these fields. This increases the viscosity of the pumped mixture and ARP deposits on the tubing of wells and pipelines, leading to equipment decommissioning. These problems arise not only in heavy oil fields in the Russian Federation, but they are also present in oil production on the artic shelf, on offshore drilling platforms in the northern part of the globe, as well as in the northern regions of the United States (Alaska) and the Channels [22–26] and in the operation of oil wells in the southern part of Patagonia (Argentina).

To solve these problems, it is necessary to introduce new methods and instruments to control the properties of ODS and their flow rates in pipelines under changed difficult conditions. An analysis of various studies and authors' work experience has shown that for a comprehensive solution to the problems noted, it is necessary to control the following parameters in ODS in real time. These are the concentrations of water and ARP, the viscosity and molecular weight of oil, the pour point, and the dispersed distribution of water droplets in emulsions, crude oil, and salt-contaminated waters. In addition, when extracting heavy oil grades, it is necessary to use centrifugal pumps. When changing the composition in the ODS, a problem arises associated with measuring the moment of resistance M_C on the shaft of the submersible pump motor, caused by the need to overcome surface tension forces at the interface between water drops and oil in a water–oil emulsion; the presence of water W and gas factor G, increased viscosity η and density ρ. This leads to a change in flow rates during well operation to an increase in electrical energy consumption. The availability of data on the ODS parameters and the G value allows for optimizing the production process and removing the unnecessary load from the equipment.

Many technologies and devices have been developed to measure these parameters, which have advantages and disadvantages [27–37]. Combining them into a single system up to a certain point allows you to successfully solve the problems noted until the quality of the ODS deteriorates significantly. This increases the measurement error in these devices and the wear of the measuring sensors. Therefore, in the world, preference for measurements under these conditions is given to non-contact devices [38–47]. There are many of these devices to ensure the measurement of all the noted parameters, which creates difficulties with their integration into a single system at the drilling rig, as well as the economic feasibility of use (cost, the need for maintenance by various companies). Therefore, it is most expedient to use methods and devices based on them, which would be less affected by these factors.

The non-contact, non-destructive, and express nuclear (proton) magnetic resonance method has such possibilities. This method is actively used in various systems for express control of condensed matter [48–52]. Despite a large number of developments and studies, its possibilities have not yet been fully disclosed, especially in the field of advanced petroleum engineering technologies [53–55]. The wide possibilities of a variety of PMR relaxometry (PMRR) in the control of ODS were demonstrated in [56–58], and it was found that PMRR is unique for the express control of oil emulsions [59]. The unique properties of the PMRR approach are associated with the possibility of quantum mechanical analysis of the chemical and physical properties of substances at different structural levels as a single complex. Based on several fundamental PMR parameters, the dynamics of molecules, phase compositions, and diffusion processes in oil and oil aggregates can be studied [55].

To control the three-component ODS of liquids in a flow, the authors of [60] studied the influence of a flowing liquid on PMR signals, determined the possibilities for measuring the distribution of flow velocities, and developed a method and instrument for studying them.

A feature of the PMRR method is that it allows for information to be obtained about the relaxation parameters of protons in their three phases $i = A, B, C$ in ODS (in particular, in emulsions): spin-lattice T_{1i} and spin-spin T_{2i} relaxation times characterizing the proton groups with different molecular mobility in two fractions of oil; the populations of the proton phases P_{1i} and P_{2i} corresponding to these times; and interproton distances R_{ij}. The method practically does not require sample preparation and reagents, and multiple accumulations of signals minimize error values during measurements and when using calibration curves built using standard samples with high correlation coefficients R^2 and with minimal deviations [56–59,61–67].

In their work, the authors of the proposed AMC tried to realize all the advantages of this method, as well as consider the advantages of several studies by scientists in the development of laboratory relaxometers and analyzers for multiphase flow measurements [60–62]. Particular attention was paid to the methodology for determining the ODS flow in a pipeline in real time, considering previous studies [64,68–73]. Electromagnetic and Coriolis flowmeters [74–78] currently used to determine the ODS flow cannot provide factory measurement accuracy with deterioration in oil quality. In addition, it is planned to implement the function of ecological monitoring of flowing liquids in the developed AMC [79–83].

2. Methodology and Apparatus for Monitoring of the Oil Disperse Systems

Currently, monitoring the quality of ODS in real time is difficult due to high pressures, temperatures, distributions of different flows, and phase separations in pipelines. In this case, it is necessary to use non-contact methods of control. This possibility is provided by a method based on the phenomenon of nuclear (proton) magnetic resonance [10,48,59,60,71,72,78].

To implement the technology of the method, an AMC was developed for automatic flow measurement of ODS properties using PMRR. Figure 1 shows a block diagram of the test bench of the complex.

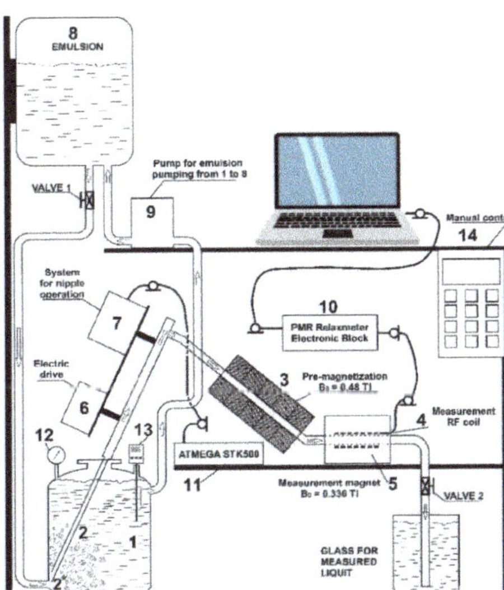

Figure 1. Structural diagram of the testing ground of automated measurement complex.

In the AMC, the NMR NP relaxometer [58,59,84] (which has no analogs) is the main module for determining the ODS properties. The relaxometer can be powered by a 12 V battery or the mains. The measurement time is less than 2 min. Its sensitivity is $K = \nu^2 D^3$ = 2700–4150 $MHz^2 cm^3$. Using a laptop program, the envelopes of the spin-echo (SE) signals are decomposed into three exponential components to determine the PMR parameters, by which the ODS characteristics are calculated.

Figure 2 shows the electric principle diagram of the sampling system for controlling the sampling from a stream.

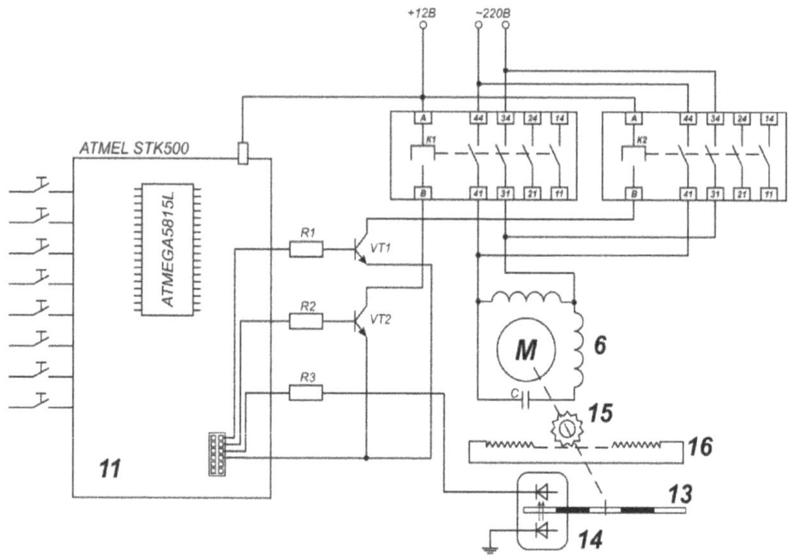

Figure 2. Electric principle diagram of the sampling system.

The principle of sample sampling is based on the Bernoulli equation, according to which, with a continuous flow, the change in pressure P_i in different sections S of the measuring tank 1 for flow velocity v_i is described by the equation:

$$P_i/\rho g + v_i^2/2g = \text{const}, \tag{1}$$

If flow rate $Q_i = S_i v_I$ is constant, then the pressures P_1 and P_2 in different sections S_1 and S_2 of the pipe will be connected by the equation:

$$P_I/\rho + \text{const}/S_I^2 = P_2/\rho + \text{const}/S_2^2, \tag{2}$$

The *ATMEGA* 8515*L* microcontroller with the *STK*500 kit for Atmel AVR flash controllers on the *SCKT*3000*D*3 panel automatically controls the sampling systems of the AMC. A sampling of emulsions (a complex multi-component ODS) is carried out in small portions according to the ISO 3171 Code of Practice. The fluid flow entering tank 1 reduces the velocity v at an increased pressure P (pressure and temperature are controlled by sensors 12 and 13) in proportion to the square root $\sqrt{S}$ of the cross section according to Equation (2). Under the action of the pressure difference ($P_P - P_B$) between the pipeline (on the bench—between tanks 8 and 1), the pressure P_P in the position of the nozzle 2* and P_2 in any position 2 in tank 1, all three components (water, oil, and gas) are intensely turbulent are mixed and homogenized in tank 1, and then the sample is delivered through the pre-polarizing Halbach magnets 3 with magnetic induction B_o = 0.48 Tl to the measuring radio frequency (RF) coil 4, located between the pole pieces of magnet 5 made of an alloy based on the rare earth element *NdFeB-37* with magnetic induction B_o = 0.336 Tl (resonant frequency on protons $\nu_o = 14.32 \times 10^6$ Hz) and field inhomogeneity $\delta B_o = 10^{-3}$ in 1 cm^3.

Due to the coiled sensor coil, the RF field B_1 inhomogeneity is less than 2% in 75% of the sample volume. To measure flow rates, the nozzle is located in position 2, for which the dependence of the spin-spin relaxation times T_2* in the flow at the maximum steepness of the dependence is pre-calibrated and entered into the laptop database. For measurements of emulsion properties, the nozzle is placed in position 2*, at which the pressure difference $(P_P - P_2) = 0$, and thus the liquid in coil 4 is stationary. This eliminates the need for explosion-proof valves. In addition, the branch pipe can be moved with the help of an electric drive 6 controlled by *ATMEGA 8515L 11* to any section of tank 1 at a distance from 1 to 150 mm with a step of 5 mm and, accordingly, sampling can be done from any section; it is possible to average over all sections. The movement of the branch pipe connected to the band 16 is controlled by the obturator 13 on the axis with the rotor of the electric drive 6, the rack-wheel 15, the number of rotations of which is counted by photodiodes 14 according to the rotations of the obturator 13 (see Figure 2). In coil 4 of sensor magnet 5, the sample is irradiated with the known Carr–Purcell–Meiboom–Gill (CPMG) sequence $90° - \tau_o - (180° - 2\tau_o)_N - T$, where N is the number of 80°-pulses, T = 9 s—series start period, the time between pulses τ_o = 200 µs. Between 180°-pulses, spin-echo (SE) signals with amplitudes A_i are formed and transmitted via cable to the receiver amplifier of the relaxometer 10 (Figure 3).

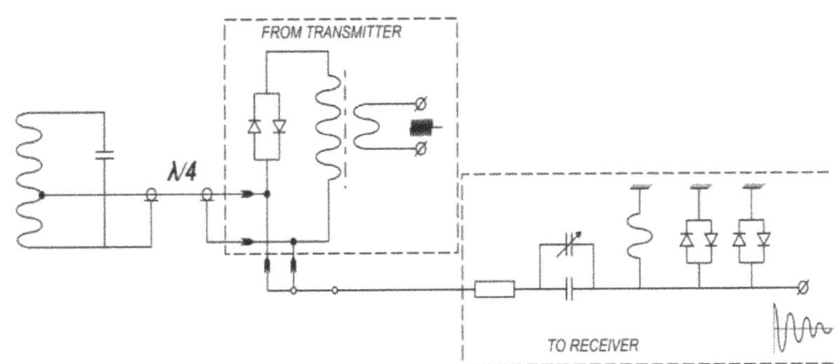

Figure 3. Principle electric scheme of the spin-echo signals transitions to/from RF coil.

PMR parameters are generated from measurements with a relaxometer, the electrical circuit shown in Figure 4. The measured liquid is squeezed out of coil 4 by the next portion from tank 1 and poured into a separate container to confirm the analysis result with a PMR laboratory relaxometer or alternative methods. The remaining liquid not selected for analysis is pumped into tank 8.

The process of measurements in the deposit is organized according to explosives and fire safety requirements. The cable length must be $l = \lambda_o/4$ or $3\lambda_o/4$ = 5.5 or 16 m (antinode of the standing wave), long enough for fire safety, where $\lambda_o = c/\nu_o$ is the resonant wavelength, *c*—light velocity. Then, analog signal data are transferred to an analog-to-digital converter (ADC) 11 and notebook 12, using the program for SE exponential envelope decomposition by equation $A = \sum A_i \exp(-t/T_{2i})$, and equations, correlating PMR-parameters with the oil characteristics, oil properties are obtained. Relaxation times T_{2i} and proton phase concentrations are attributed to i = *A*, *B*, *C* phases in water, light (benzene), and heavy (oil residues) oil fractions.

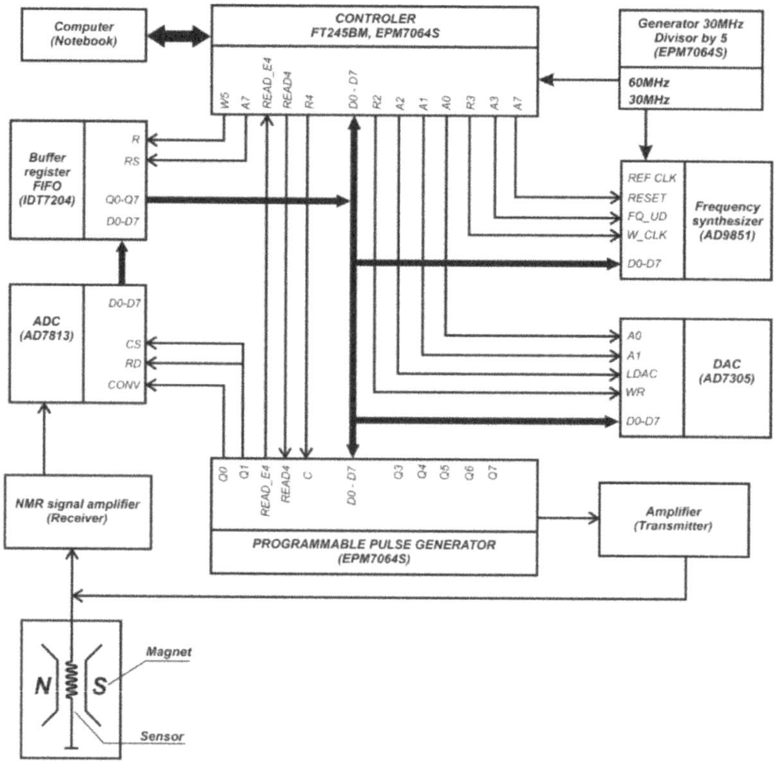

Figure 4. Principle electric scheme of Relaxometer PMR-NP.

To measure the values of magnetic induction, the developed sensor on the AD22151YZX chip that implements the Hall effect was used. A special program of the Arduino Uno/Nano microcontroller calculates the field value using a 10-bit ADC with a frequency of 10 kB/s. The data are calculated and displayed via the Arduino USB port on the laptop monitor. With a 5 V supply and an output ratio of 0.4 mV/G, the maximum measurement range is Bo = 1.25 Tl.

The developed AMC has the following advantages:

1. Versatility and ease of installation in production lines, and control of opaque, aggressive liquids in real time.
2. A wide range of measured ODS characteristics in the entire range of their changes: velocities υ_i of the fluid component flows; concentrations of water *W* and oil *O*, gas factor *G*, density ρ, viscosity η, and concentrations of ARP; and molecular weight and pour points. Multi-component analysis by a single complex, and selection of homogenized samples from pipes of any diameter in the bypass mode.
3. Lack of contact with the measured liquid and, therefore, the absence of its destruction and destructive effect on the equipment. No moving parts for measurements.

3. Results and Discussion

The possibilities of online flow measurements using PMR relaxation are determined by the following: the liquid flowing into the RF coil of the sensor in the gap of the magnet of the PMR relaxometer has the magnetization M_{IN}, and the fluid flowing out of the coil has the magnetization M_{OUT}, then the magnetization M of the liquid flowing in the coil through its volume V at an average flow rate Q will change at a rate of:

$$dM/dt = (M_{IN} - M_{OUT})Q/V, \quad (3)$$

The equation describes the rate of magnetization M change due to relaxation processes:

$$dM/\ dt = (M_0 - M)/T_2, \quad (4)$$

If the conditions $M_{IN} = M_0$ and $M_{OUT} = M$ are true, then the change in the M rate will be:

$$dM/dt = (M_0 - M)(1/T_2 + 1/T_2{}^{/}) = (M_0 - M)/T_2{}^{*}, \quad (5)$$

where $T_2{}^{/}$—is the time of liquid being in the RF-coil, at which the 1/e part of depolarized protons is substituted by the polarized. If the probe head liquid is fully mixed, then $T_2{}^{/} = V/Q$. However, it is true only for one-phase liquid.

For two-phase oil–water emulsions, the correctness of Equation (5) is confirmed by us experimentally. For 100% water, 90%, 75%, and 25% emulsions were received dependences of effective spin-spin relaxation rates $(T_{2eff})^{-1}$ from the flow velocity v in the range $v = 0 \div 0.7$ m/s, presented in Figure 5.

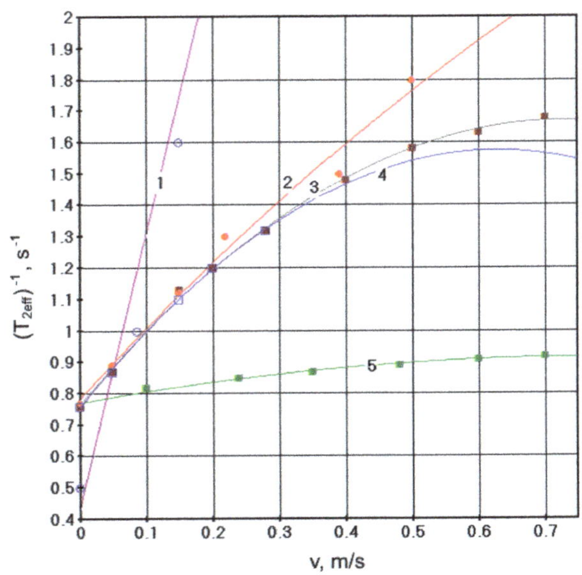

Figure 5. Relaxation rate $(T_{2eff})^{-1}$ from oil-well liquid flow velocity v for 1 is water, 2 is 90% water emulsion, 3 is 85% water, 4 is 25% water, and 5 is 20% water emulsion flow.

For used emulsions, the dependences are two component, and with correlation coefficients, $R^2 = 0.93$–0.99 and mean quadratic error $S = 0.01$–0.08 for $v > 0.2$ m/s are described by the equation [56–59,63,70,79]:

$$v\ (\text{m/s}) = k_1 \exp(-k_2 \cdot T_2), \quad (6)$$

where k_1 (m/s) = 6.2; 6.1 and 24; k_2 (s^{-1}) = 3.2; 3.7 and 13.8 for 90%, 75%, and 25% emulsions. For $v < 0.2$ m/s for the same emulsions, the equation is:

$$v\ (\text{m/s}) = k_3/T_2 - k_4, \quad (7)$$

where k_3 (m) = 0.45; 0.76 and 1.77; k_4 (m/s) = 0; 0; 1.355.

Coefficients k_1, k_2, and k_3 depend on water concentration W in emulsions [56–59,63,70,79]:

$$k_1 = 40.2\exp(-0.022W), \tag{8}$$

$$k_2 = 24.1\exp(-0.023W), \tag{9}$$

$$k_3 = 3\exp(-0.02W), \tag{10}$$

Equations (8)–(10) are necessary for the choice by the computer program of grade curves for flowrates v_{PMR} from $T_{2eff}{}^{-1}$ measurements. In addition, as the spin-echo amplitude envelope can be decomposed on the several (usually three) components with proper relaxation times and proton phases concentrations, an opportunity appeared for the determination of flow viscosities v_i and yields of emulsions components Q_i using the equation [56–59,63,70,79]:

$$Q_i = v_i \cdot S = Q \cdot A_{oi} / \sum A_{oi}, \tag{11}$$

where S—pipeline cross section, and $A_{oi}/\sum A_{oi}$—protons concentration of the i-th component in emulsion (water or oil fractions), determined from the spin-echo amplitude envelope.

For the trustworthy proper appreciation of Equation (5) and precision of the curves in Figure 5, relaxation times T_{2oW} and T_{2oO} in the immobile water and oil are compared, estimated from the curves with the relaxation times, and calculated from Equation (5). From curve 1 in Figure 5 for immobile water a relaxation rate $(T_{2oW})^{-1}$ = 0.43 s^{-1} is received, which corresponds to T_{2oW} = 2.32 s and differs from T_{2W} = 2.26 s received from a direct measurement in immobile water on 0.06 s. From curve 3, the most complicated for measurements, emulsion (due to phase inversion at $W \sim 75\%$) was received $(T_{2o75})^{-1}$ = 0.654 s^{-1}. Considering the different contributions of the components in this 75% emulsion, a value of $(T_{2o75})^{-1}$ is calculated considering the percentage of 75% and 25% of the phases by the equation [56–59,63,70,79]:

$$(T_{2o75})^{-1} = 0.75(T_{2oW})^{-1} + 0.25(T_{2oO})^{-1} = 1.375, \tag{12}$$

where $(T_{2oO})^{-1}$ = 1.37. The result from Equation (12) calculation corresponds to T_{2oO} = 0.727 s, which differs from T_{2O}= 0.7 s of direct measurement on 0.027 s. These trustworthy appreciations confirm the accuracy of the measurements in the range of error limits.

For 90% and 25% emulsions, the dependences can be described with correlation coefficient R^2 = 0.95 by equations [56–59,63,70,79]:

$$v\ (\text{m/s}) = 1.77/T_2{}^*(c) - 1.35, \tag{13}$$

$$v\ (\text{m/s}) = 6 \cdot \exp(-3.7 \cdot T_2{}^*(c)), \tag{14}$$

and the dependences are a monocomponent.

The dependences of SE amplitudes A (a.u.) from flow velocity v (m/s) for the same water and emulsions were also established. They are presented in Figure 6.

So, the flow rate can alternatively be determined from relaxation rates and echo amplitudes with error $\delta < \pm 2.3\%$.

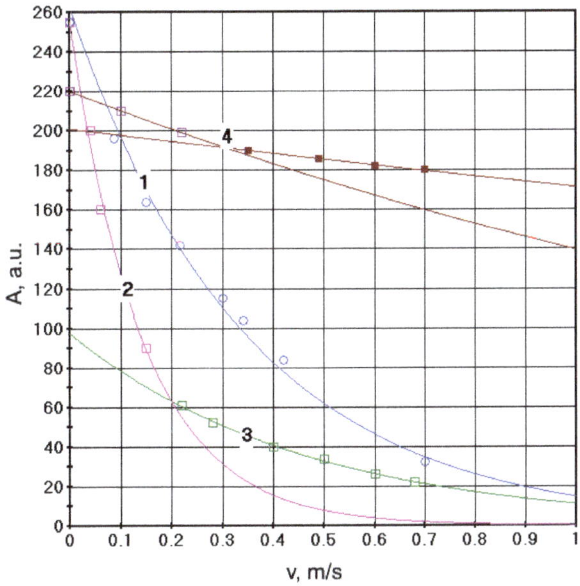

Figure 6. PMR-signal SE amplitudes A (a.u.) from flow rate$v$ (m/s). Curves: 1—100% water, 2—90% water in emulsion ($v < 0.2$ m/s), 3—90% water in emulsion ($v > 0.2$ m/s), 4—25% water in emulsion ($v < 0.3$ m/s), and 5—oil ($v > 0.3$ m/s).

Elaborated methods for the express control of crude oil properties have the following algorithm:

- Measurement of the spin-spin relaxation times and SE amplitudes A_i in immobile water T_{2W} and oil/oil product T_{2O} in the time range $t = 2N\tau$, where N—number of RF-pulses in the CPMG-sequence $90° - T\tau_o - T[180° - 2\tau_o]_N - T$;
- Online measurement of the effective spin-spin relaxation times $T_2{}^*$ in flow emulsion and using them for:

1. Determination of water concentration in emulsion by the relation [56–59,63,70,79]:

$$W_{\text{ПМР}} = T_{2W}\,(T_2{}^* - T_{2O})100\%/T_2{}^*\,(T_{2W} - T_{2O}), \tag{15}$$

The accuracy of single measurement in the range $0.5 \div 100\%$ $\delta \approx \pm 1\%$ in the immobile sample and $\delta \approx \pm 3\%$ in flowing liquid, which is better than for the nearest analog MERA-MIG with $\delta \approx \pm 10\%$ in the range 70–95%. The measurement time is three times shorter.

2. Determination of gas saturation of the oil-well liquid G_{PMR} in the range G_{PMR} = 0–250 with error $\delta \approx \pm 3.8\%$ using the equation [56–59,63,70,79]:

$$G_{PMR} = K_G(A_0 - A_G)/A_0, \tag{16}$$

where A_0 and A_G—initial SE amplitudes in the filled by liquid probe head and filled oil-well liquid, containing gas, K_G—correction coefficient. It should be mentioned that PMR parameter G_{PMR} allows for the control of specific energy consumption (SEC) at pipeline transportation because for SEC and G_{PMR}, the following equation is valid:

$$SEC = 155.3 - 0.796G_{PMR}, \tag{17}$$

3. Measurement of oil density ρ_o in the expanded range 700–1200 kg/m^3 with main reduced error $\Delta\rho/\rho_{max}$~±1%:

$$\rho_o = \kappa_1 - \kappa_2\,(T_{2A}) - \kappa_3\,(T_{2A})^2 \text{ for } \rho_o = 700\text{–}900 \text{ kg/m}^3, \quad (18)$$

$$\rho_0 = \kappa_4 \exp[-\kappa_5\,(T_{2A})] \text{ for } \rho_0 = 900\text{–}1100 \text{ kg/m}^3, \quad (19)$$

That is more precise than the inflow densitometer PLOT-3B-1P with a main reduced error = 1.3%. The measurement time is six times shorter.

4. Measurement of viscosity with a main reduced error of about ±1.5%, by [56–59,63,70,79]:

$$\nu = \eta/\rho = (1.12/\rho)\cdot(T/298\cdot T_{2A})^{1.25}, \quad (20)$$

which is more precise than the inflow viscometer *Viscosite* with δ = ±2%.

5. Measurement of integral characteristics of disperse size distribution of water droplets in emulsions by spin-lattice relaxation times T_{1W} using equations from [56–59,63,70,79]:

$$D_{CA}\,(\mu m) = 0.164 \exp(2.84\cdot T_{1A}\,(s)), \quad (21)$$

$$D_{max} = 0.32\cdot\exp(1.37\cdot T_{1A}), \quad (22)$$

$$r_{3/2} = D_{3/2}/2 = 2.40\cdot(T_{1A})^{4,27}, \quad (23)$$

6. Measurement of oil mean molecular mass with the error $\delta \approx 2.1\%$ in the expanded range MM = 50–1000 a.u.m. using the equation:

$$MM\,(\text{a.u.m.}) = 3011 + 3871.3 \exp(-5.585 T_{2O}), \quad (24)$$

7. Measurement of temperatures of freezing in high paraffinic oils in the range $T_{FR} = -16 \div +56$ °C using the equation [56–59,63,70,79]:

$$T_{FR}\,(^\circ C) = 275 - 0.62 T_{2A} + 2.8 \times 10^{-4}\,(T_{2A})^2, \quad (25)$$

8. Measurement of salts concentrations C in water using the equation:

$$C\,(M) = 24.35 T_{1W}{}^{-1} - 5.8, \quad (26)$$

9. Measurement of asphaltene–resin (AR) concentrations in the whole range with error $\delta \approx \pm 1\%$ using equations [56–59,63,70,79]:

$$AR(\%) = -3.76 \ln(T_{1A}) + 25.8, \quad (27)$$

$$AR(\%) = -2.76 \ln(T_{2A}) + 14.6, \quad (28)$$

Instrumental methods for AR determination are not found.

The component of the strength moment M_C of the pump electric drive, considering the formation of stresses at water drops/oil boarders, is determined by the resistance of the oil-well liquid in which the pump works. It originates additional resistance to the pump's electric drive shaft, caused by the necessity to overcome the forces of the surface tension between the droplets of water and oil border. Resistance depends on gas factor and water W content because the density and viscosity depend on these characteristics.

$$M_C = HQ/\omega\eta_H + M_{C0}(1 - G)\,(1 + \rho_H/\rho), \quad (29)$$

where H—is the pressure head, ω—is the rotatory rate of the electric drive shaft, and ηh—the coefficient of efficiency (CE). Introducing in Equation (29) the dependences of oil-well properties (Q, G, W, ρ_O) from PMR-parameters Equations (11), (15)–(18), (20), and M_C is received at mean oil-well liquid temperature T = 50 °C [56–59,63,70,79]:

$$M_C = HK_C S[(T_2{}^*)^{-1} + (\tau)^{-1}]/\omega\eta_H + M_{CO}[1 - (A_0 - A_G)/A_0][1 + (896.7 - 18.557(T_{1O}) - 130.8(T_{1O})^2)]/[65\ln[T_{2W}(T_2{}^* - T_{2O})100\%/T_2{}^*(T_{2W} - T_{2O})] + 830], \quad (30)$$

where K_C—pipeline reduction coefficient, S—pipeline section, τ—time of liquid presence in the coil.

Figure 7 presented the dependencies $H = f(Q)$ for oils with ρ = 882 kg/m^3 (curve 2) and ρ = 888 kg/m^3 (curve 3). Curve 1—for water.

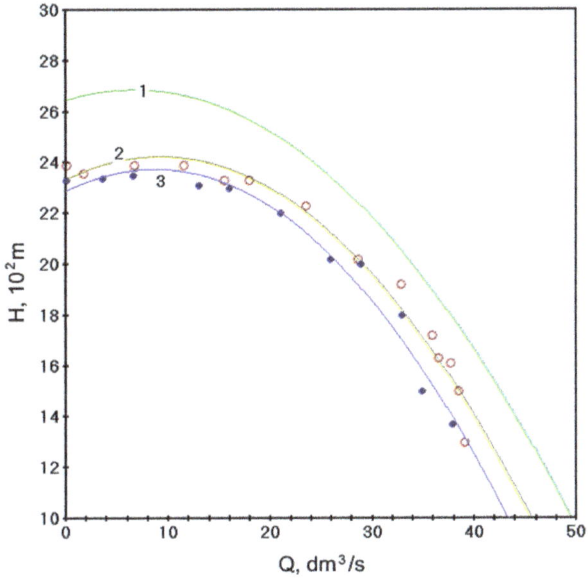

Figure 7. Curves $H = f(Q)$ for water (curve 1) and oils with ρ = 882 kg/m^3 (curve 2) and ρ = 888 kg/m^3 (curve 3).

From Figure 7, the experimental parameters of the borehole are the following: H_o = 2340 m; H = 1800 m; and Q = 30 l/s = 0.03 m^3/s. So, $C = (H_0 - H)/Q^2 = 6000$. At the pump rotation rate ω = 1450 r/min = 151.8 rad/s $A = H_0/\omega_H{}^2 = 0.001$ м·с^2/rad. If $\eta_H = 0.51$, then using equation $\eta_H = HQ/P_{mec}$, we can calculate the mechanical power $P_{mec} = HQ/\eta_H = 1.06$ kW, and the value of the additional resistance moment M'_C for the flow rates in the range 33–36 l/s will be $M'_C = \Delta H \cdot \Delta Q/\omega\eta_H = 1.7 \cdot 3/151.84 \cdot 0.5 = 0.68$ N·m. To overcome this M'_C and convert the oil mining productivity curve $H(Q)$ for oil with density ρ_o = 888 kg/m^3 to a productivity curve for more light oil, the pump power consumption must be increased on $\Delta P = \Delta H \cdot \Delta Q/\eta_H = 10.2$ kW, which can be done by increasing of rotary rate of electric drive on $\Delta\omega = \Delta P/\Delta M = 15$ Hz by frequency inverter. Flow rate control, instead of the usually used turbine flowmeters, having a great error on multiphase liquids, can be performed by AMC using the PMR-relaxometry method and Equations (1) and (2) for determination of the different phases flow velocities v_i in the sections S_i of pipes of any diameters and by calculating $Q_i = v_i \cdot S_i$.

For estimation of electric energy consumption $W_{p.c}$ at pipelines oil transport proposed to use the equation [56–59,63,70,79]:

$$W_{p.c} = 0.496\rho^{1.22}(Sv/L)^{2.75} \cdot \nu^{0.28} L/\rho^{1.25} d^{4.75} \eta_{pp} + 2.726 \cdot 10^4 \cdot \rho(\Delta z S v/L\eta_{pp})(\eta_p/\eta_{ed}), \quad (31)$$

where V—the volume of the transported oil, м^3 (flowrate is V/t); t—measurement time, ν—kinematic viscosity of oil, м^2/c; L—longitude of the pipeline, m; d—equivalent diameter of the pipe, m; η_{pp}—efficiency factor for the part of a pipeline; Δz—static head; and η_{ed} and η_p—efficiency factors of electric drive and pump. An equation must be considered that [56–59,63,70,79]:

$$\nu\ (\text{mm}^2/\text{s}) = 10^{-6}\nu\ (\text{m}^2/\text{s})10^{-3}\eta(\text{Pa·s})/\rho\ (\text{kg}/\text{m}^3), \quad (32)$$

and density and viscosity depend on relaxation time T_{2A} via Equations (18) and (20). Inserting them into Equation (31), the equation for energy consumption $W_{p.c}$ from PMR-parameters is constructed [56–59,63,70,79]:

$$W_{p.c} = 7.6V^{2.75}L^{-1.5}d^{4.75}\eta_{np}\cdot\exp(7T_{2A})/(T_{2A} - 0.017)^{1.25} + 2.873\cdot10^7(\Delta zSv/L\eta_{np})(\eta_{nas}/\eta_{ed}), \quad (33)$$

So, using AMC for permanent control of oil PMR-parameters, the energy consumption for oil transport by pipelines can be estimated.

4. Conclusions

The research results showed that using PMR as part of the AMC makes it possible to control most of the ODS properties in real time more efficiently than previously developed multifunctional complexes consisting of devices of different types. It provides more efficient process control in digital smart fields. In addition, the safety of oil production, treatment, and transportation facilities is increased, and the negative impact on the environment is reduced.

Preliminary assessments based on the results of the studies showed that the use of the methodology for monitoring the quality of ODS and wastewater could increase the lifecycle of oil production and treatment facilities by 2–3 times and reduce the number of accidents at the final stages of field operation. It should also be noted that the developed AMC makes it possible to control and manage oil purification processes from such impurities as salts, sulfur, asphaltene, resins, and paraffins. This prevents ARP deposits in the pipes.

It is essential to note the possibility of estimating ODS flowing through the pipes using the developed AMC in real time. It allows automatic control of the pump operating modes, reducing electrical energy consumption. In the standard mode, the work is based on the maximum amount of gas in ODS with the maximum electrical energy consumption. Evaluation of the additional moment of resistance on the shaft of a submersible electric motor of a centrifugal oil pump makes it possible to avoid the overload mode and increase the service life of the equipment.

The authors plan to continue research to adapt the AMC to severe operating conditions in automatic flow mode in cooperation with PJSC Tatneft for ODS with different properties and temperatures.

Author Contributions: Conceptualization, R.K. and O.K.; methodology, V.K.; software, N.D.A.; validation, R.K., V.D., N.D.A. and O.K.; formal analysis, V.K. and O.K.; investigation, R.K.; resources, R.K. and O.K.; data curation, V.K.; writing—original draft preparation, R.K.; writing—review and editing, R.K. and V.D.; visualization, N.D.A. and V.D.; supervision, R.K.; project administration, O.K.; funding acquisition, O.K. All authors have read and agreed to the published version of the manuscript.

Funding: The research is partially funded by the Ministry of Science and Higher Education of the Russian Federation under the strategic academic leadership program 'Priority 2030'.

Data Availability Statement: Not applicable.

Conflicts of Interest: The authors declare no conflict of interest.

Nomenclature

ODS	oil dispersed system
AMC	automatic measuring complex
PMR	nuclear (proton) magnetic resonance
DSF	digital smart field
AR	asphaltene–resins
ARP	asphaltene–resins–paraffins
PMRR	PMR relaxometry
RF	radio frequency
SE	spin-echo
ADC	analog-to-digital converter
SEC	specific energy consumption

References

1. Liu, Q.; Zhao, Z.; Liu, Y.; He, Y. Natural resources commodity prices volatility, economic performance and environment: Evaluating the role of oil rents. *Resour. Policy* **2022**, *76*, 102548. [CrossRef]
2. Cengiz, E.; Babagiray, M.; Emre Aysal, F.; Aksoy, F. Kinematic viscosity estimation of fuel oil with comparison of machine learning methods. *Fuel* **2022**, *316*, 123422. [CrossRef]
3. Xu, Y.; Lun, Z.; Pan, Z.; Wang, H.; Zhou, X.; Zhao, C.; Zhang, D. Occurrence space and state of shale oil: A review. *J. Pet. Sci. Eng.* **2022**, *211*, 110183. [CrossRef]
4. Chen, Z.; Wang, L.; Wei, Z.; Wang, Y.; Deng, J. Effect of components on the emulsification characteristic of glucose solution emulsified heavy fuel oil. *Energy* **2022**, *244*, 123147. [CrossRef]
5. Gizatullin, B.; Gafurov, M.; Murzakhanov, F.; Vakhin, A.; Mattea, C.; Stapf, S. Molecular Dynamics and Proton Hyperpolarization via Synthetic and Crude Oil Porphyrin Complexes in Solid and Solution States. *Langmuir* **2021**, *37*, 6783–6791. [CrossRef]
6. De Robbio, R.; Cameretti, M.C.; Mancaruso, E. Investigation by modelling of a plug-in hybrid electric commercial vehicle with diesel engine on WLTC. *Fuel* **2022**, *317*, 123519. [CrossRef]
7. Luján, J.M.; García, A.; Monsalve-Serrano, J.; Martínez-Boggio, S. Effectiveness of hybrid powertrains to reduce the fuel consumption and NOx emissions of a Euro 6d-temp diesel engine under real-life driving conditions. *Energy Convers. Manag.* **2019**, *199*, 111987. [CrossRef]
8. Davydov, R.; Dudkin, V. The Nuclear Magnetic Flowmeter for Monitoring the Consumption and Composition of Oil and Its Complex Mixtures in Real-Time. *Energies* **2022**, *15*, 3259. [CrossRef]
9. Nascimento da Silva, G.; Rochedo, P.R.R.; Szklo, A. Renewable hydrogen production to deal with wind power surpluses and mitigate carbon dioxide emissions from oil refineries. *Appl. Energy* **2022**, *311*, 118631. [CrossRef]
10. Davydov, V.V.; Myazin, N.S.; Davydov, R.V. Multiphase NMR Flowmeter-Relaxometer for Controlling the State and Rapidly Changing Flow Rates of Oil Mixtures. *Meas. Tech.* **2022**, *65*, 444–452. [CrossRef]
11. Davydov, V.V.; Myazin, N.S.; Grebenikova, N.M. Determination of the Composition and Concentrations of the Components of Mixtures of Hydrocarbon Media in the Course of its Express Analysis. *Meas. Tech.* **2020**, *62*, 1090–1098. [CrossRef]
12. Nourelfath, M.; Lababidi, H.M.S.; Aldowaisan, T. Socio-economic impacts of strategic oil and gas megaprojects: A case study in Kuwait. *Int. J. Prod. Econ.* **2022**, *246*, 108416. [CrossRef]
13. Faksness, L.-G.; Leirvik, F.; Taban, I.C.; Engen, F.; Vensen, H.V.; Holbu, J.W.; Dolva, H.; Bråtveit, M. Offshore field experiments with in-situ burning of oil: Emissions and burn efficiency. *Environ. Res.* **2022**, *205*, 112419. [CrossRef] [PubMed]
14. Chen, Y.; Zhang, Y.; Wang, J.; Lu, Z. Optimal Operation for Integrated Electricity–Heat System with Improved Heat Pump and Storage Model to Enhance Local Energy Utilization. *Energies* **2020**, *13*, 6729. [CrossRef]
15. Rosendahl, T.; Hepso, V. *Integrated Operations in the Oil and Gas Industry: Sustanability and Capabability Development*; IGI Global: Hershey, PA, USA, 2013.
16. Tihomirnov, L.I. Digitalization: Key factors of success. *Oil Gas* **2019**, *1–2*, 142–143.
17. Marchenko, I. Features of geotechnical surveys and leg penetration analysis for drilling platforms in the Arctic seas. *IOP Conf. Ser. Mater. Sci. Eng.* **2019**, *700*, 012047. [CrossRef]
18. Tarasenko, M.Y.; Lenets, V.A.; Akulich, N.V.; Yalunina, T.R. Features of use direct and external modulation in fiber optical simulators of a false target for testing radar station. In *Lecture Notes in Computer Science*; Springer: Cham, Switzerland, 2017; Volume 10531, pp. 227–232.
19. Petrov, A.A. Digital Frequency Synthesizer for ^{133}Cs-Vapor Atomic Clock. *J. Commun. Technol. Electron.* **2017**, *62*, 289–293. [CrossRef]
20. Zhang, S.; Lei, Q.; Wu, L.; Wang, Y.; Zheng, L.; Chen, X. Supply chain design and integration for the Co-Processing of bio-oil and vacuum gas oil in a refinery. *Energy* **2022**, *241*, 122912. [CrossRef]
21. Ocampo Batlle, E.A.; Escobar Palacio, J.C.; Silva Lora, E.E.; Bortoni, E.D.C.; Horta Nogueira, L.A.; Carrillo Caballero, G.E.; Vitoriano Julio, A.A.; Escorcia, Y.C. Energy, economic, and environmental assessment of the integrated production of palm oil biodiesel and sugarcane ethanol. *J. Clean. Prod.* **2021**, *311*, 127638. [CrossRef]
22. Alekseeva, M.N.; Svarovskaya, L.I.; Yashchenko, I.G. Risks of pollution of Arctic territories by oil and oil products. *AIP Conf. Proc.* **2019**, *2167*, 02000. [CrossRef]

23. Oliveira-Pinto, S.; Rosa-Santos, P.; Taveira-Pinto, F. Electricity supply to offshore oil and gas platforms from renewable ocean wave energy: Overview and case study analysis. *Energy Convers. Manag.* **2019**, *186*, 556–569. [CrossRef]
24. Pye, M. The discovery and development of the Brae Area Fields, U.K. South Viking Graben. *AAPG Mem.* **2018**, *115*, 155–161. [CrossRef]
25. Jones, D.W.; Van Bergen, P. Reservoir geology of the upper jurassic brae sandstone member, kingfisher field, South Viking Graben, U.K. North Sea. *AAPG Mem.* **2018**, *115*, 283–305. [CrossRef]
26. Krueger, S.; Sharpe, R.; Attridge, W.; Ruszka, J. Introduction of new drilling technology provides continuous high build rate capability in complex corkscrew well trajectory, accessing unswept reserves in the north sea. In Proceedings of the SPE/IADC Drilling Conference and Exhibition, Hague, The Netherlands, 14–16 March 2017. [CrossRef]
27. Feng, Q.; Xu, L.; Liu, C.; Hao, Y.; Yang, Z.; Zhang, T.; Liu, Y.; Su, H. Enhancing the anode performance of microbial fuel cells in the treatment of oil-based drill sludge by adjusting the stirring rate and supplementing oil-based drill cuttings. *Sustain. Energy Fuels* **2021**, *5*, 5773–5788. [CrossRef]
28. Coto, B.; Suárez, I.; Tenorio, M.J.; Huerga, I. Extraction of aromatic and polyaromatic compounds with NMP: Experimental and model description. *Fluid Phase Equilibria* **2022**, *554*, 113293. [CrossRef]
29. Li, Y.; Wang, D.; Xu, G.; Li, Q.; Yong, L.; Hongyu, G.; Lei, S.; Dongwei, L.; Meng, G.; Guoran, L.; et al. ZIF-8/PI Nanofibrous Membranes with High-Temperature Resistance for Highly Efficient PM0.3 Air Filtration and Oil-Water Separation. *Front. Chem.* **2021**, *9*, 810861. [CrossRef] [PubMed]
30. Zhao, L.; Zeng, X.; Zhao, W.; Zhu, F.; Hou, M.; Fan, G. Structural optimization for an axial oil-water separator with multi-stage separation. *Heat Mass Transf.* **2021**, *57*, 1949–1963. [CrossRef]
31. Jafarinejad, S.; Esfahani, M.R. A review on the nanofiltration process for treating wastewaters from the petroleum industry. *Separations* **2021**, *8*, 206. [CrossRef]
32. Li, Z.; Zhu, Y.; Xi, J.; Ye, D.; Hu, W.; Song, L.; Hu, Y.; Cai, W.; Gui, Z. Scalable production of hydrophobic and photo-thermal conversion bio-based 3D scaffold: Towards oil-water separation and continuous oil collection. *J. Clean. Prod.* **2021**, *319*, 128567. [CrossRef]
33. Erith, M.; Alfonso, Z.; Erik, L. A Multi-Sensor Approach to Separate Palm Oil Plantations from Forest Cover Using NDFI and a Modified Pauli Decomposition Technique. In Proceedings of the International Geoscience and Remote Sensing Symposium (IGARSS 2020), Waikoloa, HI, USA, 26 September—2 October 2020. [CrossRef]
34. Shafi, H.; Velswamy, K.; Ibrahim, F.; Huang, B. A hierarchical constrained reinforcement learning for optimization of bitumen recovery rate in a primary separation vessel. *Comput. Chem. Eng.* **2020**, *140*, 106939. [CrossRef]
35. Acharya, T.; Casimiro, L. Evaluation of flow characteristics in an onshore horizontal separator using computational fluid dynamics. *J. Ocean Eng. Sci.* **2020**, *5*, 261–268. [CrossRef]
36. Liu, M.; Zhu, Z.; Zhang, Z.; Chu, Y.; Yuan, B.; Wei, Z. Development of highly porous mullite whisker ceramic membranes for oil-in-water separation and resource utilization of coal gangue. *Sep. Purif. Technol.* **2020**, *237*, 116483. [CrossRef]
37. Mahmoud, M.; Tariq, Z.; Kamal, M.S.; Al-Naser, M. Intelligent prediction of optimum separation parameters in the multistage crude oil production facilities. *J. Pet. Explor. Prod. Technol.* **2019**, *9*, 2979–2995. [CrossRef]
38. Patrone, P.N.; Cooksey, G.; Kearsley, A. Dynamic Measurement of Nanoflows: Analysis and Theory of an Optofluidic Flowmeter. *Phys. Rev. Appl.* **2019**, *11*, 034025. [CrossRef]
39. Gu, Y.; Zhao, Y.; Lv, R.; Yang, Y. Theory and structure of a modified optical fiber turbine flowmeter. *Flow Meas. Instrum.* **2016**, *50*, 178–184. [CrossRef]
40. Shaaban, S. Design and optimization of a novel flowmeter for liquid hydrogen. *Int. J. Hydrogen Energy* **2017**, *42*, 14621–14632. [CrossRef]
41. Klingensmith, W.C.; Mays, D.C. Information Content of Wastewater Flowmeter Data before and during a Surcharge. *J. Environ. Eng.* **2018**, *144*, 05018004. [CrossRef]
42. Yang, Y.; Ha, W.; Zhang, C.; Liu, M.; Zhang, X.; Wang, D. Measurement of high-water-content oil-water two-phase flow by electromagnetic flowmeter and differential pressure based on phase-isolation. *Flow Meas. Instrum.* **2022**, *84*, 102142. [CrossRef]
43. Jin, N.; Yu, C.; Han, Y.; Yang, Q.; Ren, Y.; Zhai, L. The Performance Characteristics of Electromagnetic Flowmeter in Vertical Low-Velocity Oil-Water Two-Phase Flow. *IEEE Sens. J.* **2021**, *21*, 464–475. [CrossRef]
44. Wang, Y.; Li, H.; Liu, X.; Chen, L.; Xie, R.; Han, L. A Novel Method for Measuring the Flow Rate of High Viscous Fluid in Polymer Injection Well by Nonuniform Magnetic Field Electromagnetic Flowmeter. In Proceedings of the 8th Annual IEEE International Conference on Cyber Technology in Automation, Control and Intelligent Systems, Tianjin, China, 19–23 July 2018. [CrossRef]
45. Davydov, V.V. Control of the longitudinal relaxation time T_1 of a flowing liquid in NMR flowmeters. *Russ. Phys. J.* **1999**, *42*, 822–825. [CrossRef]
46. Han, Y.; Jin, N.; He, Y.; Ren, Y. Flow Measurement of Oil-in-Water Emulsions Using Arc-Type Conductivity Probes and Electromagnetic Flowmeter. *IEEE Trans. Instrum. Meas.* **2018**, *67*, 667–677. [CrossRef]
47. Wang, Y.; Li, H.; Liu, X.; Zhang, Y.; Xie, R.; Huang, C.; Hu, J.; Deng, G. Novel downhole electromagnetic flowmeter for oil-water two-phase flow in high-water-cut oil-producingwells. *Sensors* **2016**, *16*, 1703. [CrossRef] [PubMed]
48. Davydov, V.V.; Myazin, N.S.; Makeev, S.S. Method for Monitoring the Longitudinal Relaxation Time of 1Flowing Liquids Over the Entire Range of Flow Rate. *Meas. Tech.* **2020**, *63*, 368–374. [CrossRef]

49. Myazin, N.S.; Yushkova, V.V.; Davydova, T.I. New nondestructive method for determining the composition of components in biological objects in express mode. *J. Phys. Conf. Ser.* **2017**, *917*, 042017. [CrossRef]
50. Davydov, V.V.; Grebenikova, N.M. On the Possibility of Express Recording of Nuclear Magnetic Resonance Spectra of Liquid Media in Weak Fields. *Tech. Phys.* **2018**, *63*, 1845–1850. [CrossRef]
51. Myazin, N.S. Peculiarities of magnetic resonance signals processing during the express control of the liquid media state. *J. Phys. Conf. Ser.* **2019**, *1236*, 012026. [CrossRef]
52. Davydov, V.V.; Dudkin, V.I.; Karseev, A.Y. Feasibility of Using Nuclear Magnetic Spectroscopy for Rapid Monitoring of Liquid Media. *J. Appl. Spectrosc.* **2015**, *82*, 794–800. [CrossRef]
53. Shikhof, I.; Arns, C.H. Temperature-Dependent Oxygen Effect on NMR D-T_2 Relaxation-Diffusion Correlation of *n*-Alkanes. *Appl. Magn. Res.* **2016**, *47*, 1391. [CrossRef]
54. Zalesskiy, S.S.; Danieli, E.; Blumich, B.; Ananikov, V.P. Miniaturization of nmr systems: Desktop spectrometers, microcoil spectroscopy, and "nmr on a chip" for chemistry, biochemistry, and industry. *Chem. Rev.* **2014**, *114*, 5641–5677. [CrossRef]
55. Haratt, W.N.; Lee, V.E.; Priestley, R.D.; Cabral, J.T. Precision polymer particles by flash nanomicrofluidic droplet extraction. *ACS Appl. Polym. Mater.* **2021**, *3*, 4746–4768.
56. Kashaev, R.S.; Kozelkov, O.V. On-line express-analyzer and method of water monitoring on the base of proton magnetic resonance. *IOP Conf.Ser. Earth Environ. Sci.* **2019**, *288*, 012035. [CrossRef]
57. Kashaev, R.S.H. Oil Disperse Systems Study Using Nuclear Magnetic Resonance Relaxometry (NMRR). In *Advances in Energy Research*; Nova Science Publishers: Hauppauge, NY, USA, 2013; Volume 16.
58. Kashaev, R.; Faskchiev, N. Nuclear (proton) magnetic resonance relaxometry study of the effect of rotating magnetic field on the emulsion structure. *Appl. Magn. Reson.* **2011**, *41*, 31–43. [CrossRef]
59. Kashaev, R.S.; Kien, N.C.; Tung, T.V.; Kozelkov, O.V. Fast Proton Magnetic Resonance Relaxometry Methods for Determining Viscosity and Concentration of Asphaltenes in Crude Oils. *J. Appl.Spectrosc.* **2019**, *86*, 890–895. [CrossRef]
60. Callaghan, P.T. Principles Reo-NMR: Nuclear Magnetic Resonance and the Rheology of Complex Fluids. *Rep. Prog. Phys.* **1999**, *62*, 599–668. [CrossRef]
61. Davydov, V.V.; Dudkin, V.I.; Karseev, A.Y. Formation of the nutation line in NMR measuring system with flowing samples. *Tech. Phys. Lett.* **2015**, *41*, 355–358. [CrossRef]
62. Khasanova, N.; Sakharov, B.; Volkov, V.; Nurgaliev, D. Low field method for analysis of heavy oils without extraction. In Proceedings of the 17th International Multidisciplinary Scientific Geoconference SGEM2017, Viena, Austria, 27–29 November 2017; pp. 297–304.
63. Kashaev, R.S.; Masiab, A.N. Study of phase transitions in some phase changing organic matters by use of NMR relaxometry. *Chem. Mater. Eng.* **2013**, *1*, 78–84. [CrossRef]
64. Davydov, V.V.; Dudkin, V.I.; Karseev, A.Y. A Compact Nuclear Magnetic Relaxometer for the Express Monitoring of the State of Liquid and Viscous Media. *Meas. Tech.* **2014**, *57*, 912–918. [CrossRef]
65. Aleksandrov, A.S.; Ivanov, A.A.; Archipov, R.V.; Gafurov, M.R. TagirovMS Pulsed NMR spectrometer with dynamic nuclear polarization for weak magnetic fields. *Magn. Reson. Solids* **2019**, *21*, 19203.
66. Davydov, V.V.; Dudkin, V.I.; Velichko, E.N. Method for Measuring the Sensitivity of Nuclear-Resonance Magnetometers with Flowing Liquid. *Meas. Tech.* **2016**, *59*, 176–182. [CrossRef]
67. Hogendoorn, J.; Boer, A.; Appel, M.; de Jong, H.; de Leeuw, R. Magnetic Resonance Technology. A New Concept for Multiphase Flow Measurement. In Proceedings of the 31th International North Sea Flow Measurement Workshop, Tonsberg, Norway, 22–25 October 2013.
68. Deng, F.; Xiao, L.; Wanget, M.; Tao, Y.; Kong, L.; Zhang, X.; Liu, X.; Geng, D. Online NMR Flowing Fluid Measurements. *Appl. Magn. Reson.* **2016**, *47*, 1239–1253. [CrossRef]
69. Davydov, V.V. Some specific features of the NMR study of fluid flows. *Opt. Spectrosc.* **2016**, *121*, 18–24. [CrossRef]
70. Safieva, R.Z.; Mishin, V.D. Systems Analysis of the Evolutiom of Views on Oil Systems: From Petroleum Chemistry to Petroinformatics. *Pet. Chem.* **2021**, *61*, 539–554. [CrossRef]
71. Davydov, V.V.; Dudkin, V.I.; Karseev, A.Y. A Compact Marked Nuclear-Magnetic Flowmeter for Measurement of Rapidly Varying Flow Rates of Liquid. *Meas. Tech.* **2015**, *58*, 317–322. [CrossRef]
72. Davydov, V.V.; Dudkin, V.I.; Myazin, N.S. Nutation line shape for the nonstationary regime of magnetic resonance flowmeter–relaxometerю. *J. Commun. Technol. Electron.* **2016**, *61*, 1159–1165. [CrossRef]
73. Uskova, E.I.; Doroginitski, M.M.; Skirda, V.D.; Fatkhutdinov, I.H. New Approach to Analyze 2D Map T_1-T_2. *Appl. Magn. Res.* **2020**, *51*, 183–193. [CrossRef]
74. Mills, C. Process effects on coriolis flowmeter calibration. *Hydrocarb. Process.* **2020**, *99*, 8.
75. Prakhova, M.Y.; Krasnov, A.N. Coriolis flowmeter for commercial accounting of crude oil. *J. Phys. Conf. Ser.* **2020**, *1582*, 012071. [CrossRef]
76. Rondon, J.; Barrufet, M.A.; Falcone, G. A novel downhole sensor to determine fluid viscosity. *Flow Meas. Instrum.* **2012**, *23*, 9–18. [CrossRef]
77. Falcone, G.; Hewitt, G.F.; Alimonti, C.; Harrison, B. Multiphase Flow Metering: Current Trends and Future Developments. *J. Pet Technol.* **2002**, *54*, 77–84. [CrossRef]

78. Davydov, V.V.; Myazin, N.S.; Kiryukhin, A.V. Nuclear-Magnetic Flowmeter-Relaxometers for Monitoring Coolant and Feed water Flow and Status in Npp. *Atom. Energy* **2020**, *127*, 274–279. [CrossRef]
79. Kashaev, R.S.; Kozelkov, O.V.; Temnikov, A.N.; Van Tung, T.; Kein, N.T. Relaxometer of proton magnetic resonance. *Instrum. Exp. Tech.* **2019**, *2*, 145–148.
80. Davydov, V.V.; Dudkin, V.I.; Karseev, A.Y.; Vologdin, V.A. Special Features in Application of Nuclear Magnetic Spectroscopy to Study Flows of Liquid Media. *J. Appl. Spectrosc.* **2016**, *82*, 1013–1019. [CrossRef]
81. Davydov, V.V.; Myazin, N.S.; Dudkin, V.I.; Davydov, R.V. Determination of the Longitudinal Relaxation Time of a Flowing Liquid Using a Differential Nuclear Magnetic Spectrometer. *Tech. Phys. Lett.* **2020**, *46*, 1147–1151. [CrossRef]
82. Khan, N.S.; Kumam, P.; Thounthong, P. Computational Approach to Dynamic Systems through Similarity Measure and Homotopy Analysis Method for Renewable Energy. *Crystals* **2020**, *10*, 1086. [CrossRef]
83. Khan, N.S.; Gul, T.; Kumam, P.; Shah, Z.; Islam, S.; Khan, W.; Zuhra, S.; Sohail, A. Influence of Inclined Magnetic Field on Carreau Nanoliquid Thin Film Flow and Heat Transfer with Graphene Nanoparticles. *Energies* **2019**, *12*, 1459. [CrossRef]
84. Kashaev, R.S.; Idiyatullin, Z.S.; Temnikov, A.N.; Dautov, I.R. Design for Measurement of Composition and Flow Rate by NMR-Method. RF Patent № 74710, 24 April 2008.

MDPI
St. Alban-Anlage 66
4052 Basel
Switzerland
www.mdpi.com

Energies Editorial Office
E-mail: energies@mdpi.com
www.mdpi.com/journal/energies

www.ingramcontent.com/pod-product-compliance
Lightning Source LLC
Chambersburg PA
CBHW061518210926
51850CB00008B/1958
* 9 7 8 3 7 2 5 8 0 4 9 6 2 *